AF566764

METAL OPTICS AND SUPERCONDUCTIVITY

Proceedings of the Lebedev Physics Institute Academy of Sciences of the USSR

Volume 165	**Solitons and Instantons, Operator Quantization,** Edited by V.L. Ginzburg
Volume 166	**The Nonlinear Optics of Semiconductor Lasers,** Edited by N.G. Basov
Volume 167	**Group Theory, Gravitation and Elementary Particle Physics,** Edited by A.A. Komar
Volume 168	**Issues in Intense-Field Quantum Electrodynamics,** Edited by V.L. Ginzburg
Volume 169	**The Physical Effects in the Gravitational Field Black Holes,** Edited by M.A. Markov
Volume 170	**The Theory of Target Compression by Longwave Laser Emission,** Edited by G.V. Sklizkov
Volume 171	**Research on Laser Theory,** Edited by A.N. Orayevskiy
Volume 172	**Phase Conjugation of Laser Emission,** Edited by N.G. Basov
Volume 173	**Quantum Mechanics and Statistical Methods,** Edited by M.M. Sushchinskiy
Volume 174	**Nonequilibrium Superconductivity,** Edited by V.L.Ginzburg
Volume 174 Supplemental Volume 1	**Thermodynamics and Electrodynamics of Superconductors,** Edited by V.L. Ginzburg
Volume 174 Supplemental Volume 2	**Electron Liquid Theory of Normal Metals,** Edited by V.P. Silin
Volume 175	**Luminescence Centers of Rare Earth Ions in Crystal Phosphors,** Edited by M.D. Galanin
Volume 175 Supplemental Volume	**Luminescence and Anisotropy of Zinc Sulfide Crystals,** Edited by M.D. Galanin

Volume 176	**Classical and Quantum Effects in Electrodynamics,** Edited by A.A. Komar
Volume 176 Supplemental Volume	**Quantization, Gravitation and Group Methods in Physics,** Edited by A.A. Komar
Volume 177	**Stoichiometry in Crystal Compounds and Its Influence on Their Physical Properties,** Edited by Yu. M. Popov
Volume 178	**The Delfin Laser Thermonuclear Installation: Operational Complex and Future Directions,** Edited by G.V. Sklizkov
Volume 179	**X-Ray Plasma Spectroscopy and the Properties of Multiply-Charged Ions,** Edited by I.I. Sobel'man
Volume 179 Supplemental Volume 1	**Interaction of Ultrashort Pulses with Matter,** Edited by M.D. Galanin
Volume 179 Supplemental Volume 2	**Electron-Excited Molecules in Nonequilibrium Plasma,** Edited by N.N. Sobolev
Volume 180	**Inelastic Light Scattering in Crystals,** Edited by M.M. Sushchinskiy
Volume 181	**Pulsed Metal and Metal Halogenide Vapor Lasers,** Edited by G.G. Petrash
Volume 182	**Luminescence of Wideband Semiconductors,** Edited by M.D. Galanin
Volume 183	**Invariants and the Evolution of Nonstationary Quantum Systems,** Edited by M.A. Markov
Volume 184	**Electron Processes in MIS-Structure Memories,** Edited by A.F. Plotnikov
Volume 185	**Injection Lasers in Optical Communication and Information Processing Systems,** Edited by Yu. M. Popov
Volume 186	**Photoproduction of Pions on Nucleons and Nuclei,** Edited by A.A. Komar

Proceedings of the Lebedev Physics Institute
Academy of Sciences of the USSR
Series Editor N.G. Basov
Volume 190

METAL OPTICS AND SUPERCONDUCTIVITY

Edited by A.I. Golovashkin

Translated by Al Peabody

Nova Science Publishers
New York

Deputy Series Editor: M.A. Man'ko

Nova Science Publishers, Inc.
283 Commack Road
Suite 300
Commack, New York 11725

This book is being published under exclusive English language rights granted to Nova Science Publishers, Inc. by the All-Union Copyright Agency of the USSR (VAAP).

Library of Congress Cataloging-in-Publication Data available upon request

ISBN 0-941743-80-2

The original Russian-language version of this book was published by Nauka Publishing House in 1988

Graphic Design by Elenor Kallberg and Peggy Harvey

Printed in the United States of America

CONTENTS

FOREWORD

The articles contained in this collection are dedicated to the study of the electron structure of transition metals and superconducting alloys and compounds based on them. The study of the electron structure of materials is one of the central problems of solid-state physics and defines the solution of a number of problems. One of them is the problem of high-temperature superconductivity which has attracted exceptional attention from physicists in connection with the discovery of new classes of ceramic oxides which are superconducting at liquid-nitrogen temperature.

The electron structure is one of the "three whales" on which all of superconductivity rests. It is frequently our ignorance of the electronic properties of a metal, alloy or compound in its normal state which makes it impossible to predict superconductivity in the material, preventing us from calculating the parameters of the superconducting state. There are now a number of effective methods for investigation of the electron structure of the metals and allows. This collection discusses metal optics, tunneling and magnetic measurements in superconductors. These methods are quite informative and allow us to obtain many important electron characteristics and temperature relations. Various characteristics of the superconducting compounds Nb_3Ge, Nb_3Al, Nb_3Sn and Nb_3Ga with A15 structure and NbN with B1 structure, having rather high critical temperatures, are experimentally studied. The optical method of determining the dispersion of complex dielectric constant in the visible and IR spectral areas and its variation as a function of temperature, as well as the determination of the variation of critical magnetic field as a function of temperature and mean free path length, allowed estimation of the values of various electron characteristics of the compounds studied. The tunnel method and the method of studying the current-voltage characteristic of junctions with direct conductivity allowed determination of the superconducting state parameters. A number of singularities and anomalies obtained experimentally in the relations studied are discussed. The experimental results are supplemented by calculations based on the first principles of the band structure, as well as such characteristics as the dielectric constant, electron energy loss function and reflectivity over a broad spectral range for many transition metals.

Although significant efforts have been directed toward the investigation of the new classes of high-temperature superconductors recently discovered — the lanthanum and yttrium superconductors — studies of superconductors with A15 and B1 structure (as well as other classes) remain timely. There is still no clear answer to the question, why do these superconductors have such high critical temperatures. The results obtained for these materials deepen our conception of superconductivity, allowing us to draw certain conclusions as to the direction of future research. We should note that superconductors with A15 and B1 lattice are superior in a number of parameters even to ceramic superconductors. These parameters include significantly higher critical current density, significantly greater stability and independence of external conditions. Further progress in our understanding of the entire phenomenon of superconduction, increases in critical parameters of superconductors, the creation of new superconducting materials, the problem of stabilizing their properties require us to study superconductors of a wide variety of types. This collection responds to this need.

A. I. Golovashkin

MICROSCOPIC CALCULATION OF THE DIELECTRIC RESPONSE FUNCTION OF METALS

I. I. Mazin, Ye. G. Maksimov, S. N. Rashkeyev, Yu. A. Uspenskiy

Results are presented from calculations of dielectric constant for a large number of metals. The electronic structure of the metals was computed within the framework of the density functional theory using the LMTO method for band calculations. The theoretical results obtained are compared with the experimental data. There is seen to be a reasonable agreement between theory and experiment both for intraband and for interband contributions to the frequency-dependent conductivity. The electron-energy loss spectrum is analyzed for fast electrons in metals, directly expressed through the collective excitation spectrum of the metals. It is noted that there are specific collective excitations in the transition metals, related to intensive interband transitions. Optical spectra and electron energy loss spectra are analyzed in detail in $4d$ metals, as is the change in these spectra in the $4d$ series.

INTRODUCTION

Metal optics is perhaps one of the oldest branches in all metal physics. Experimental and theoretical studies of the optical properties of metals have been conducted for a long time. In recent years, additional stimuli have appeared for the conduct of such studies. On the one hand, there are the practical demands of laser and optical instrument design requiring, in particular, materials for mirrors with high reflectivity in the extreme areas of the spectrum — the infrared and ultraviolet bands, with minimum losses at high light flux power. The creation of such materials requires careful experimental investigation over a broad frequency range, including with synchrotron radiation, as well as strict quantitative computation of the optical characteristics of metals.

Another stimulus for increasing interest in metal optics has been the creation and broad development of methods for precise, quantitative calculation of the physical characteristics of solids, using only the number of the element in the periodic table as input data. In 1964/65, Hohenberg, Kohn and Sham suggested the so-called density functional theory for systems of interacting electrons allowing, in principle, the problem of the behavior of systems of strongly interacting particles to be reduced to the solution of a single-frequency problem of the behavior of one electron in an effective self-consistent field. The development of this method served as an impetus to the development of band calculation methods, since the possibility appeared of constructing self-consistent calculation systems. In constructing such systems, the exchange-correlation energy functional is usually selected in the local approximation, i.e., it is considered that the density of this energy depends on the electron density at the given point. In combination with such powerful band calculation methods as the augmented plane waves (APW) method, the Korringi-Kohn-Rostocker (KKR) method, the linear muffin-tin orbitals (LMTO) method, this allows the construction of single-frequency potentials, as well as the determination of single-electron band spectra for a wide variety of substances.

Previously, information on the band structure of substances was used in metal optics only to interpret the available experimental data based on the electron energy level structure. At the present time, thanks to the development of band calculation methods, it has become possible to reverse the problem, i.e., to produce optical properties from the band structure of a material. For example, this can be used to predict metallic materials with desired optical properties.

Another important application of these calculations is the possibility of proper interpretation of experimental optical data. The data of various authors differ significantly, which is apparently related to the sensitivity of the experiments performed to the methods used in conducting them as well as the methods of preparation of specimens. These differences are particularly great in the transition metals. This is especially true of the low-frequency optical characteristics of such metals in the infrared ($\hbar\omega \leqslant 0.5$ eV) area (cf. Chapter 5). Frequently, experimenters have interpreted the peculiarities of the optical characteristics of metals in this area as experimental defects, which can arise, for example, due to poor quality of preparation of the surface of specimens studied. The true nature of the situation can be determined, either by significantly increasing the accuracy of the experiments (which will hardly be possible in the near future), or by careful theoretical analysis, requiring the ability to calculate optical characteristics accurately from "first principles," i.e., without direct utilization of experimental data. The same can also be said of the nature of the plasma peaks in the electron energy-loss spectra of the transition metals, the interpretation of which is frequently a far from trivial task.

Some dozens of works have now been published in which the optical properties of metals are computed on the basis of a wide variety of band structure calculation methods. In spite of this variety of methods, all of these calculations share common difficulties. The first problem which arises is to determine how, based on the known single-electron spectrum and known wave functions, to construct the macroscopic dielectric constant function $\epsilon_M(\omega, \mathbf{q}\to 0)$. Most authors do this using the random-phase approximation.

This approach does not consider local field effects, related to the presence of rapidly oscillating coulomb fields in the crystalline lattice, and also ignores exchange and correlation effects (which are partially considered in the single-electron spectrum calculated by the density functional method) completely. However, even with this simplified analysis of the dielectric constant $\epsilon_M(\omega, \mathbf{q}\to 0)$, all such calculations encounter one significant difficulty. The problem is that in computing optical properties, accuracy of calculation of the current operator matrix element on the Bloch wave functions is very important. Since all modern band calculations are variational in nature, the wave functions are determined with less accuracy than the electron energies. Consequently, the matrix elements of the current operator based on such functions have lower accuracy than the energy spectrum. For this reason, the quality of the band calculations (i.e., the calculation of electron energies) still cannot guarantee quality of calculation of optical properties. As we stated earlier, an incorrect approach to the calculation of current matrix elements can in some cases make the calculation of optical constants quite senseless and lead to clearly inaccurate results.

However, in spite of these difficulties, the available calculations of the optical properties of metals in the single-frequency approximation sometimes provide good agreement with experimental data both in terms of the position of the major singularities on the frequency scale and their amplitude. Since this agreement has been repeatedly observed for various simple, transition and noble metals, this leads us to the thought that it is not a result of chance, i.e., apparently, multifrequency effects are not very significant for the computation of optical properties. Clearly, the suitability of the single-frequency approximation can only be determined numerically. Agreement with experimental results is not the only method of verifying the quality of calculations. It is possible to verify the accuracy of calculations by the agreement of the results with a number of precise sum rules for $\epsilon_M(\omega, \mathbf{q}\to 0)$, for example the f-sum rule.

The purpose of this work is to calculate the optical properties of metals in the single-frequency approximation, and to provide a comprehensive analysis of the applicability of this approximation. We have attempted to analyze the broadest possible class of phenomena for a large number of simple, transition and noble metals. Numerical calculations were based on the LMTO method of computing band structure. Calculations are performed for 15 simple, transition and noble metals and comparisons are made with experimental data on dielectric constant $\epsilon(\omega)$, optical conductivity $\sigma(\omega)$, reflectivity $R(\omega)$ and electron energy-loss function $L(\omega)$. Analysis of the results allows us to determine the singularities of the optical characteristics which are reproduced well by single-frequency calculations, as well as those which contain significant errors. We have also undertaken a theoretical analysis of this problem. It has been shown that the effects of exchange and correlation and local-field effects largely compensate each other. It is therefore probable that the single-frequency approximation will yield better results than would be expected a priori.

Numerical calculations in the single-frequency calculation were also used to study two special questions — the mechanism of formation of the electron-energy loss spectrum in metals with d electrons and the optical properties of metals in the infrared area of the spectrum. Here we did not limit ourselves to the assumption of agreement or disagreement of the calculations with experimental data, but rather attempted to use them to explain the regularities observed.

CHAPTER 1
THE DIELECTRIC CONSTANT AND OPTICAL PROPERTIES OF METALS

1.1. Relationship of Dielectric Constant with the Optical Characteristics of Metals

The response of a system to external electromagnetic radiation with small wave vector $(|\mathbf{q}|\sim\omega/c)$ is determined by the long-wave macroscopic transverse dielectric constant $\epsilon_{\perp}(\omega,\mathbf{q}\to 0)$. In cubic crystals, the only type we will analyze in this work, within limits of small $\mathbf{q}\to 0$ the transverse dielectric constant coincides with the longitudinal $\epsilon_{\|}(\omega,\mathbf{q}\to 0)$, the response of the system to an external scalar potential.

Therefore, in what follows we should discuss the longitudinal dielectric constant, omitting the $\perp$ and $\|$ with $\epsilon(\omega)$. The dielectric constant of $\epsilon(\omega)$ has real and imaginary parts which are other than zero,

$$\epsilon(\omega)=\epsilon_1(\omega)+i\epsilon_2(\omega) \tag{1.1}$$

and satisfies the Kramers–Kronig relation:

$$\varepsilon(\omega) = 1 + \frac{1}{\pi}\int_0^\infty \frac{d\omega'^2\varepsilon_2(\omega')}{\omega'^2-\omega^2-i\delta}. \tag{1.2}$$

The optical characteristics of a metal, such as the complex index of refraction

$$\tilde{n}(\omega)=n(\omega)+ik(\omega),$$

as well as the reflectivity for a normally incident electromagnetic wave $R(\omega)$ are expressed directly through the dielectric constant $\epsilon(\omega)$. Here

$$n(\omega) = \sqrt{{}^1/_2\left[\varepsilon_1(\omega) + \sqrt{\varepsilon_1^2(\omega) + \varepsilon_2^2(\omega)}\right]}, \tag{1.3}$$

$$k(\omega) = \sqrt{{}^1/_2\left[-\varepsilon_1(\omega) + \sqrt{\varepsilon_1^2(\omega) + \varepsilon_2^2(\omega)}\right]} \tag{1.4}$$

and

$$R(\omega) = \left|\left(\frac{\sqrt{\varepsilon(\omega)} - 1}{\sqrt{\varepsilon(\omega)} + 1}\right)^2\right|. \tag{1.5}$$

The dielectric constant $\epsilon(\omega)$ also expresses the fast electron energy loss function $L(\omega)$

$$L(\omega) = \mathrm{Im}\left[-1/\varepsilon(\omega)\right] = \frac{\varepsilon_2(\omega)}{\varepsilon_1^2(\omega) + \varepsilon_2^2(\omega)}, \tag{1.6}$$

which in the first Born approximation is proportional to the twice differential scattering cross sections of electrons at zero angle with energy loss $\Delta E = \hbar\omega$[1].

The f–sum rule is correct for the dielectric constant $\epsilon(\omega)$[1, 2]

$$\int_0^\infty \omega\varepsilon_2(\omega)\, d\omega = \frac{\pi}{2}\omega_p^2, \tag{1.7}$$

where ω_p is the electron plasma frequency

$$\omega_p^2 = 4\pi ne^2/m \tag{1.8}$$

and n is the electron density of the metal. This relation is independent of the nature of the interaction of the electrons with each other and with ions, reflecting only the requirement that the number of particles be conserved. When equation (1.7) is integrated to infinite energies in the literal sense, n must be understood to be the full density of all electrons, including core electrons. Much more reasonable for crystals is the f–sum rule for finite energies[3]

$$\int_0^E \omega\varepsilon_2(\omega)\, d\omega = \frac{2\pi^2 e^2}{m} n_{\mathrm{eff}}(E), \tag{1.9}$$

which determines the effective number of electrons taking part in optical transitions up to a given energy E. In many metals the transitions related to excitation of core electrons are in the area of very high energies, significantly greater than the energy of plasma oscillations of conductivity electrons $\omega_p=\sqrt{4\pi n_v e^2/m}$, where n_v is the density of the valent electrons of the metal. In such metals integration in equation (1.9) to $E<E_g$, where E_g is the lower excitation energy of core electrons, yields the f–sum rule for valent electrons. There are, however, metals in which the area of optical excitation of conductivity electrons and the area of core electron excitation energies overlap. The f–sum rule can be applied to such metals only with caution[3].

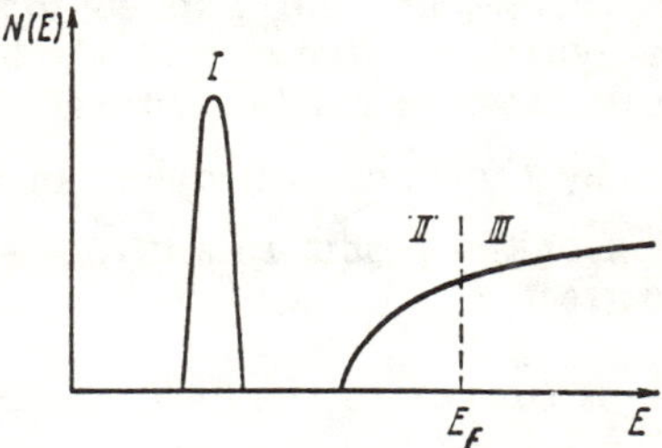

Figure 1. Typical Electron–State–Density Curve of a Metal Considering the Nearest Core Level.

Let us study a metal, the electron state density in which is schematically shown in Figure 1. Area *I* here describes the states occupied by the nearest core level expanded, however, by an overlap to the band. Area *II* consists of the states of the conductivity electrons and *III* is the empty state area, in which only real transitions from area *II* or area *I* are possible.

A similar situation obtains, for example, in certain alkali metals (Rb, Cs) at high pressures[4]. Even if we ignore the real transitions from states in area *I*, it is easy to see that the quantity n_{eff}, corresponding to valent electron transitions, will differ from the real quantity n_{II}. We must first write the quantity $n^{\text{tot}}_{\text{eff}}$, corresponding to consideration of all transitions, both *II–III* and *I–III*:

$$n^{\text{tot}}_{\text{eff}} = n^{I-III}_{\text{eff}} + n^{II-III}_{\text{eff}} = n_I + n_{II}.$$

Here n_I is the full number of electrons in area *I*, n_{II} is the same for area *II*. Further, placing the Fermi level in the area of the gap between areas *I* and *II*, we can write the following relation

$$n_I = n_{\rm eff}^{I-II} + n_{\rm eff}^{I-III}.$$

Now subtracting the second relation from the first, we obtain

$$n_{\rm eff}^{II-III} = n_{II} + n_{\rm eff}^{I-II}.$$

This equation means that the f–sum rule, applied to the transitions involving only conductivity electrons, yields a value of $n_{\rm eff}$ which is greater than the true value. The excess in the number of conductivity electrons yields the value of transitions $n_{\rm eff}^{I-II}$ forbidden by the Pauli principle. In actuality, of course, this means only that application of the f–sum rule in this situation for conductivity electrons alone is quite incorrect.

Yet another sum rule is correct for the dielectric constant $\epsilon(\omega)$[1]

$$\int_0^\infty \omega \operatorname{Im} \frac{1}{\varepsilon(\omega)}\, d\omega = -\frac{\pi}{2}\omega_p^2, \tag{1.10}$$

a particular case of the Thomas–Reich–Kuhn–sum rule. In our calculations, it will also be analyzed for finite energies

$$\int_0^E \omega \operatorname{Im} \frac{1}{\varepsilon(\omega)}\, d\omega = -\frac{2\pi^2 e^2}{m} \tilde{n}_{\rm eff}(\omega) \tag{1.11}$$

and used in comparison of the oscillator forces of various plasma peaks in the electron energy loss spectra.

1.2. Microscopic Determination of Dielectric Constant. Single–Frequency Approach

Numerical computation of the function $\epsilon(\omega)$ requires microscopic determination of this quantity. For a system of strongly interacting particles exposed to an external periodic potential, this is a very difficult task. We shall return to discussion of this problem in Chapter 4. In most works on metal optics, both the interpretation of experimental data and microscopic calculations of function $\epsilon(\omega)$ are based on a simple single–frequency approach to the system of metal electrons within the framework of the theory of Bloch. In this approach, the electrons follow the simple equation of Schroedinger

$$\left(-\frac{\hbar^2\nabla^2}{2m}+V_{\rm eff}(r)\right)\psi_\lambda(\mathbf{r})=E_\lambda\psi_\lambda(\mathbf{r}) \qquad (1.12)$$

with periodic effective potential $V_{\rm eff}(\mathbf{r})$. The corresponding Hamiltonian equation is

$$\widehat{H}=-\frac{\hbar^2\nabla^2}{2m}+V_{\rm eff}(\mathbf{r}). \qquad (1.13)$$

Adding to this Hamiltonian the components corresponding to the interaction of an electron with an electromagnetic wave, and considering the interaction of electrons in the random-phase approximation, it is easy to obtain the following expression for the dielectric constant system $\epsilon(\omega)$[5]:

$$\varepsilon(\omega)=1-\frac{8\pi e^2}{3\Omega q^2}\sum_{\mathbf{k},\lambda,\lambda'}\frac{f_{\mathbf{k}\lambda}-f_{\mathbf{k}+\mathbf{q},\lambda'}}{E_{\mathbf{k}\lambda}-E_{\mathbf{k}+\mathbf{q},\lambda'}+\hbar\omega+i\delta}|\langle\mathbf{k}+\mathbf{q},\lambda'|e^{i\mathbf{q}\mathbf{r}}|\mathbf{k}\lambda\rangle|^2.$$
$$(\mathbf{q}\to 0) \qquad (1.14)$$

Here $|\mathbf{k}\lambda\rangle$ is the Bloch electron wave function with quasi-momentum **k** in band λ with energy $E_{\mathbf{k}\lambda}$; $f_{\mathbf{k}\lambda}$ is the Fermi filling function of single-electron states. Summation with respect to **k** is performed within the Brillouine zone, Ω is the volume of the crystal.

At the limit $\mathbf{q}\to 0$ we can, by isolating contributions with $\lambda=\lambda'$ (intraband portion) and considering that where $\lambda\neq\lambda'$ and $\mathbf{q}\to 0$

$$\langle\mathbf{k}\lambda|e^{-i\mathbf{q}\mathbf{r}}|\mathbf{k}+\mathbf{q},\lambda'\rangle=\frac{\hbar}{e}\frac{\langle\mathbf{k}\lambda|\mathbf{q}\mathbf{j}|\mathbf{k}\lambda'\rangle}{E_{\mathbf{k}\lambda'}-E_{\mathbf{k}\lambda}},$$

where $\mathbf{j}=(e\hbar/im)\nabla$ is the longitudinal current operator, obtain an expression for the dielectric constant

$$\varepsilon(\omega)=1-\frac{\tilde{\omega}_p^2}{\omega^2}-\frac{8\pi\hbar^2}{3\Omega}\sum_{\mathbf{k},\lambda'\neq\lambda}\frac{f_{\mathbf{k}\lambda}-f_{\mathbf{k}\lambda'}}{(E_{\mathbf{k}\lambda}-E_{\mathbf{k}\lambda'})^2}\frac{|\langle\mathbf{k}\lambda'|\mathbf{j}|\mathbf{k}\lambda\rangle|^2}{\hbar\omega-E_{\mathbf{k}\lambda'}+E_{\mathbf{k}\lambda}+i\delta}. \qquad (1.15)$$

Here

$$\epsilon^{\rm intra}(\omega,\mathbf{q}\to 0)=1-\tilde{\omega}_p^2/\omega^2 \qquad (1.16)$$

is the intraband (Drude) contribution to the dielectric constant, while the last component in the right portion of (1.15) is $\epsilon^{\text{inter}}(\omega, \mathbf{q}\to 0)$ — the contribution of interband transitions.

In expression (1.15)

$$\tilde{\omega}_p^2 = \frac{8\pi e^2}{3\Omega} \sum_{\mathbf{k},\lambda} |\mathbf{v}_{\mathbf{k}\lambda}|^2 \, \delta(E_{\mathbf{k}\lambda} - E_F), \tag{1.17}$$

E_F is the Fermi energy, while

$$\mathbf{v}_{\mathbf{k}\lambda} = \left\langle \mathbf{k}\lambda \left| \frac{\hbar \nabla}{im} \right| \mathbf{k}\lambda \right\rangle = \frac{1}{\hbar} \frac{\partial E_{\mathbf{k}\lambda}}{\partial \mathbf{k}} \tag{1.18}$$

is the mean velocity of an electron in state $|\mathbf{k}\lambda\rangle$. Description of intraband transitions within the framework of our single-frequency approach can be refined by considering electron–phonon interaction processes, as well as other electron relaxation processes by introducing the corresponding collision frequency γ.

As a result of this, we arrive at the ordinary Drude–Ziener formula for the intrazone contribution to the dielectric constant $\epsilon(\omega)$

$$\varepsilon^{\text{intra}}(\omega) = 1 - \frac{\tilde{\omega}_p^2}{\omega(\omega + i\gamma)}. \tag{1.19}$$

In metals, the collision frequency is determined basically by electron–phonon interaction processes and can be computed from the electric resistance ρ at temperatures T above the Debye temperature by the following equation[6]:

$$\gamma = \tilde{\omega}_p^2 \cdot \rho \cdot 4\pi. \tag{1.20}$$

The imaginary portion of the dielectric constant $\epsilon_2(\omega)$ can also be represented, by applying equations (1.15) and (1.19), as the sum of the intraband and interband contributions

$$\epsilon_2(\omega) = \epsilon_2^{\text{intra}}(\omega) + \epsilon_2^{\text{inter}}(\omega), \tag{1.21}$$

where

$$\varepsilon_2^{\text{intra}}(\omega) = \frac{\gamma}{\omega} \frac{\tilde{\omega}_p^2}{\omega^2 + \gamma^2} \tag{1.22}$$

and

$$\varepsilon_2^{\text{inter}}(\omega) = \frac{4\pi^2}{3\omega^2\Omega} \sum_{\mathbf{k},\lambda' \neq \lambda} |\langle \mathbf{k}\lambda' | \mathbf{j} | \mathbf{k}\lambda \rangle|^2 f_{\mathbf{k}\lambda} (1 - f_{\mathbf{k}\lambda'})\, \delta(E_{\mathbf{k}\lambda'} - E_{\mathbf{k}\lambda} - \hbar\omega). \tag{1.23}$$

Substituting equations (1.21)–(1.23) into the f–sum rule, we obtain

$$\tilde{\omega}_p^2 + \frac{8\pi}{3\Omega} \sum_{\mathbf{k},\lambda' \neq \lambda} \frac{f_{\mathbf{k}\lambda} - f_{\mathbf{k}\lambda'}}{E_{\mathbf{k}\lambda'} - E_{\mathbf{k}\lambda}} |\langle \mathbf{k}\lambda' | \mathbf{j} | \mathbf{k}\lambda \rangle|^2 \approx \omega_p^2. \tag{1.24}$$

Here ω_p^2 is determined by the density of those electrons which are considered in the system of filled bands. If only valent electrons are included in the number of filled bands, then ω_p^2 is determined by their density. If we include in our analysis also the core electrons (for example, the $4d$ electrons we analyzed in Cd), then ω_p^2 must be determined considering these electrons as well.

1.3. Band Computations and Determination of Optical Properties of Metals

It follows from (1.14) that calculation of the dielectric constant in the single-frequency approximation requires that we be able to calculate both the electron band energies at an arbitrary $\mathbf{k}$ point of the Brillouine zones, and the matrix elements of the current operator with respect to the wave functions $|\mathbf{k}\lambda\rangle$ and $|\mathbf{k}\lambda'\rangle$.

At present there are various methods of calculating the band structure $E_{\mathbf{k}\lambda}$ of metals, differing from each other both in the selection of the basis of the wave functions, and in the speed of numerical computer calculations. For simple metals, the pseudopotential method works, analyzing the periodic potential of the crystalline lattice as a perturbation. More complex methods, such as linear combination of atomic orbitals, the KKR method and, finally, the LMTO method, allow electron energies to be calculated both in the bands, which can be considered to consist of nearly free electrons, as well as the energy of electrons with a high degree of localization. It has therefore become possible to calculate the band structure not only of simple metals, but also of substances with more complex energy spectra, such as the transition and noble metals. The various methods usually yield results which agree rather well. Up to the present time, the improvement of all methods for calculating the band structure has been directed toward increasing the accuracy of computation of the energy spectrum $E_{\mathbf{k}\lambda}$.

The status of calculating current matrix elements is not as satisfactory. Due to the variational nature of most modern band calculations, the approximate wave functions $|\mathbf{k}\lambda\rangle$ obtained as a result of such calculations always have lower accuracy than the calculated energies $E_{\mathbf{k}\lambda}$. For this reason, the matrix current element in all such calculations is computed with lower accuracy than $E_{\mathbf{k}\lambda}$. Furthermore, we see that in some approximations, yielding reasonable results for the band structure (for example, the canonical d–zone model), the matrix current elements may be equal to zero, making the interband component of the dielectric constant also equal to zero. These examples show how necessary caution is in computing the matrix current elements.

Some authors have attempted to avoid this difficulty by assuming the matrix element to be a constant independent of **k**. For example, Christensen[7-10] used the method of augmented plane waves considering relativistic corrections to compute the optical and photoemission properties of Ag, Au, Rh and Pd, assuming the matrix element to be constant. Pickett and Allen[11] used this same method to calculate the so–called optical state density (proportional to the imaginary portion of the dielectric constant in the constant matrix element approximation) for Nb and Mo. In particular, they noted the presence of a low–frequency peak in the optical state density of niobium $\hbar\omega{\sim}0.2$ eV, the presence of which was confirmed later by us in the framework of a more accurate computation.

However, the constant matrix element approximation, first of all, does not allow determination of the absolute values of function $\epsilon(\omega)$ and, secondly, as has been repeatedly noted[12, 13], in actuality the matrix elements are highly dependent on **k**. Therefore, they must be considered.

One example of the calculation of optical properties considering the matrix element is the work of Kubo, et al.[14, 15]. They calculated the optical characteristics of the $3d$–transition metals Cr, Fe and Ni and the $4d$ metals Nb and Mo. In general, the dielectric constants which they obtained agree rather well with the experimental data, but the number of singularities on the calculated curves is significantly greater than are observed experimentally. This is apparently a result of the need to consider electron–phonon and electron–scattering, leading to a finite survival time of electron levels. The augmented plane wave method, considering the matrix element, was also applied by other authors[16] to compute photoelectron emission in the metals W, Mo and Ta.

The pseudopotential methods have been repeatedly used to compute the optical properties of various metals (cf. reviews [17, 18]). In addition to the traditional use of the pseudopotential method to calculate the properties of simple metals, attempts have been made to use the nonlocal pseudopotential to calculate the optical characteristics of metals with more complex energy spectrum. For example, in [19] the optical properties of V, Nb and Ta were calculated. Kasovski[20] attempted to calculate the optical conductivity of Zn and Cd. He constructed a nonlocal pseudopotential from experimental data on the Fermi surface and obtained fair agreement with experimental data in the area $\hbar\omega \lesssim 0.5$ eV. However, the band structure of Zn and Cd which he computed differs significantly both from the energy spectrum of the elements obtained by other methods[21], and from experimental data on photoemissions[22]. The agreement of optical conductivity with the experimental data in the low-frequency area resulted from the fact that the optical transitions involving *d* bands in zinc and cadmium, which are poorly reproduced by the pseudopotential method, occur only at rather high energy ($\hbar\omega \sim 10$ eV), while in the low-frequency area these metals can be considered simple bivalent metals.

Smith, et al., used a model Hamiltonian method to calculate the optical properties of metals[23-28]. The matrix parameters of the model Hamiltonian $\hat{H}(\mathbf{k})$ were adjusted so that the band spectrum $E_{\mathbf{k}\lambda}$ obtained by diagonalizing these matrices corresponded to the results of more accurate first-principle band calculations. The following equation was used for the current operator

$$\hat{\mathbf{j}} = \frac{e}{\hbar} \frac{\partial \widehat{H}(\mathbf{k})}{\partial \mathbf{k}}. \qquad (1.25)$$

The advantage of this calculation plan is that it makes it possible to construct a model Hamiltonian matrix with comparatively few dimensions, yielding a true energy spectrum. This allows all operations with these matrices to be performed rapidly by computers. This calculation planned from "second principles" (i.e., with initial adjustment of the matrix) was used to calculate the optical dielectric constants of Cu, Ag, Au, Rh, Pd, as well as the alloys $Ag_{1-x}Cd_x$, ferromagnetic Ni, etc. The results of these calculations agree with the experimental data. Photoemission spectra, magnetooptic properties, etc. were also studied.

The KKR method is presently one of the most common methods used to calculate band structures. The nonself-consistent formulation of this method has been used, for example, to calculate the optical properties of copper[29] with the Chodorov potential and palladium with the Gaspar potential[30]. Calculation of the imaginary portion of the dielectric constant of copper[29] agreed rather well with experimental data restored by the Kramers-Kronig method in the energy area $\hbar\omega \lesssim 6$ eV.

Recently, increasing attention has been given to the so-called self-consistent band calculations. Self-consistent systems can be constructed by various methods. In particular, Callaway, et al., [12, 31, 32] constructed a self-consistent band calculation based on the classical LCAO method, and used this method to compute the optical properties of Na, Fe, Ni and paramagnetic Cr. Exchange interaction was considered within the framework of the Slater $X\alpha$ approximation, and the basis functions used were the 11-*s*, 7-*p* and 5-*d* gaussian orbitals. It was shown, in particular, that the matrix elements of momentum vary strongly with **k**.

An attempt to construct a self-consistent KKR system was undertaken in 1975 by Janak, Williams and Moruzzi[33]. Self-consistency was introduced in the $X\alpha$ approximation, and renormalization of the spectrum due to multiple-electronic effects was considered by introducing the parameter λ

$$\widetilde{E}_n(\mathbf{k}) = E_n(\mathbf{k}) + \lambda \cdot (E_n(\mathbf{k}) - E_F).$$

The parameter α was selected from the requirement of best agreement of calculated values with experimental data on the Fermi surface, while λ was selected on the basis of best agreement of data on photoemissions. The resulting $\epsilon_2(\omega)$ curve of copper in the $\hbar\omega \sim 0$–6 eV agrees with the experimental data better than the curve calculated with the Chodorov potential. The matrix elements of the momentum operator were selected in the form of integrals within a muffin-tin sphere

$$\mathbf{p}_{nm} = -i\hbar \int \frac{\Psi_n^*(r)\, \nabla V(r)\, \Psi_m(r)}{E_m - E_n}\, d\mathbf{r}, \tag{1.26}$$

where $V(\mathbf{r})$ is the self-consistent effective muffin-tin potential. Expression (1.26) is accurate in Bloch functions within the framework of the muffin-tin approximation. Expression (1.26) for the momentum operator matrix element was extended by Chen[34] to the case of variation of effective potential as a function of energy E angular momentum modulus l.

Better KKR self-consistency was obtained by Janak, Williams and Moruzzi[21] based on the electron-density functional of Hohenberg, Kohn and Sham[35, 36]. Using the muffin-tin approximation and approximating the local density functional in the form of Lundquist and Hedin[37], they calculated the self-consistent single-frequency potential of 32 metals, as well as their band structures, position of Fermi level, bond energy and other characteristics. Self-consistency was based on density. The optical properties were not analyzed.

However, in spite of all the advantages of the KKR method, it is rather cumbersome and requires a great deal of machine time for numerical calculations of the band structure, since the structural constants included in the equations used to determine the energy spectrum depend on energy and must be constantly recomputed.

The search for simpler band calculation systems, required for self-sufficiency, led to the creation of the atomic sphere approximation method[38] and the LMTO method, suggested by Andersen in 1975[39], which has been used repeatedly to calculate self-consistent potentials, band structures, Fermi surfaces and magnetic properties of both cubic and hexagonal metals[40-42]. The LMTO method has advantages over the KKR method in that it utilizes structural constants which do not depend on energy. This allows the band calculation problem to be reduced to computation of the natural values of a matrix. The time necessary for numerical band structure calculation is thus decreased by 10^2–10^3 times. The narrow *d* bands are somewhat better reproduced by this method than the more "blurred" *p* and *f* bands. However, concrete calculations have shown that this accuracy is quite sufficient for determination of many properties of metals.

The numerical calculations of the band structure of metals which we performed utilize the LMTO method. The single-frequency muffin-tin potential was calculated in a local approximation of the electron-density functional method by a self-consistent method. The procedure used in these calculations is described in detail in monograph[21]. 10–15 iterations were usually sufficient to yield a self-consistent potential. The wave functions and energy spectrum found were used to compute the optical properties. We shall discuss this in more detail in the next chapter.

CHAPTER 2
NUMERICAL CALCULATION OF OPTICAL PROPERTIES OF METALS

2.1. Determination of the Current Operator Matrix Element

Before going over to a description of the specific procedure for calculating the dielectric constant, we should like to discuss one important and rather basic segment of the calculation — computation of the current operator matrix elements, which were included in the expression for $\epsilon_2(\omega)$ in the single-frequency approximation. As is well known, a number of equivalent relations can be described for a current operator matrix element in block functions[43], but this equivalence is true only for precise wave functions. However, since all band calculations deal with approximate wave functions, the question arises as to which relation to select. In our opinion, the most logical path is to determine this matrix element by using a continuity equation relating the change of the **q** component of electron density operator with time to the **q** component of current

$$-i\dot{\rho}_{\mathbf{q}}=\mathbf{q}\mathbf{j}_{\mathbf{q}}. \tag{2.1}$$

The change in electron density is determined by the equation of motion

$$i\hbar\dot{\rho}_{\mathbf{q}}=[H,\ \rho_{\mathbf{q}}], \tag{2.2}$$

where

$$H=-\frac{\hbar^2\nabla^2}{2m}+V_{\mathrm{eff}}(\mathbf{r})$$

The current operator can, according to (2.1) and (2.2), be defined as follows:

$$\lim_{\mathbf{q}\to 0} \mathbf{j}_{\mathbf{q}} = \lim_{\mathbf{q}\to 0}\left(-\frac{\mathbf{q}}{\hbar\,|\,\mathbf{q}\,|^2}\right)[H,\ \rho_{\mathbf{q}}] = \frac{e\hbar\nabla}{im}. \tag{2.3}$$

Expression (2.3) for the current operator, which satisfies the continuity equation, which is quite important from the standpoint of the basic principles of the theory, is also generally used to calculate the matrix elements $j^{\mathbf{k}}_{\lambda\lambda'}$

$$\mathbf{j}^{\mathbf{k}}_{\lambda\lambda'} = \left\langle \mathbf{k}\lambda \left| \frac{e\hbar\nabla}{im} \right| \mathbf{k}\lambda' \right\rangle. \tag{2.4}$$

Here the quantity $|\mathbf{k}\lambda\rangle$ is the approximate wave function of an electron in band λ, found within the framework of some method of band structure computation. How correct is this approach?

In the overwhelming majority of band calculations, the wave function is taken as a linear combination of basis functions $|\mathbf{k}n\rangle$, which we shall consider orthonormalized for simplicity

$$|\mathbf{k}\lambda\rangle = \sum_n a_n^{\mathbf{k}\lambda}\,|\mathbf{k}n\rangle. \tag{2.5}$$

The coefficients $a_n^{\mathbf{k}\lambda}$ and the eigenvalues $E_{\mathbf{k}\lambda}$ are determined by the system of matrix equations

$$\sum_{n'} \langle \mathbf{k}n\,|\,H\,|\,\mathbf{k}n'\rangle\, a_{n'}^{\mathbf{k}\lambda} = E_{\mathbf{k}\lambda} a_n^{\mathbf{k}\lambda}, \tag{2.6}$$

where summation with respect to n' is limited by the finite number of basis functions. This approach to calculation of the wave functions $|\mathbf{k}\lambda\rangle$ and energies $E_{\mathbf{k}\lambda}$ means that these quantities are sought not throughout Hilbert space, but in the subspace of a finite number of basis functions $|\mathbf{k}n\rangle$. This subspace can be obtained by projection with the operator

$$P = \sum_{\mathbf{k}n} |\,\mathbf{k}n\rangle\,\langle\mathbf{k}n\,|,$$

which differs from one as the incompleteness of the basis.

The wave functions $|\mathbf{k}\lambda\rangle$ are the eigenfunctions of the projection of Hamiltonian $\tilde{H}=PHP$, not of the Hamiltonian H itself. The electron density operator in this subspace is $\tilde{\rho}_{\mathbf{q}}=P\rho_{\mathbf{q}}P$. If we now define the effective current operator as in (2.3), we obtain

$$\tilde{\mathbf{j}}_{\mathbf{q}} = \lim_{\mathbf{q}\to 0}\left(-\frac{\mathbf{q}}{\hbar\,|\,\mathbf{q}\,|^2}\right)[\tilde{H},\, \tilde{\rho}_{\mathbf{q}}]. \tag{2.7}$$

Due to the nonlinearity of the commutation operation with respect to the projection operator, the current $\tilde{\mathbf{j}}$ determined according to (2.7) is generally not equal to the projection of the operator $e\hbar\nabla/im$ and, accordingly,

$$\tilde{\mathbf{j}}^{\mathbf{k}}_{\lambda\lambda'} \neq \left\langle \mathbf{k}\lambda \left| \frac{e\hbar\nabla}{im} \right| \mathbf{k}\lambda' \right\rangle.$$

Let us study this problem in greater detail and calculate the commutator

$$\lim_{\mathbf{q}\to 0} \frac{\mathbf{q}}{\hbar\,|\,\mathbf{q}\,|^2}[\tilde{H},\, \tilde{\rho}_{\mathbf{q}}] = \lim_{\mathbf{q}\to 0} \frac{\mathbf{q}}{\hbar\,|\,\mathbf{q}\,|^2} \sum_{\substack{\mathbf{k}_1,\mathbf{k}_2\\ n_1,\dots,n_4}} \Big[|\,\mathbf{k}_1 n_1\rangle\, H^{\mathbf{k}_1}_{n_1 n_2}\, \langle \mathbf{k}_1 n_2\,|,$$

$$|\,\mathbf{k}_2+\mathbf{q},\, n_3\rangle\, \rho^{\mathbf{k}_2}_{\mathbf{q},n_3 n_4}\, \langle \mathbf{k}_2 n_4\,|\Big] = \frac{1}{\hbar}\sum_{\mathbf{k},nn'} |\,\mathbf{k}n\rangle \left\{ \frac{\partial H^{\mathbf{k}}_{nn'}}{\partial \mathbf{k}} + \right. \tag{2.8}$$

$$\left. + \lim_{\mathbf{q}\to 0} \sum_{n_1} \left[H^{\mathbf{k}}_{nn_1} \frac{\partial \rho_{\mathbf{q},n_1 n'}}{\partial \mathbf{q}} - \frac{\partial \rho_{\mathbf{q},nn_1}}{\partial \mathbf{q}} H^{\mathbf{k}}_{n_1 n'} \right] \right\} \langle \mathbf{k}n'\,|.$$

Here we have introduce the symbols

$$H^{\mathbf{k}}_{nn'} = \langle \mathbf{k}n\,|\,H\,|\,\mathbf{k}n'\rangle$$

and

$$\rho^{\mathbf{k}}_{\mathbf{q},nn'} = \langle \mathbf{k}+\mathbf{q},\, n\,|\,\rho_{\mathbf{q}}\,|\,\mathbf{k}n'\rangle,$$

where $\rho_{\mathbf{q}}=e^{i\mathbf{q}\mathbf{r}}$, and considered that $\rho^{\mathbf{k}}_{\mathbf{q}=0,\,nn'}=\langle \mathbf{k}n\,|\,\mathbf{k}n'\rangle=\delta_{nn'}$. An expression just like (2.8) is obtained for the effective current if we define the current as follows::

$$\tilde{\mathbf{j}} = -\frac{e}{\hbar}\frac{\delta\tilde{H}\,(\mathbf{A})}{\delta\mathbf{A}}, \tag{2.9}$$

where $\mathbf{A}$ is the vector potential, introduced to the Hamiltonian by a calibration-invariant method.

Computing the matrix elements of current $\tilde{\mathbf{j}}$, we obtain

$$\tilde{\mathbf{j}}^{\mathbf{k}}_{\lambda\lambda'} = \langle \mathbf{k}\lambda | \tilde{\mathbf{j}} | \mathbf{k}\lambda' \rangle = -\frac{1}{\hbar} \sum_{nn'} (a_n^{\mathbf{k}\lambda})^* \times \times a_{n'}^{\mathbf{k}\lambda'} \left\{ \frac{\partial H^{\mathbf{k}}_{nn'}}{\partial \mathbf{k}} + (E_{\mathbf{k}\lambda} - E_{\mathbf{k}\lambda'}) \lim_{\mathbf{q}\to 0} \frac{\partial \rho^{\mathbf{k}}_{\mathbf{q},nn'}}{\partial \mathbf{q}} \right\}. \tag{2.10}$$

We note immediately that the work of Smith[26] utilized determination of the matrix element of the current operator without considering the second component in expression (2.10). This approach, as we can see from our previous discussion, cannot be considered internally consistent. Furthermore, without additional analysis of accuracy, it may lead to significant factual errors.

It is convenient for further transformation of matrix element $\mathbf{j}^{\mathbf{k}}_{\lambda\lambda'}$ to isolate the factor $e^{i\mathbf{kr}}$ explicitly in the wave functions $|\mathbf{k}\lambda\rangle$ and the basis $|\mathbf{k}n\rangle$

$$|\mathbf{k}\lambda\rangle = e^{i\mathbf{kr}} | u_{\mathbf{k}\lambda} \rangle,$$
$$|\mathbf{k}n\rangle = e^{i\mathbf{kr}} | u_{\mathbf{k}n} \rangle.$$

We then easily see that

$$\frac{\partial H^{\mathbf{k}}_{nn'}}{\partial \mathbf{k}} = \left\langle \mathbf{k}n \left| \frac{i\hbar^2 \nabla}{m} \right| \mathbf{k}n' \right\rangle + \left\langle \frac{\partial u_{\mathbf{k}n}}{\partial \mathbf{k}} \right| e^{-i\mathbf{kr}} H | \mathbf{k}n' \Big\rangle + \Big\langle \mathbf{k}n | H e^{i\mathbf{kr}} \left| \frac{\partial u_{\mathbf{k}n'}}{\partial \mathbf{k}} \right\rangle \tag{2.11}$$

and

$$\lim_{\mathbf{q}\to 0} \frac{\partial \rho^{\mathbf{k}}_{\mathbf{q},nn'}}{\partial \mathbf{q}} = \left\langle \frac{\partial u_{\mathbf{k}n}}{\partial \mathbf{k}} \middle| u_{\mathbf{k}n'} \right\rangle = -\left\langle u_{\mathbf{k}n} \middle| \frac{\partial u_{\mathbf{k}n'}}{\partial \mathbf{k}} \right\rangle. \tag{2.12}$$

Using (2.11) and (2.12), we obtain the final expression

$$\tilde{\mathbf{j}}^{\mathbf{k}}_{\lambda\lambda'} = \mathbf{j}^{\mathbf{k}}_{\lambda\lambda'} + \delta\mathbf{j}^{\mathbf{k}}_{\lambda\lambda'} = \left\langle \mathbf{k}\lambda \left| \frac{e\hbar\nabla}{im} \right| \mathbf{k}\lambda' \right\rangle + \left(-\frac{e}{\hbar}\right) \sum_n (a_n^{\mathbf{k}\lambda})^* \times \times \left\{ \left\langle \frac{\partial u_{\mathbf{k}n}}{\partial \mathbf{k}} \right| e^{-i\mathbf{kr}} (H - E_{\mathbf{k}\lambda'}) | \mathbf{k}\lambda' \Big\rangle + \Big\langle \mathbf{k}\lambda | (H - E_{\mathbf{k}\lambda}) e^{i\mathbf{kr}} \left| \frac{\partial u_{\mathbf{k}n}}{\partial \mathbf{k}} \right\rangle \right\} a_n^{\mathbf{k}\lambda'}. \tag{2.13}$$

The last two components in (2.13) are corrections to the normally used expression (2.4) for operator current matrix elements. They can be written as

$$\delta \mathbf{j}^{\mathbf{k}}_{\lambda\lambda'} = \left(-\frac{e}{\hbar}\right) \sum_n (a_n^{\mathbf{k}\lambda})^* \left\{ \left\langle \frac{\partial u_{\mathbf{k}n}}{\partial \mathbf{k}} \right| e^{-i\mathbf{k}\mathbf{r}} (I - P)(H - E_{\mathbf{k}\lambda'}) \left| \mathbf{k}\lambda' \right\rangle + \right.$$

$$\left. + \left\langle \mathbf{k}\lambda \right| (H - E_{\mathbf{k}\lambda}) (I - P) e^{i\mathbf{k}\mathbf{r}} \left| \frac{\partial u_{\mathbf{k}n}}{\partial \mathbf{k}} \right\rangle \right\} a_n^{\mathbf{k}\lambda'}, \tag{2.14}$$

where I is a unit operator, while P is a projection operator.

Let us analyze the results obtained. Obviously, the corrections $\delta \mathbf{j}^{\mathbf{k}}_{\lambda\lambda'}$ disappear if the system of functions $|\mathbf{k}n\rangle$ is full (projection operator equal to unit operator) or if the functions $|\mathbf{k}\lambda\rangle$ and $|\mathbf{k}\lambda'\rangle$ are precise eigenfunctions of the Hamiltonian H. The corrections also disappear if functions $u_{\mathbf{k}\lambda}(\mathbf{r})$ are independent of $\mathbf{k}$. This occurs in the case of a plane wave basis, where

$$|\mathbf{k}n\rangle = \frac{1}{\sqrt{V}} e^{i(\mathbf{k}+\mathbf{G}_n)\mathbf{r}},$$

where $\mathbf{G}_n$ is the vector of the reciprocal lattice. However, this case is an exception. In particular, the corrections are not equal to zero for bases constructed in the localized orbitals

$$|\mathbf{k}n\rangle = \sum_{\mathbf{R}} e^{i\mathbf{k}\mathbf{R}} \Phi_n(\mathbf{r} - \mathbf{R}),$$

used in many band structure calculation methods.

The presence in expression (2.13) of corrections for wave–function inaccuracy should not be surprising. Possibly more unexpected is the fact that the accuracy of calculation of matrix current elements can in many cases be quite poor (without considering correcting terms $\delta \mathbf{j}^{\mathbf{k}}_{\lambda\lambda'}$) in spite of very accurate calculation of the energy spectrum. Usually this disagreement is caused by the discontinuous nature of the wave function or its derivative. Let us study two characteristic examples.

It is well known that the KKR method where $l_{max}=2$ (l is the orbital moment) allows highly accurate determination of the band structure of a metal. In[34], the matrix elements $\left\langle \mathbf{k}\lambda \left| \frac{\hbar\nabla}{im} \right| \mathbf{k}\lambda' \right\rangle$ were computed for copper with $l_{max}=2$ and $l_{max}=4$ at points of high symmetry. It was demonstrated that these matrix elements differ in the two cases in question by a factor of 2–3. Since the continuous periodic function $|\mathbf{k}\lambda\rangle$ should contain spherical harmonics with arbitrarily large l, clearly where $l_{max}=2$ the wave function will have significantly greater discontinuities than where $l_{max}=4$. This factor leads to a significant change (increase) in the matrix elements $\left\langle \mathbf{k}\lambda \left| \frac{\hbar\nabla}{im} \right| \mathbf{k}\lambda' \right\rangle$ where $l_{max}=4$ in comparison with $l_{max}=2$. A still clearer example is the model of a canonical d band, in which the wave function contains only spherical harmonics with $l=2$. We know[39] that this model describes the energy spectrum of a metal in the area of the d band rather well. As the same time, the matrix elements $\left\langle \mathbf{k}\lambda \left| \frac{\hbar\nabla}{im} \right| \mathbf{k}\lambda' \right\rangle$ for it are precisely equal to zero, and without considering corrections $\delta \mathbf{j}^{\mathbf{k}}_{\lambda\lambda'}$, in this case it is simply impossible to describe the optical characteristics of the metal.

Let us briefly analyze other possible representations for nondiagonal current operator matrix elements, used in many works to compute the contribution of interband transitions to $\epsilon_2(\omega)$, namely

$$\mathbf{j}^{\mathbf{k}}_{\lambda\lambda'} = \langle \mathbf{k}\lambda | \mathbf{r} | \mathbf{k}\lambda' \rangle (E_{\mathbf{k}\lambda} - E_{\mathbf{k}\lambda'}) \frac{1}{\hbar i}, \tag{2.15}$$

$$\mathbf{j}^{\mathbf{k}}_{\lambda\lambda'} = \left\langle \mathbf{k}\lambda \left| \frac{\nabla V(\mathbf{r})}{\hbar i} \right| \mathbf{k}\lambda' \right\rangle \frac{1}{E_{\mathbf{k}\lambda} - E_{\mathbf{k}\lambda'}}. \tag{2.16}$$

It is not difficult to show, using the commutator derivation of the corresponding relations for the current operator, that (2.15) and (2.16), in addition to corrections of the same type as in equation (2.14), actually do not contain the additional corrections

$$\frac{1}{\hbar i} \{\langle \mathbf{k}\lambda | (H - E_{\mathbf{k}\lambda}) (I - P) \mathbf{r} | \mathbf{k}\lambda' \rangle - \langle \mathbf{k}\lambda | \mathbf{r} (I - P) (H - E_{\mathbf{k}\lambda'}) | \mathbf{k}\lambda' \rangle\}$$

for formula (2.15) and

$$\frac{1}{\hbar i} \{\langle \mathbf{k}\lambda | (H - E_{\mathbf{k}\lambda}) (I - P) \nabla V(\mathbf{r}) | \mathbf{k}\lambda' \rangle - \langle \mathbf{k}\lambda | \nabla V(\mathbf{r}) (I - P) (H - E_{\mathbf{k}\lambda'}) | \mathbf{k}\lambda' \rangle\}$$

for (2.16). Thus, equations (2.15) and (2.16) are not more accurate than (2.4).

In conclusion to this section we note briefly that the problem of self-consistent determination of the current operator and its matrix elements does not arise just in the calculation of the optical properties of metals. The same problems also exist in calculating the forces of oscillators in atoms[44], the probability of absorption of x-rays in metals[45] and in a number of other problems. Thus, for example, in a work by Volkov et al., [46], in an analysis of the possibility of existence of spontaneous currents in the system, a self-consistent definition of the current operator was suggested within the Kohn-Latinger representation for a finite number of energy bands.

2.2. Computation of Current Operator Matrix Element by LMTO Method

As we discussed in the previous section, the corrections $\delta\mathbf{j}^{\mathbf{k}}_{\lambda\lambda'}$ to the matrix element of the current operator $\mathbf{j}^{\mathbf{k}}_{\lambda\lambda'}=\left\langle\mathbf{k}\lambda\left|\frac{e\hbar\nabla}{im}\right|\mathbf{k}\lambda'\right\rangle$ disappear when precise wave functions of the Hamiltonian $H=-\frac{\hbar^2\nabla^2}{2m}+V(\mathbf{r})$ are used as $|\mathbf{k}\lambda\rangle$. The value of these corrections in approximate calculations of functions $|\mathbf{k}\lambda\rangle$ can be computed directly by using equation (2.14). We shall use a different procedure to estimate the value of these corrections. We note first of all that it follows from the definition of a current operator matrix element (2.14) that the quantity $\tilde{\mathbf{j}}^{\mathbf{k}}_{\lambda\lambda}$ satisfies the well-known equation

$$\tilde{\mathbf{j}}^{\mathbf{k}}_{\lambda\lambda} = \frac{e}{\hbar}\frac{\partial E_{\mathbf{k}\lambda}}{\partial \mathbf{k}}. \tag{2.17}$$

This means that the difference of matrix element $\left\langle\mathbf{k}\lambda\left|\frac{e\hbar\nabla}{im}\right|\mathbf{k}\lambda\right\rangle$ from $\frac{e}{\hbar}\frac{\partial E_{\mathbf{k}\lambda}}{\partial \mathbf{k}}$ results from the presence of the correction $\delta\mathbf{j}^{\mathbf{k}}_{\lambda\lambda'}$. The value of the ratio

$$\left|\left\langle\mathbf{k}\lambda\left|\frac{e\hbar\nabla}{im}\right|\mathbf{k}\lambda\right\rangle\right| \Big/ \left|\frac{e}{\hbar}\frac{\partial E_{\mathbf{k}\lambda}}{\partial \mathbf{k}}\right|$$

in this case gives us the value of these corrections.

Let us briefly study the computation of the matrix element $\mathbf{j}^{\mathbf{k}}_{\lambda\lambda'}=\left\langle\mathbf{k}\lambda\left|\frac{e\hbar\nabla}{im}\right|\mathbf{k}\lambda'\right\rangle$ by the LMTO method. The LMTO wave function is[39]

$$\psi_{\mathbf{k}\lambda}(\mathbf{r}) = \sum_L a_L^{\mathbf{k}\lambda}\chi_L^{\mathbf{k}}(\mathbf{r}), \tag{2.18}$$

where $\chi_L^{\mathbf{k}}(\mathbf{r})=\sum_{\mathbf{R}} e^{i\mathbf{k}\mathbf{R}}\chi_L(\mathbf{r}-\mathbf{R})$ is the Bloch sum of the MT orbitals. In the atomic sphere approximation

$$\chi_L^{\mathbf{k}(\mathrm{ASA})}(\mathbf{r}) = \sum_{L'}^{l'\leqslant l_{\max}} \left[\pi^{\mathbf{k}}_{LL'}\Phi_{L'}(\mathbf{r}, E_\gamma) + \Omega^{k}_{LL'}\mathring{\Phi}_{L'}(\mathbf{r}, E_\gamma)\right], \tag{2.19}$$

the coefficients $\pi^{k}_{LL'}$ and $\Omega^{k}_{LL'}$ are presented in a work by Andersen[39] (equations 4.21*a*, 4.21*b*), while $\Phi_L(\mathbf{r}, E_\gamma)$ and $\dot{\Phi}_L(\mathbf{r}, E_\gamma)$ are the solution of the rating of the Schroedinger equation in one location with energy E_γ and its derivative with respect to energy. The matrix elements $\mathbf{j}^{\mathbf{k}}_{\lambda\lambda'}$ in the atomic sphere approximation are equal to

$$\mathbf{j}^{\mathbf{k}(ASA)}_{\lambda\lambda'} = \sum_{LL'} (a_L^{\mathbf{k}\lambda})^* a_{L'}^{\mathbf{k}\lambda'} \left\langle \chi_L^{\mathbf{k}(ASA)} \left| \frac{e\hbar\nabla}{im} \right| \chi_{L'}^{\mathbf{k}(ASA)} \right\rangle, \tag{2.20}$$

where the angle brackets represent integration over the atomic sphere. They are not difficult to compute, using equations defining the action of operator ∇ on the function $f(|\mathbf{r}|)Y_L(\hat{\mathbf{r}})$ (for example, cf. [47]). We should note that the function $\chi_L^{\mathbf{k}(ASA)}$ is well defined only in an atomic sphere of radius s (Wigner–Setiz sphere). If we extend this function over the entire space, then, due to the break in the sum in equation (2.19) at l_{max}, this function is discontinuous, which essentially influences the satisfaction of the continuity equation.

Table 1 presents the results of calculating the equation

$$\xi_\lambda = (\mathbf{j}^{\mathbf{k}}_{\lambda\lambda})_x \Big/ \left| \frac{e}{\hbar} \frac{\partial E_{\mathbf{k}\lambda}}{\partial \mathbf{k}_x} \right|$$

representing satisfaction of the continuity equation for intraband transitions. Calculations were performed for copper with a potential taken from [21] for $\mathbf{k}$=(o.2, 0.1, 0.05)$\cdot 4\pi/a$.

As we can see from the table, the dispersion of $E_{\mathbf{k}\lambda}$ in the lower six bands for l_{max}=2 and l_{max}=3 remains almost identical, and the values of $\frac{e}{\hbar}\frac{\partial E_{\mathbf{k}\lambda}}{\partial \mathbf{k}}$ also differ very little, while the difference in $\mathbf{j}^{\mathbf{k}}_{\lambda\lambda}$ where l_{max}=2 and l_{max}=3 is significant. In higher bands where l_{max}=3, the quantity $\mathbf{j}^{\mathbf{k}}_{\lambda\lambda'}$, as before, differs significantly from $\frac{e}{\hbar}\frac{\partial E_{\mathbf{k}\lambda}}{\partial \mathbf{k}}$.

Within the framework of the LMTO method, continuity of the wave function and its derivative can be achieved if, following the lead of Andersen[39], we replace the MT orbital $\chi_L^{\mathbf{k}(ASA)}$ by the orbital $\chi_L^{\mathbf{k}(LMTO)}$

$$\chi_L^{\mathbf{k}(\mathrm{LMTO})}(\mathbf{r}) = \left[\chi_L^{\mathbf{k}(\mathrm{ASA})}(\mathbf{r}) - \tilde{\chi}_L^{\mathbf{k}(\mathrm{ASA})}(\mathbf{r})\right]\theta(s - |\mathbf{r}|) + \chi_L^{\mathbf{k}(\mathrm{ASA})}(|\mathbf{r}| = s) \times$$
$$\times \sum_{\mathbf{G}} e^{i(\mathbf{k}+\mathbf{G})\mathbf{r}} F_L(\mathbf{k}+\mathbf{G}). \tag{2.21}$$

Here

$$F_L(\mathbf{k}+\mathbf{G}) = 3(2l+1)(2l+3)\frac{j_{l+1}(|\mathbf{k}+\mathbf{G}|s)}{(|\mathbf{k}+\mathbf{G}|s)^3} Y_L(\widehat{\mathbf{k}+\mathbf{G}}) \tag{2.22}$$

and $\tilde{\chi}_L^{\mathbf{k}(\mathrm{ASA})}$ is the Bloch sum (2.19) for the crystalline potential, which is equal to zero, formalized so that $\chi_L^{\mathbf{k}(\mathrm{ASA})}(|\mathbf{r}|=s) = \tilde{\chi}_L^{\mathbf{k}(\mathrm{ASA})}(|\mathbf{r}|=s)$.

Table 1

Test for Satisfaction of Continuity Equation in Atomic Sphere Approximation

λ	E_λ, Reed	ξ_λ	E_λ, Reed	ξ_λ	λ	E_λ, Reed	ξ_λ	E_λ, Reed	ξ_λ
	$l_{max}=2$		$l_{max}=3$			$l_{max}=2$		$l_{max}=3$	
1	0.107	0.987	0.107	1.011	9	2.826	0.506	2.679	0.498
2	0.335	−0.233	0.331	0.984	10			2.780	−0.277
3	0.381	−0.524	0.377	0.776	11			3.253	−0.088
4	0.397	0.439	0.393	1.101	12			3.737	0.072
5	0.428	0.863	0.424	1.040	13			4.083	0.279
6	0.456	0.486	0.425	1.010	14			4.241	0.307
7	1.981	0.644	1.810	0.659	15			4.785	−1.575
8	2.435	0.580	2.158	0.499	16			5.480	0.079

Table 2

Test for Satisfaction of Continuity Equation in LMTO Method

λ	E_λ, Reed	ξ_λ	E_λ, Reed	ξ_λ	λ	E_λ, Reed	ξ_λ	E_λ, Reed	ξ_λ
	$l_{max}=2$		$l_{max}=3$			$l_{max}=2$		$l_{max}=3$	
1	0.107	0.989	0.107	0.998	9	2.629	0.645	2.583	0.708
2	0.336	0.354	0.335	1.008	10			2.616	0.565
3	0.379	0.273	0.378	1.028	11			2.965	0.188
4	0.396	0.628	0.395	0.907	12			3.480	0.404
5	0.426	0.858	0.425	0.923	13			3.664	0.443
6	0.454	0.636	0.453	0.922	14			3.922	0.876
7	1.880	0.714	1.800	0.746	15			4.409	0.391
8	2.240	0.574	2.150	0.656	16			4.769	0.462

Using (2.21) and (2.22), it is not difficult to see that

$$\mathbf{j}_{\lambda\lambda'}^{\mathbf{k}(\mathrm{LMTO})} = \sum_{L,L'} \Big\{ \Big\langle \chi_L^{\mathbf{k}(\mathrm{ASA})} \Big| \frac{e\hbar\nabla}{im} \Big| \chi_{L'}^{\mathbf{k}(\mathrm{ASA})} \Big\rangle - \Big\langle \tilde{\chi}_L^{\mathbf{k}(\mathrm{ASA})} \Big| \frac{e\hbar\nabla}{im} \Big| \tilde{\chi}_{L'}^{\mathbf{k}(\mathrm{ASA})} \Big\rangle +$$

$$+ \frac{e\hbar}{m} (\chi_L^{\mathbf{k}(\mathrm{ASA})}(|\mathbf{r}| = s))^* \chi_{L'}^{\mathbf{k}(\mathrm{ASA})}(|\mathbf{r}| = s)\,\Omega \sum_{\mathbf{G}} F_L^*(\mathbf{k}+\mathbf{G})\,(\mathbf{k}+\mathbf{G}) \times$$

$$\times F_{L'}(\mathbf{k}+\mathbf{G}) \Big\} (a_L^{\mathbf{k}\lambda})^* a_{L'}^{\mathbf{k}\lambda'} . \tag{2.23}$$

The last sum in equation (2.23) is easily computed by a slight modification of the program for computing F–structured constants in the LMTO method[39]. Since the continuous wave function is used with this calculation method, the relation ξ_λ is much closer to one than in the atomic sphere approximation method, which is clearly seen from Table 2.

These results apparently provide a maximum estimate for the error in nondiagonal matrix elements $\mathbf{j}_{\lambda\lambda'}^{\mathbf{k}}(E_\lambda < E_F,\ E_{\lambda'} > E_F)$. For a numerical test of this assumption, we compared the matrix elements of optical transitions which we computed by the LMTO method $\mathbf{j}_{\lambda\lambda'}^{(\mathrm{LMTO})}$ ($l_{\max}=3$) for copper (with the potential from [21]) as points of high symmetry with results of KKR calculation ($l_{\max}=4$)[34]. This confirms our estimate (Table 3).

Table 3

Comparison of Matrix Elements of Momentum $\mathbf{p}_{\lambda\lambda'}^{\mathbf{k}}$ for Copper Computed by LMTO and KKR Methods[34].

Method	Transition						
	$L_1^l \to L_2'$	$L_3^l \to L_2'$	$L_3^u \to L_2'$	$L_2' \to L_1^u$	$X_1^l \to X_4'$	$X_5 \to X_4'$	$X_4' \to X_1^u$
$E_\lambda - E_{\lambda'}$, Reed	0.354	0.224	0.113	0.341	0.555	0.298	0.394
$\lvert\mathbf{p}_{\lambda\lambda'}\rvert$ (LMTO) ($l_{\max}=3$)	0.240	0.108	0.322	0.820	0.218	0.472	0.911
$\lvert\mathbf{p}_{\lambda\lambda'}\rvert$ (KKR) ($l_{\max}=4$)	0.214–0.220	0.081–0.104	0.302–0.384	0.857–0.843	0.201–0.203	0.350–0.350	0.977–0.963
Note. Results given for KKR method for various calculation equations.							

Thus, we have shown that, due to the approximate nature of computation of the wave function in band calculations, significant difficulties arise in determination of a current operator matrix element, required for calculation of the optical properties of metals. Possible means of eliminating these difficulties involve either direct consideration of the corrections $\delta \mathbf{j}^{\mathbf{k}}_{\lambda\lambda'}$ during calculation of optical constants, or improvement of the wave function, bringing it closer to the precise eigenfunction of the Hamiltonian of the system. In our opinion, the second method is preferable, since it allows us to consider multiple-frequency effects and the lifetime of electrons more simply. As comparison of the calculated quantities, even those calculated with incompletely accurate functions $|\mathbf{k}\lambda\rangle$, with experimental data shows us (cf. following sections), there is reasonable qualitative and even quantitative agreement in the position of most of the singularities of these curves. At the moment, unfortunately, it is still difficult to say what causes the difference in the amplitudes — inaccuracy in the wave functions or multiple-frequency effects. This will require further investigation both of the role of corrections $\delta \mathbf{j}^{\mathbf{k}}_{\lambda\lambda'}$, and of possible refinement of the functions $|\mathbf{k}\lambda\rangle$.

2.3. Description of Calculation Method

To provide a broad test of the capabilities of first-principle calculation of optical properties, interpretation of experimental data and a search for the relationships between electron structure and optical characteristics, we undertook calculation of $\epsilon(\omega)$, $R(\omega)$ and $L(\omega)$ over a broad frequency range $\hbar\omega \sim 0$–35 eV for 15 simple, noble and transition metals. The results of the calculations are shown below in Figures 2-16.

The optical characteristics were computed as follows. The interband component of the imaginary portion of the dielectric constant $\epsilon_2^{inter}(\omega)$ was calculated by equation (1.23). The energy structure $E_{\mathbf{k}\lambda}$ and wave functions $|\mathbf{k}\lambda\rangle$ were calculated within the framework of the LMTO method (with correcting term) with $l_{max}=3$, while the current operator matrix elements were calculated by equation (2.23). The single-frequency muffin-tin potentials included in Schroedingers equation (1.12) were computed within the framework of the local approximation of the density functional method in a self-consistent manner. The method of these calculations is described in detail in [21]. This method of computing current operator matrix elements provides a calculation accuracy of $|\mathbf{j}^{\mathbf{k}}_{\lambda\lambda'}|^2$ on the order of 10% in the frequency range $\hbar\omega \leqslant 15$ eV[13].

Expression (1.15) is generally correct only for metals with cubic lattices (simple cubic, body-centered cubic and face-centered cubic). For all these lattices it is sufficient to form integration with respect to the irreducible 1/48 portion of the Brillouine zone. In our case, integration was performed by the method of tetrahedrons, using 204 points in the 1/48 Brillouine zone for the bcc lattice and 175 points for the fcc lattice.

The real portion of $\epsilon_1^{inter}(\omega)$ was calculated by the Kramers-Kronig relation (1.2). Integration was performed at an interval of 0.01 Reed, with the upper limit the value of $\hbar\omega_{max}$, above with which the accuracy achieved by the machine $\epsilon_2^{inter}(\omega)=0$. Since in the LMTO method only a finite number of bands is computed (16 bands for $l_{max}=3$), this value of $\hbar\omega_{max}$ always exists (Table 4). The accuracy of integration by the Kramers-Kronig relation is no worse than 0.5% in this case. The parameters of the intraband component of dielectric constants $\tilde{\omega}_p$ and γ were calculated by the method of [6] (Table 4). Then, the full dielectric constant $\epsilon(\omega)$ was used to find the reflectivity $R(\omega)$ and electron energy-loss function $L(\omega)$.

Table 4
Calculation of Metal Parameters

Parameters	Metals							
	Al	Cu	Ag	Zn	Cd	V	Cr	Y
N_v	3	11	11	12	12	5	6	3
$\hbar\tilde{\omega}_p$, eV	12.741	9.687	7.741	10.820	9.411	8.078	6.270	4.684
$\hbar\gamma$, eV	0.052	0.019	0.016	0.085	0.080	0.176	0.063	0.158
$\hbar\omega_{max}$, eV	25	75	60	65	50	50	55	35
n_{eff} ($\hbar\omega_{max}$)	2.776	7.776	7.751	6.457	11.075	5.493	5.495	3.620

Parameters	Metals						
	Zr	Nb	Mo	Tc	Ru	Rh	Pd
N_v	4	5	6	7	8	9	10
$\hbar\tilde{\omega}_p$, eV	7.398	9.764	9.234	11.152	10.686	9.764	6.990
$\hbar\gamma$, eV	0.283	0.173	0.056	0.279	0.085	0.055	0.063
$\hbar\omega_{max}$, eV	40	40	45	45	50	50	55
n_{eff} ($\hbar\omega_{max}$)	4.928	6.037	7.600	8.827	9.829	10.682	11.193

Note. N_v is the number of valent electrons, $\tilde{\omega}_p$, γ represents the intraband conductivity parameters, $\hbar\omega_{max}$, $n_{eff}(\hbar\omega_{max})$ are the values for the metals computed.

A few words on calculation accuracy. As we know, the band structure in the energy area $E_{\mathbf{k}\lambda} \leqslant 15$ eV is computed by the LMTO method with an error of ~0.1 eV[39]. As our experience shows, at higher energies (15 eV$\leqslant E_{\mathbf{k}\lambda} \leqslant$35 eV), the LMTO method yields dispersion curves of the correct shape, but shifted upward by as much as 1 eV. In the area of still higher energies ($E_{\mathbf{k}\lambda}$>35 eV), the picture of the spectrum may become qualitatively incorrect. The error in calculation of the square of the modulus of a matrix element is: where $\hbar\omega \leqslant 15$ eV less than 10%, at 15 eV$\leqslant \hbar\omega \leqslant$35 eV — 20–40%, at $\hbar\omega$>35 eV — even 50–100%. Here, the value of $|\langle \mathbf{k}\lambda | \mathrm{j} | \mathbf{k}\lambda' \rangle|^2$ is usually low. Still, we did not discard the values of the imaginary portion of the dielectric constant at $\hbar\omega$>35 eV, but rather used them as a sort of extrapolation of $\epsilon_2(\omega)$ into the higher frequency area, which is required for more precise restoration of $\epsilon_1(\omega)$ by the method of Kramers and Kronig. We note also that for less closely packed bcc lattices, the error in band calculation by the LMTO method is higher than for fcc lattices, as a result of which the error in calculation of $\epsilon(\omega)$ will be somewhat greater for bcc metals than for fcc lattices.

In spite of these inaccuracies, a comparison of the calculation results with experimental data indicates that our calculation is realistic for all the metals mentioned above. However, it does not seem to us quite correct to evaluate the quality of calculations only on the basis of comparison with experimental data. When we compare computed optical constants with experimental results we run into the fact that for some metals the experimental results are rather contradictory in themselves. This apparently results from the fact that all optical experiments are quite sensitive to specimen surface contamination, leading to great complications in experimental procedure. First of all, most of the metals studied have a strong tendency toward oxidation. The presence of even a very thin oxide film on the surface of a specimen can significantly change its optical properties, as was observed in Cd[48]. Therefore, measurements must be performed in a deep vacuum. Secondly, it is well known that many metals, particularly transition metals, easily interact with gaseous impurities: hydrogen, oxygen, nitrogen, etc. This sort of absorption can occur in air and in electrolyte solutions during electrochemical polishing. In the transition metals, the surface of which usually does not carry a thin oxide film acting as a barrier, impurity gases can penetrate to a significant depth — up to 100 Å or more [49]. This factor frequently causes contradictory experimental results and forces us to maintain a critical attitude toward experimental studies.

It is therefore particularly important to verify the extent to which the calculated quantity $\epsilon(\omega)$ satisfies a number of accuracy rules, including the f–sum rule (1.9) mentioned earlier. Verification of the f–sum rule is important also because the LMTO method which we used reproduces narrow d–type bands better than the broader p and f bands. The matrix elements of $d-p$ and $d-f$ transitions, the contribution of which to the f–sum rule is quite significant, are therefore not very accurately calculated. For this reason, the f–sum rule serves as a check of the applicability of a band calculation system for computation of the optical properties over a broad frequency range. The f–sum rule for specific energies also allows us to perform a comparative analysis of the role of various types of transitions. It should also be noted that our calculation considers only band electrons, completely ignoring electron transitions from core shells. But in some of the metals we studied (such as Nb), transitions involving core electrons begin at rather low energies (in Nb at $\hbar\omega \sim 34$ eV). The presence of transitions at these energies may change both the optical constants observed in the high-frequency area, and the satisfaction of the precise f-sum rule, in which the contribution of the core must also be considered. If we exclude transitions of electrons from the core state from our analysis, then at the limit as $\hbar\omega \to \infty$, the value of $n_{eff}(\hbar\omega) \approx n_v$, where $n_v \sim$ the number of valent electrons per atom.

2.4. Numerical Results. Comparison with Experiments

The method described for calculating optical properties was verified on the example of the simple metal Al, then the properties of the noble metals Cu and Ag, the $3d$-transition metals V and Cr and all the $4d$-transition metals (Y, Zr, Nb, Mo, Tc, Ru, Rh, Pd) were calculated. The IIB group metals Zn and Cd were also studied, due to certain specifics of their collective plasma excitations. The metals Al, Cu, Ag, Rh and Pd crystallize in fcc lattices, Cr, V, Nb and Mo — in bcc lattices. We also calculated the optical characteristics of the transition metals Y, Zr, Tc, Ru and the IIB group metals Zn and Cd, crystallizing in phase-centered close-packed structure, which is not a cubic-type lattice. Calculations were performed for them for the hypothetical case of crystallization in an fcc lattice. This replacement is justified only by the fact that for all these metals the ratio of the basic fccp-structure periods c/a does not differ greatly from the corresponding value of the ideal fccp lattice ($c/a \approx 1.633$). The number of close neighbors of fcc and ideal fccp lattices is the same right out to the third coordination sphere. Therefore, in self-consistent calculation, replacement of the fcc structure with the fccp structure should not greatly change such important characteristics of the band structure as the positions of the bands and their widths, which result basically from the immediate surroundings of the atom in the lattice. Due to this, this type of replacement apparently has little influence on the quantities used in the form of integrals over the Brillouine zone. This approach was successfully used for calculation of bond energies, equilibrium volume and many other parameters of fccp metals[21].

The experiments with which we compared the calculated optical properties can be divided into three classes. In the infrared and visible areas of the spectrum (0–6 eV) the method of ellipsometry was used, allowing us to measure the complex index of refraction $\tilde{n}(\omega)$ and, consequently, to determine the real $\epsilon_1(\omega)$ and imaginary $\epsilon_2(\omega)$ parts of the dielectric constant directly. In the broader frequency area, experimenters must reproduce $\epsilon(\omega)$ using the Kramers–Kronig relations from data gained by measurement of reflectivity $R(\omega)$ over a broad energy interval (usually $\hbar\omega \sim 0$–40 eV). These measurements are performed using synchrotron radiation sources. The third class of experiments consists of inelastic scattering of fast electrons ($E \gg 10$ KeV) passing through thin metal films. In the first Born approximation, all information on such scattering can be obtained from the electron energy-loss function $L(\omega)=\text{Im}[-1/\epsilon(\omega)]$. Considering these three types of experiments, we shall usually present calculated values of $\epsilon_2(\omega)$ (or the related optical conductivity $\sigma(\omega)=\epsilon_2(\omega)\omega/4\pi$), $R(\omega)$ and $\text{Im}[-1/\epsilon(\omega)]$. Let us now go over to analysis of specific curves. The optical properties of Al have been repeatedly measured experimentally (for example, cf. [50, 51]). Comparison with experimental results demonstrates that in the simple metal Al, the agreements of calculated and experimental data is good (Figure 2). Interband transitions begin at $\hbar\omega\sim0.5$ eV, close to the value of the gap at face (III) of the Brillouine zone, while $\epsilon_2^{\text{inter}}(\omega)$ has its maximum at $\hbar\omega\sim1.5$ eV, corresponding to the gap on the (200) face. At approximately this same frequency there is a dip on the curve of $R(\omega)$, in full agreement with the experimental data[50]. The electron energy loss function has the usual form for simple metals with a single peak at $\hbar\omega\sim15.4$ eV (in experiments, 15.2 eV). Near the plasma peak there is a rather sharp drop in reflectivity, i.e., a transition to the transmission band. Since $\epsilon_2(\omega)$ decreases rapidly with increasing frequency, the main contribution to the f-sum rule is that of the 0–10 eV area, in which the matrix current elements are computed with good accuracy (~10%). Therefore, the plateau $n_{\text{eff}}(\hbar\omega_{\max})\sim2.776$ reached by the value of $n_{\text{eff}}(\omega)$ in this area is close to the number of valent electrons of Al.

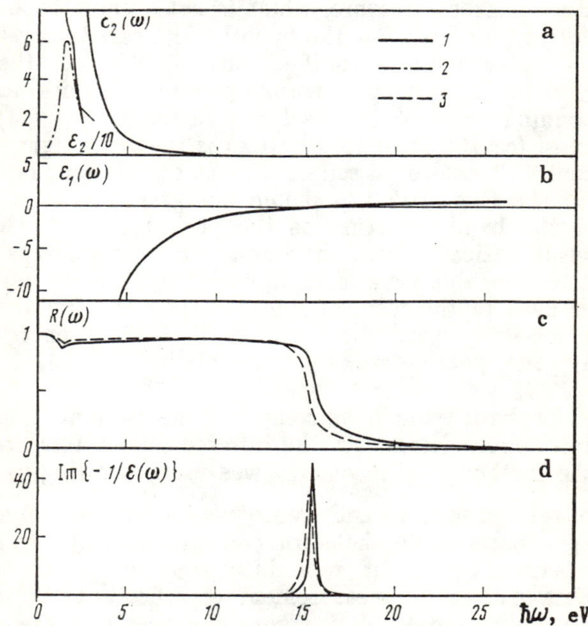

Figure 2. Optical Characteristics of Aluminum. 1 — Calculated quantities; 2 — calculated interband portion of dielectric constant; 3 — experimental values[50].

More complex are the optical properties of the noble metals: copper and silver. There are many experiments on determination of the optical constants of these metals (Cu[52-58] and Ag[56, 59-61]), the data of which agree. The optical conductivity $\sigma(\omega)$ which we calculated for Cu (Figure 3) correctly repeats all of the major characteristics of the experimental curve *a, 2:* a sharp rise at 2.1 eV, related to the beginning of interband transitions at the Fermi level, a rather rapid drop at $\hbar\omega$~6 eV, corresponding approximately to the distance from the bottom of the *d* band to E_F, and a new rise at $\hbar\omega$~15 eV due to transitions from the *d* band to the *p* band. The peaks with energy 5 eV virtually coincide, while the 3.9 eV maximum of the theoretical curve corresponds to a noticeable bend in the experimental curve. The same sort of agreement is observed between the maxima with energies of 9 and 15 eV in the theoretical curve and the peaks at 8 and 16 eV observed experimentally.

Data are also presented on the reflectivity of copper[56] and the results of our calculations (Figure 3*c*). In the 0–2.1 eV frequency interval, where there are no interband transitions, virtually total reflection is observed. The minima at 4.3 eV and 13 eV and the maxima at 5.7 eV and 16 eV on the experimental curve correspond to the calculated minima at 4.5 eV and 13.2 eV and the maxima at 6 eV and 16.5 eV. The bends on both curves at 2.6 eV are also clearly seen. The sharp drop in reflectively $R(\omega)$ at 2.1 eV results, as we know, from the reddish color of copper. We see that our calculations reproduce this characteristic of the element.

As concerns the fast electron energy loss function $\mathrm{Im}[-1/\epsilon(\omega)]$, we can see from the figure that we identify maxima at ~4.5 eV, 8 eV, 12 eV and 20.5 eV, experimentally observed in [57]. It is also notable that all the other structural specifics of the curve of $\mathrm{Im}[-1/\epsilon(\omega)]$ obtained in this experiment are reproduced well by our calculations.

A similar situation obtains in the case of silver (Figure 4), in which the calculated curve of $\epsilon_2(\omega)$ correlates well with the results of experimental studies [59]. All the remaining specifics of the theoretical curves of $R(\omega)$ and $L(\omega)$ also agree with the experimental data. The absolute values of $\epsilon_2(\omega)$ also correspond rather well to the experimental data over a broad frequency range. There is therefore good agreement in our calculation of the powerful low-frequency peak of $L(\omega)$ at $\hbar\omega$~3.8 eV in silver. We must note that the position of this singularity and the value of $L(\omega)$ at this point depend significantly on the accuracy of computation of $\epsilon_2(\omega)$. In particular, a decrease in $\tilde{\omega}_p^2$ in the intraband contribution to $\epsilon_2(\omega)$ of only 20–30% results in virtually complete disappearance of the peak of $L(\omega)$ at $\hbar\omega$~3.8 eV and degeneration of this peak to a step-type singularity. This is because $\epsilon_1(\omega)$, after a decrease in $\tilde{\omega}_p^2$, no longer reaches the value of zero in the vicinity of this singularity (Figure 4).

We should note that in the noble metals Cu and Ag the f–sum rule is not as effective as in the simple metals (cf. below). For example, for Ag our calculated value of $n_{eff}(\hbar\omega \sim 50$ eV$) \sim 7.8$, while the experimental value of n_{eff} at $\hbar\omega = 25$ eV is 5.0[56], whereas $n_v = 11$. In this case, the poor satisfaction of the f–sum rule results not from calculation or experimental error, but rather from the nature of the metal itself. In most chemical compounds Ag has a valence of 1, which agrees with the value of n_{eff} at 3.0 eV $\leqslant \hbar\omega \leqslant$ 3.9 eV, i.e., up to the start of transitions from the d band. As $\hbar\omega$ increases, the value of n_{eff} increases rather slowly. It might be thought that a significant contribution to the f–sum rule would be made by the area of intensive d–f transitions ($\hbar\omega \sim$ 40–70 eV). This is indirectly indicated by the great value of $\hbar\omega_{max}$ in Ag and Cu (cf. Table 40). Unfortunately, at such high energies our calculations do not yield a correct description of $\epsilon_2(\omega)$. Experimental studies have not yet been performed in this frequency range. A similar situation obtains in the IIB–group metal Zn.

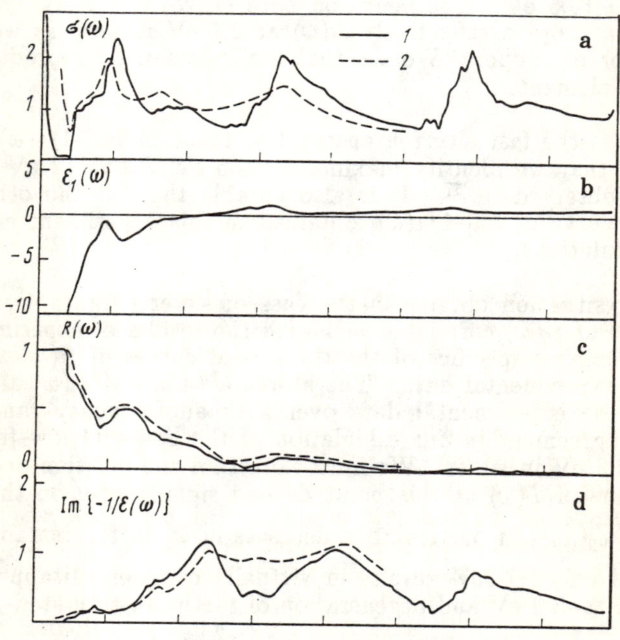

Figure 3. Optical Characteracteristics of Copper. 1 — Calculated values; 2 — experimental values; a — [55]; c — [56]; d — [57].

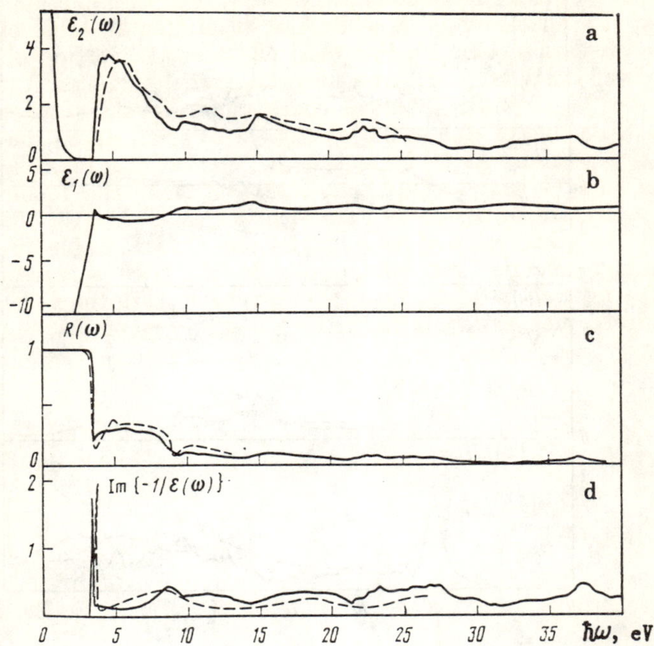

Figure 4. Optical Characteristics of Silver. 1 — Calculated values; 2 — experimental values; a — [59]; c — [60]; d — [56].

Among the $3d$ and $4d$–transition metals studied are metals with fcc, bcc and fccp lattices. Our computations fit best with the fcc metals rhodium and palladium. This is apparently because fcc metals have a more closely packed structure than bcc metals. The fact that the LMTO method which we used to compute the electron structure works better for metals with closer packing was noted by Andersen[39]. The optical properties of these metals were studied experimentally, in [61-64] for rhodium and in [59, 65, 66] for palladium. The curves of $\epsilon(\omega)$, $R(\omega)$ and $L(\omega)$ for these metals have complex shape typical for transition metals with a large number of maxima and minima. The functions $\epsilon_2(\omega)$ and $R(\omega)$ for Rh and Pd correlate well with experimental data on the position of their main singularities. Thus, comparison of the reflectivity of Rh with the experimental data of [62] (Figure 5) shows that the positions of the minimum at 3 eV and the maximum at 4 eV agree. The plateau at 2 eV, found by calculation, corresponds to a bend on the experimental curve, while the arm at 9 eV corresponds to an arm at 7.8 eV observed in [62]. The minimum at 10.5 eV and the maximum at 11.5 eV which we calculated agree with the corresponding singularities in the experimental data at 9.5 eV and 12 eV.

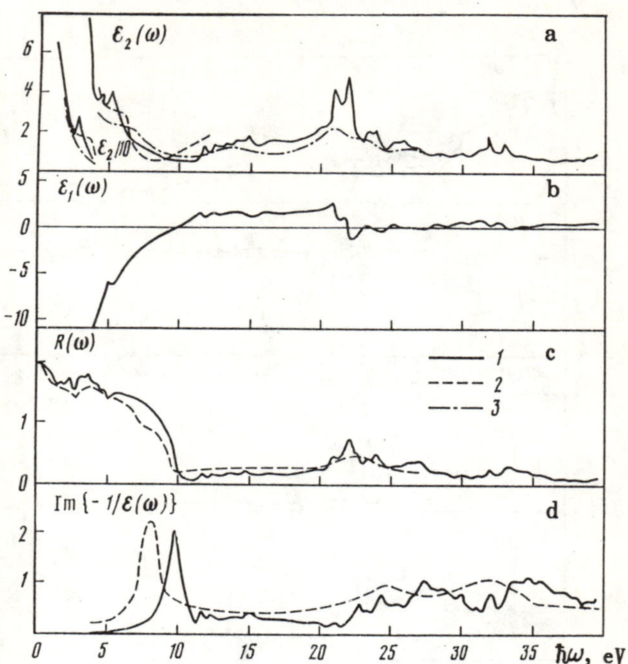

Figure 5. Optical Characteristics of Rhodium. 1 — Calculated values; 2 — experimental values; *a* — [64]; *c* — [62]; *d* — [61] (prod. units); 3 — [62].

A detailed comparison can also be performed for palladium (Figure 6). As before, we see good agreement between theory and experiment both in absolute value of $R(\omega)$ and the position of the corresponding singularities. Thus, for example, there is agreement of the minima at 4 eV, maxima at 5 eV and the extended plateau at 9–19 eV, while the peak at 2.2 eV found by calculation corresponds to a bend on the experimental curve at the same energy.

Further, the graph of $\epsilon_2(\omega)$ for Rh indicates that at $\hbar\omega \sim 11$ eV $d-p$ transitions begin, while the corresponding value for Pd does not have such a strong singularity, i.e., $d-p$ transitions are somewhat suppressed. The results of calculation of the optical properties of Cu, Ag, Rh and Pd were published by us in [13, 67, 68] [1].

[1]In previously published computations $\epsilon_2^{\text{inter}}(\omega)$ lacked the factor $\pi/2$, which led to a reduced value of $\epsilon_2(\omega)$

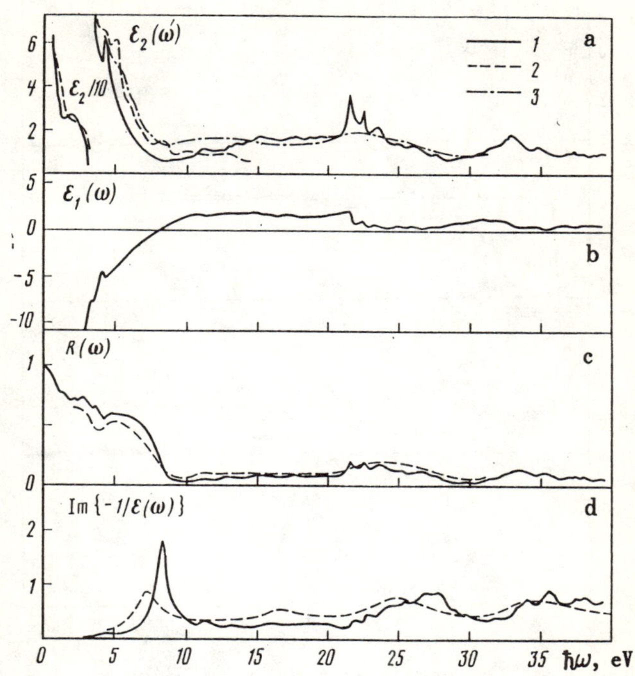

Figure 6. Optical Characteristics of Palladium. 1 — Calculated values; 2 — experimental values; *a* — [65]; *c* — [66]; *d* — [59]; 3 — [66].

Experimental studies of the 4*d* metals Nb and Mo, which crystallize as bcc lattices, have been performed for some time (cf. [61, 69, 70] on Nb and [61, 70, 72] on Mo). The calculation situation is somewhat more complex for them. The behavior of $\epsilon(\omega)$ at low energies (up to $\hbar\omega \sim 10$ eV) agrees rather well with the experimental results (Figures 7 and 8). At higher frequencies, the agreement is worse. The main reason for this is apparently that, due to the insufficient accuracy of the LMTO method, the bottom of the hybridized $p-f$ band was shifted upward by approximately 1 eV. As a result, there is a significant rise in $\epsilon_2(\omega)$, related to the beginning of transitions from the *d* band to the $p-f$ band in niobium, not at 10.5 eV, as was observed experimentally, but rather at 12 eV. For the same reason the broad minimum of $\epsilon_2(\omega)$ between 11 eV and 12 eV is significantly broader in the calculated curve, leading to an increase in the calculated reflectivity in this frequency spectrum not observed experimentally.

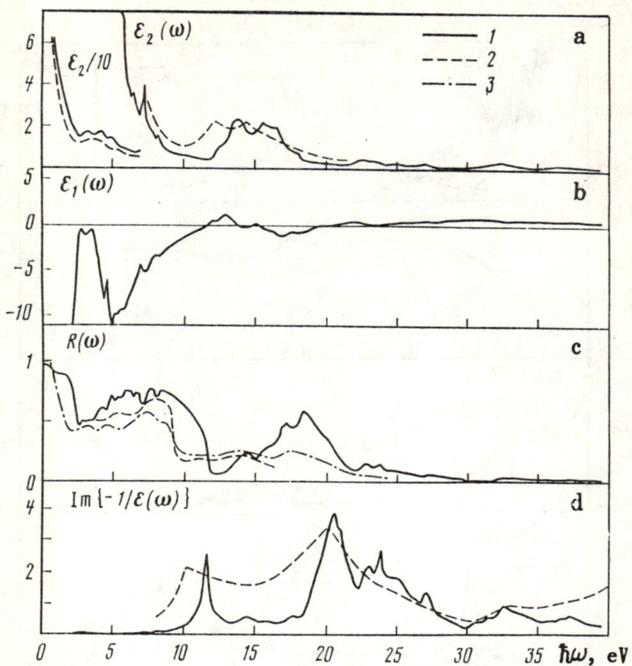

Figure 7. Optical Characteristics of Niobium. 1 — Calculated values; 2 — experimental values; *a* — [69]; *c* — [70]; *d* — [61] (prod. units); 3 — [69].

A comparison of the results of our calculation of the optical characteristics of the bcc transition metals V and Cr, in the $3d$ series, with experimental data (cf. [58, 70, 71, 73] for V and [58, 74] for Cr) shows that the nature of some singularities of the optical spectra of these metals is not quite clear. For example, some experimenters [71] did not observe the characteristic area of d–p transitions in vanadium, starting, according to our calculations, at $\hbar\omega \sim 11$ eV (Figure 9). This may result from insufficient accuracy of the experiments — correctly verifying the f–sum rule, the authors of [71] obtain for vanadium $n_{eff}(\hbar\omega) \sim 2.2$ at $\hbar\omega \sim 34$ eV (with five valent electrons), whereas the calculations yield $n_{eff}(\hbar\omega) \sim 5.493$. Possibly, for this same reason they obtained a significant decrease in the electron energy–loss curve in niobium[69]. The value of $n_{eff}(\hbar\omega_{max}) \sim 6.2$ which they present at $\hbar\omega \sim 26$ eV is clearly elevated, since it has been firmly established that in niobium transition of electrons from the core levels starts at $\hbar\omega \sim 34$ eV. Elevated values of $\epsilon_2(\omega)$ could lead to a drop in the amplitude of the electron energy loss curve, as is observed in [69]. The same authors gave a reduced amplitude of $L(\omega)$ for molybdenum[71]. Another experiment[72] gave an electron energy loss function for Mo closer in amplitude to our calculated curve (cf. Figure 8). As concerns chromium, in the low-energy area ($\hbar\omega \lessdot 6.5$ eV), the position of the basic singularities $\epsilon_2(\omega)$ agrees with our calculations (Figure 10). Unfortunately, we do not know of works in which optical constants of Cr were presented over a broad frequency range. Comparison of the calculated results for $4d$-transition metals crystallizing in fccp lattices with the experiments (for example, cf. [61, 75, 76] for Y, [61, 70] for Zr, [77] for Tc and [78] for Ru) shows that the calculated optical properties of these metals in fcc lattice allow qualitative analysis of the optical spectra of the metals. Of course, this replacement does not allow reproduction of all details of the frequency variation of $\epsilon(\omega)$. However, the coarsest specifics of the optical properties, related to the position of the Fermi level with respect to the bottom and top of the *d* band, as well as the mutual placement of *d, p* and *f* bands, as calculations indicate, will be reflected here. The electron energy loss functions are best reflected (Figure 11–14). A detailed analysis of the curve of $L(\omega)$ in Tc was undertaken by us in [79]. We should note that in all the transition metals the calculated value of $n_{\text{eff}}(\hbar\omega_{\text{max}})$ differs from the number of valent electrons by no more than 30%, while in the noble metals and Zn the difference may be almost

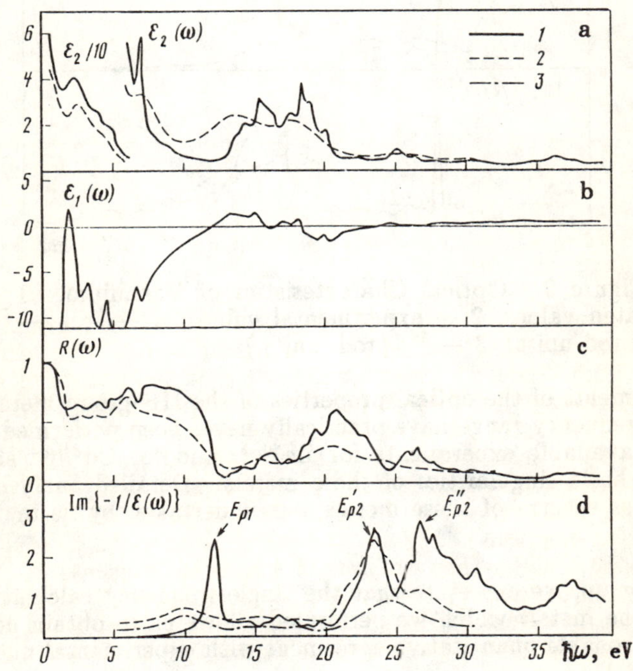

Figure 8. Optical Characteristics of Molybdenum. 1 — Calculated values; 2 — experimental values[72]; 3 — [71].

by a factor of two (cf. Table 4). This is apparently related to the fact that in the noble metals the highly energetic d–f transitions (40–70 eV) play a more significant role than in the transition metals. This is indirectly indicated by the large values of $\hbar\omega_{max}$ in the noble metals.

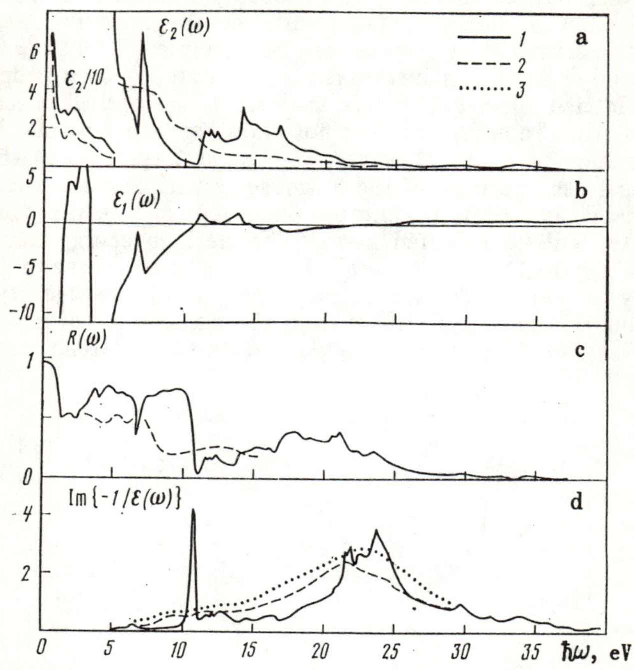

Figure 9. Optical Characteristics of Vanadium. 1 — Calculated values; 2 — experimental values; a — [71]; c — [70]; d — [73] (prod units); 3 — [58] (prod. units.)

Measurements of the optical properties of the IIB–group metals Zn and Cd over a broad frequency range have practically never been performed. A comparison with the available experiments for Zn[80, 81] and for Cd[81, 84] shows that we reproduce the basic singularities of these metals. Detailed analysis of the electron energy–loss spectra of these metals was undertaken by us in [81] (Figure 15, 16).

Summing up, we can state that the single–frequency calculation of optical properties of the metals which we performed allows us to obtain not only qualitative, but reasonable quantitative agreement with experimental data.

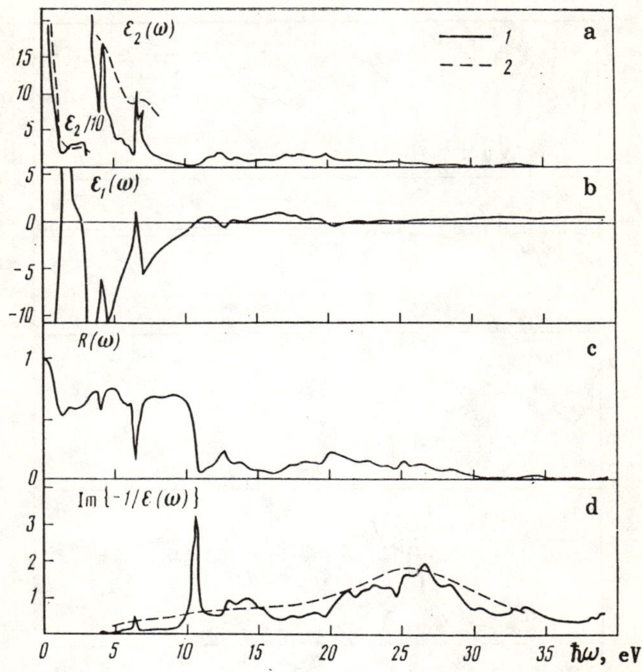

Figure 10. Optical Characteristics of Chromium. 1 — Calculated values; 2 — experimental values; *a* — [74]; *d* — [58] (prod. units).

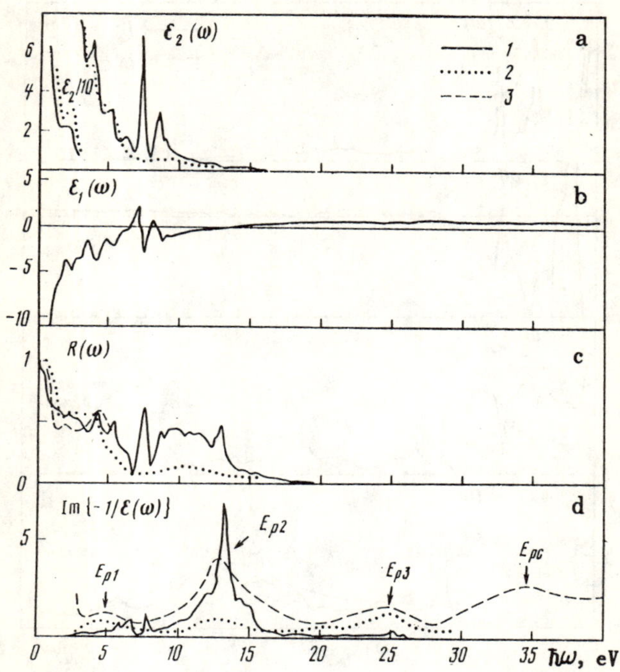

Figure 11. Optical Characteristics of Yttrium. 1 — Calculated values; 2 — experimental values[76]; 3 — *c* — [75]; *d* — [61] (prod. units).

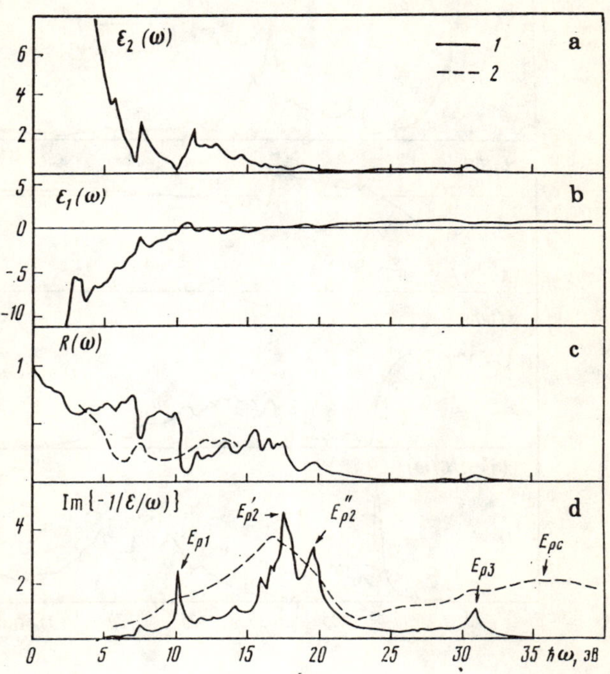

Figure 12. Optical Characteristics of Zirconium. 1 — Calculated values; 2 — experimental values; *c* — [70]; *d* — [61] (prod. units).

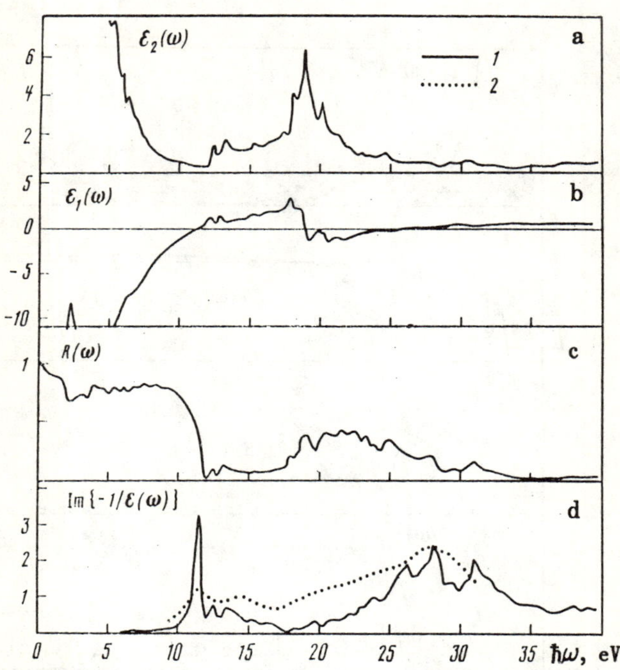

Figure 13. Optical Characteristics of Technetium. 1 — Calculated values; 2 — [77] (prod. units).

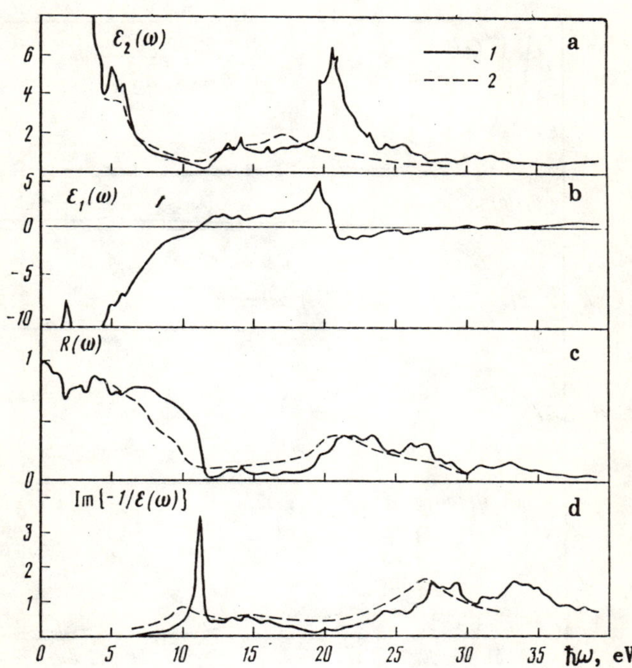

Figure 14. Optical Characteristics of Ruthenium. 1 — Calculated values; 2 — experimental values[78].

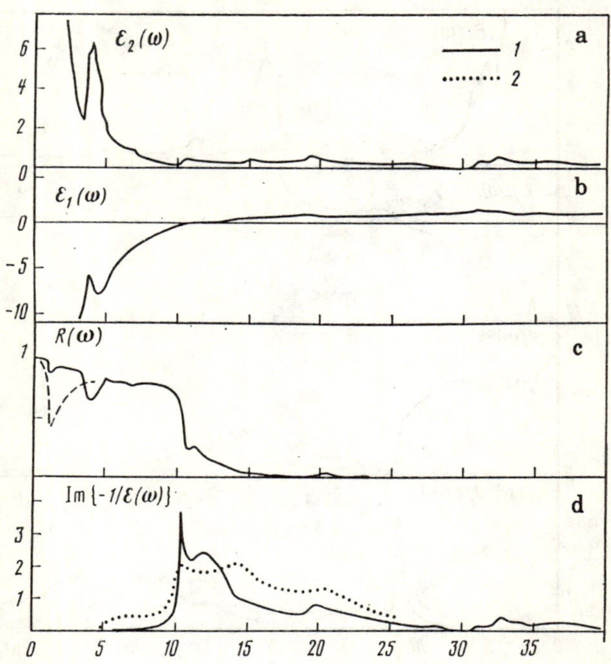

Figure 15. Optical Characteristics of Zinc. 1 — Calculated values; 2 — experimental values[81] (prod. units); dash line[80].

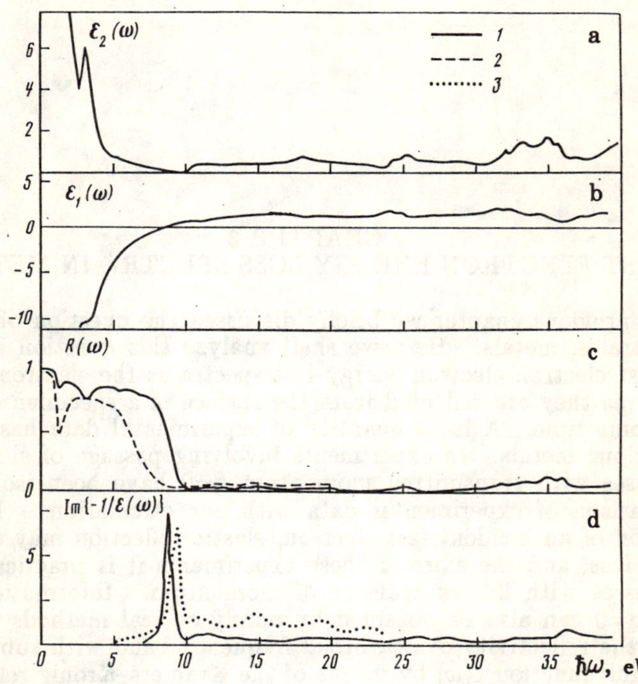

Figure 16. Optical Characteristics of Cadmium. 1 — Calculated values; 2 — [48]; 3 — [81] (prod. units).

CHAPTER 3
FAST ELECTRON ENERGY LOSS SPECTRA IN METALS

In the previous chapter we briefly discussed the question of electron energy-loss spectra in metals. Here we shall analyze this question in more detail. Studies of fast electron electron energy-loss spectra as the electrons pass through thin films or as they are reflected from the surface of a specimen have been conducted for some time. A large quantity of experimental data has been accumulated for various metals. In experiments involving passage of electrons through films, processes with transmitted momentum $\mathbf{q}=0$ have been isolated, allowing direct comparison of experimental data with our calculations. In experiments with reflection of an incident fast electron, elastic reflection may occur before or after energy loss, and therefore in these experiments it is practically possible to isolate processes with known transfer of momentum. Information on the loss function at $\mathbf{q}=0$ can also be obtained by purely optical methods involving measurement of the reflectivity over a broad frequency band with subsequent restoration of the full function $\epsilon(\omega)$ by means of the Kramers-Kronig relations.

The simplest situation occurs in simple metals, in which the electron system is similar in its properties to a homogeneous electron gas of the same density. Here the dominant singularity of the electron energy loss spectrum is the powerful plasma resonance, corresponding to excitation of long-wave electron-density oscillations. The frequency of this resonance corresponds closely to the value of ω_p at which the dielectric constant of a homogeneous electron gas $\epsilon(\omega)$ becomes equal to zero, i.e.,

$$\omega_p = \sqrt{\frac{4\pi n_v e^2}{m\Omega}}, \tag{3.1}$$

where Ω is the atomic volume. Due to the interaction with the lattice, which in a simple metal can be described by the weak electron-ion potential, the plasma peak takes on a certain nonzero width even at **q**=0. It also easily shifts in frequency. This effect may be rather well described by the theory of pseudopotential perturbations (cf. review[18]). The calculation which we performed using the LMTO method yielded the same results for the simple metal Al (cf. Figure 2): The calculated resonance width agrees both with calculations by the pseudopotential method[17] and with experiments [50], while the position of resonants $\hbar\omega$=15.4 eV is close to the experimental value of 15.2 eV. We note that in the model of a homogeneous electron gas the plasma frequency is slightly higher — 15.8 eV.

A more complex situation obtains in d–transition metals. Here, the distribution of electrons within a unit cell is quite heterogeneous, and the electron energy loss spectrum consists of several resonances of different widths and amplitudes. Additional difficulties arise due to the comparatively low threshold of excitation of core p electrons: In elements at the beginning of the d series it is ~30 eV (for 4d metals), while by the end of the d series it increases to 50–60 eV. At these and higher energies, core electron excitation processes begin to play a major role in the formation of the electron energy loss spectrum.

The functions we calculated $\mathrm{Im}[-1/\epsilon(\omega)]$, as well as the experimental data on electron energy–loss spectra for all 4d metals in the 0–40 eV range are shown in Chapter 1 on Figures 5–8 and 11–14. Good agreement of calculated and experimental data is observed both in terms of the number and position of the peaks of the function $\mathrm{Im}[-1/\epsilon(\omega)]$, and in terms of the absolute value of the electron energy–loss function (results of optical studies[72, 73] normalized). Let us briefly list the major singularities of the electron energy–loss spectra, marked on Figures 11 and 12 by the arrows. The tall, narrow maximum at E_{p1}≈10 eV exists in all 4d transition metals. On the experimental curve it is greatly smoothed and in some cases looks more like a bulge than a maximum. The main maximum is seen as a well defined single peak only in yttrium (Figure 11). In zirconium (Figure 12) on the calculated curve this peak is split into two closely located maxima with energies E'_{p2} and E''_{p2}. By the end of the period, the picture is more complex. Here with characteristic energies E_{p2}~20–30 eV we usually see a whole group of such maxima (cf. Figures 5, 6). On the experimental curves as the atomic number increases the main plasma peak significantly expands. However, the effect of splitting the main plasmon is less clearly expressed than in the calculated curves. Still, in Mo in the immediate vicinity of energy E''_{p2} there is a significant bulge on the experimental electron energy–loss function[72] (cf. Figure 8). Further, in Rh[61] instead of one plasma peak the experiments show two blurred bulges (cf. Figure 5). The maximum at energy E_{p3} (frequently seen as a bulge on the experimental curves) matches the calculated value rather well, although the accuracy of calculations in this area of energies is not great. At higher energies, the experimental curve shows a peak related to excitation of the core 4p electrons, not considered in our calculations (E_{pc}, cf. Figure 11, 12).

The greatest difference between our calculations and the experiments consists of the smoother nature of the experimental curves. The smooth nature of the curves, obtained by reflecting fast electrons from the surface of a specimen in[61], is not particularly surprising, since in this case the experiment determined the electron energy-loss function as averaged with respect to moment **q**. However, the data of [72, 77, 78], obtained for transmission of fast electrons through a thin film of metal and reflection of light, i.e., corresponding to $\mathbf{q}=0$, also do not repeat the fine structure of the electron energy-loss spectrum seen in the calculated curves. As was shown in [79], this difference cannot be explained by insufficient resolution of the instruments. There are two main reasons for this difference between calculated and experimental data. The first is inaccuracy in the calculation of the electron zone of the structure and the momentum operator matrix elements, leading to a systematic decrease in the value of $\mathrm{Im}[\epsilon(\omega)]$. This can be particularly clearly seen in the example of the first peak in the electron energy-loss spectrum, the position of which with respect to energy in all $4d$ metals is located in the area of the drop on the curves of $\epsilon_2(\omega)$ or, which amounts to the same thing, of $\sigma(\omega)$. A comparison of the $\sigma(\omega)$ we calculated with the experimental data of [13, 82] shows that our values of $\sigma(\omega)$ in the area of the drop are significantly reduced in comparison to the experimental values. This factor may be related to the systematic shift in our calculations in comparison to the experiment, showing the second peak in $\sigma(\omega)$ at energies on the order of 0.5–1 eV higher. A more significant cause, which may also result in smoothing of singularities on the curves of $\sigma(\omega)$ and $\mathrm{Im}[-1/\epsilon(\omega)]$ is the influence of the finite lifetime of electrons in excited states on the dielectric constant. The corresponding effects are completely ignored in the density functional method we used. These effects are not difficult to reproduce artificially, by averaging the calculated curves of $\sigma(\omega)$ over an energy interval, the width of which increases with increasing energy $\hbar\omega$. We did not undertake this operation, since its result is rather obvious, and it is easier to establish the relationship of the electron energy-loss spectrum with electron structure using the unsmoothed curve.

Now, utilizing the good agreement between calculation and experiment, let us attempt to understand what determines the complex nature of the electron energy-loss spectrum in transition metals. The $4d$ metals, which we calculated fully, form a convenient object for investigation. As Payns[1] noted, an understanding of the electron energy-loss spectrum of the transition metals requires knowledge of the energy and force of the basic interband transition oscillators. The interband transitions, in turn, are determined by the electron structure of the metal, which we shall analyze first.

As an example, let us discuss the electron structure of a typical $4d$ metal — niobium, the state density $N(E)$ of which is shown in Figure 17. The Fermi level is slightly below the middle of the $4d$ band, occupying the energy interval from 6–15 eV. The curve of $N(E)$ at high energies is quite jagged and differs greatly from the form of $N(E)$ postulated in the frequently used $s-d$ model. Analysis of the curves of partial state densities (Figure 17) shows that the area from 21.5 to 50.0 eV is occupied by the greatly hybridized $5p$–$4f$ band. In the lower portion (from 21.5 to 32.5 eV) a significant contribution to *N(E)* is made by *p* electrons (more precisely, states with l=1), whereas its upper portion (from 32.5 to 50.0 eV) consists basically of *f* electrons (states with l=3). Below, for brevity we shall refer to the lower portion of the *p–f* band as the *p* band, the upper portion as the *f* band. The valent 5*s* electrons are concentrated mainly in the area below the bottom of the 4*d* band and in its lower portion, as well as the lower portion of the *p* band. One important feature of the electron structure of 4*d*–transition metals is the broad gap in electron state density separating the *d* and *p* bands. There if no such sharp gap between the *p* and *f* bands. The break in the spectrum at E≈51 eV is artificial. It is related to the limited basis ($l \leqslant 3$) used in our calculations of electron structure. Speaking of the electron structure of the 4*d* metals, we must note the comparatively small depth of the 4*p* levels in the core (in Nb, 35 eV below E_F). Their excitation makes a significant contribution to the optical spectra of the metals and the electron energy–loss function at high energies. In our calculations, this factor was not considered, and therefore at these energies our calculations differ from the experimental data. The basic electron structure parameters of the remaining 4*d* metals — the width and mutual placement of *p, d* and *f* bands — are shown in Figure 18.

Let us now see how the electron transitions form the function $\epsilon(\omega)$ in the 4*d* metals. To do this, we shall analyze the frequency variation of interband conductivity

$$\sigma(\omega)=\frac{\omega}{4\pi}\mathrm{Im}[\epsilon^{\mathrm{inter}}(\omega)]$$

for all 4*d* metals (Figure 19). This value directly determines the contribution of each frequency interval to the *f*–sum rule[79]. The numerical data on the basic electron transitions are presented in Table 5. Let us discuss them in more detail.

In the low–frequency area ($\hbar\omega \leqslant \hbar\gamma \sim 1$ eV), the main role is played by intraband transitions. In contrast to the simple metals, this contribution to the 4*d* metals is hardly dominant in the *f*–sum rule, representing only 0.6–1.3 electrons per atom. The value of the contribution varies from metal to metal quite irregularly, reminiscent of the change in $N(E_F)$ in 4*d*–series metals (Table 5, Figure 20).

In the optical area of the spectrum (1 eV$<\hbar\omega \leqslant 5$ eV), transitions between occupied and free *d*–band subbands are most significant. Since the momentum operator connects states with $\Delta l=\pm 1$, the matrix elements of interband *d–d* transitions are different from zero only due to mixing with the pure *d* orbitals of states with l=1 and l=3 or, speaking in the language of strong bond approximation, due to overlapping of *d* orbitals localized in neighboring atoms. The small value of matrix elements for interband *d–d* transitions is compensated, however, by the large phase volume. For this reason, the contribution of these transitions to the *f*–sum rule is greater than the contribution of intraband transitions.

Table 5
Energies and Forces of Electron Transition Oscillators;
Energy and Contribution to *f*–Sum Rule of Basic Plasmons,
Experimental Data on Plasma Oscillations in 4*d*–Transition Metals

Metal $\hbar\omega_p$, eV	Calculated					Experimental	
	Transition Type	E_{trans}, eV	N_{trans}	E_{plasm}, eV	N_{plasm}	E_{plasm}, eV	Type of Singularity
Y,11.1	*d–d*	3.1*	2.24	6.5	0.34	4.0	Maximum
	d–p	7.2	0.45	7.7	0.15		
		8.5	0.90	13.0	2.95	12.4	Maximum
	d–f	25.2	0.23	25.2	0.37	25.4	Maximum
	d–d	3.9*	3.86	10.2	0.48	8.0	Bulge
Zr,15.4		11.3	0.57				
	d–p	14.7	0.19	17.0	2.4	15.6	Maximum
		16.3	0.31	19.5	1.45		
	d–f	30.7	0.40	30.7	0.46	28.7	Bulge
	d–d	4.7*	4.2	11.5	0.72	9.5	Maximum
Nb,19.6		13.6	0.48	14.5	0.48		
	d–p	15.4	0.72	20.2	1.2	19.6	Maximum
		22.5	0.51	23.7	1.5		
	d–f	32.3	0.54	32.3	1.14	32.4	Bulge
	d–d	4.8*	4.25	11.8	0.75	9.9	Maximum
Mo,23.3		15.2	0.75	16.5	0.30	14.3	Bulge
	d–p	18.2	1.17	23.3	1.51	22.8	Maximum
		24.9	0.73	27.0	2.96		
	d–f	35.9	0.89	36.2	2.04	36.0	Bulge
	d–d	4.6*	4.01	11.5	0.45	11.0	Maximum
Tc,25.9		12.8	0.15	13.5	0.51	15.1	Bulge
	d–p	18.8	2.4	25.8	1.3		
		24.5	0.16	27.3	1.9		
						27.0	Maximum

Table 5, continued

Metal $\hbar\omega_p$, eV	Calculated					Experimental	
	Transition Type	E_{trans}, eV	N_{trans}	E_{plasm}, eV	N_{plasm}	E_{plasm}, eV	Type of Singularity
		30.2	0.54	30.9	1.65		
	d–f	37.6	0.64	37.9	0.40		
		40.2	0.6	42.3	0.45		
						46.0	Bulge
		45.5	0.44	46.5	1.2		
	d–d	4.2*	4.3	11.1	0.51	9.5	Maximum
Ru,28.7		13.9	0.3	14.5	0.45	14.6	Bulge
				24.0	0.9		
	d–p	20.5	2.4	27.4	1.05		
		27.2	0.9	29.0	0.75	26.1	Maximum
		31.4	0.81	33.0	2.1	30.0	Bulge
	d–f	43.3	1.65	44.3	2.7		
		47.3	0.45	49.0	1.65		
	d–d	3.8*	3.1	9.8	0.48	7.9	Maximum
Rh,29.9		14.8	0.53	15.1	0.51		
	d–p	23.8	3.06	24.3	0.9	24.6	Broad Merging Maxima
		32.2	1.4	27.0	1.53	32.2	
				34.9	1.7		
	d–f	40.8	1.63			41.1	Slight Bulge
				41.2	1.58		
		46.5	1.38	47.6	4.2	49.0	Bulge
	d–d	3.0*	1.98	8.3	0.47	6.8	Maximum
Pd,30.5		16.2	1.2	16.	0.21	16.1	Bulge
	d–p	21.8	2.5	20.	0.3	24.2	Maximum
				24.8	1.67		
		32.8	1.9	35.5	1.31	34.3	Maximum
	d–f	39.0	1.35	38.	1.27		
		43.8	1.6	45.5	3.3	46.0	Bulge
		47.8	0.75	49.4	2.7		

* Half width of *d* band given as *d–d* transition energy.

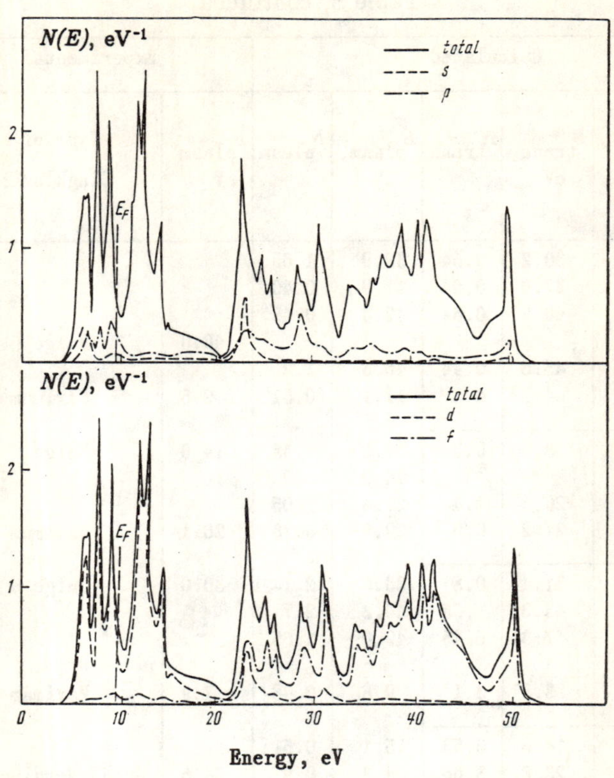

Figure 17. Full and Partial Electron State Densities in Nb.

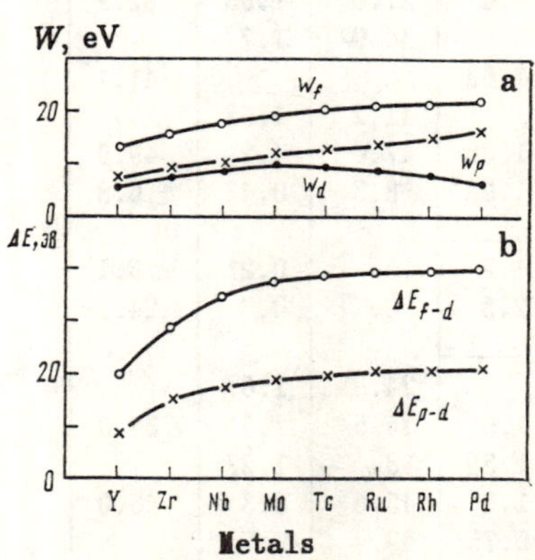

Figure 18. Basic Parameters of Electron Structure of $4d$–Transition Metals. a — Bandwidth; b — band energy difference.

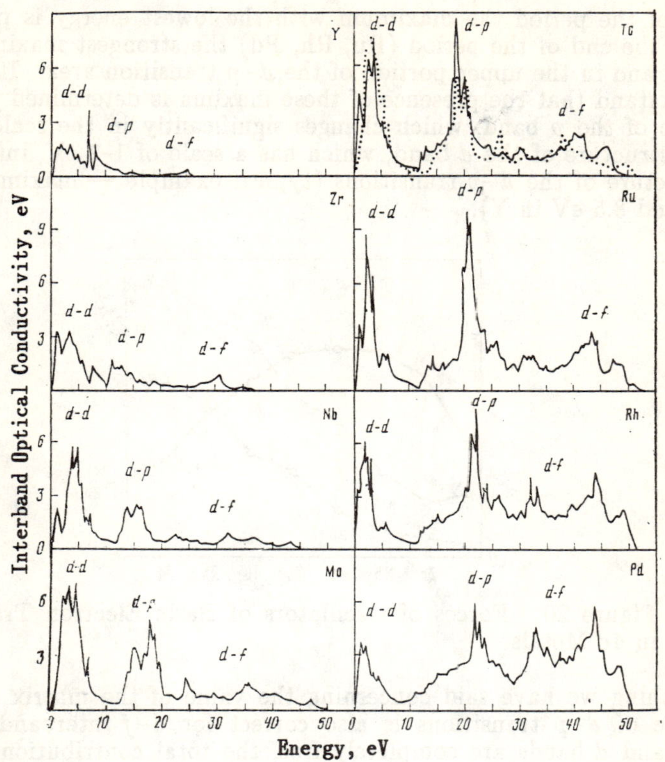

Figure 19. Results of Calculation of Optical Conductivity $\sigma(\omega)$ in 4d Metals. Points for Tc show calculated values for bcc lattice.

Both of the types of transitions studied occur within the d band and have energy less than $\hbar\omega_p$. In our analysis of electron energy-loss, it is convenient to combine them to a single group of d–d transitions. The contribution of d–d transitions changes regularly across the 4d series: in metals at the beginning and end of this period, where the number of filled and free d–states, respectively, is small, this contribution is small, while in metals in the middle of the period it reaches its maximum (Table 5, Figure 20).

The next type of transition consists of transitions from the occupied portion of the d band into the unfilled p band. Since states with l=1 and l=3 predominate in the p band, the matrix elements of d–p transitions are several times greater in magnitude than the interband d–d transitions. The minimum excitation energy of d–p transitions is greater than the width of the d band, and therefore in all 4d metals the areas of intensive d–d and d–p transitions are clearly separated. The area of d–p transitions is hardly homogeneous. The function $\sigma(\omega)$ has a number of maxima, and whereas in metals at the beginning and middle of the period the maximum with the lowest energy is predominant, in metals at the end of the period (Ru, Rh, Pd) the strongest maxima are those in the center and in the upper portion of the $d-p$ transition area. It is not difficult to understand that the presence of these maxima is determined primarily by the structure of the p band, which changes significantly in the scale of 4–7 eV. The energy structure of the d band, which has a scale of 1–2 eV, influences only the fine structure of the $d-p$ transitions (typical example — maxima of $\sigma(\omega)$ at $\hbar\omega$=7.5 eV and 8.5 eV in Y).

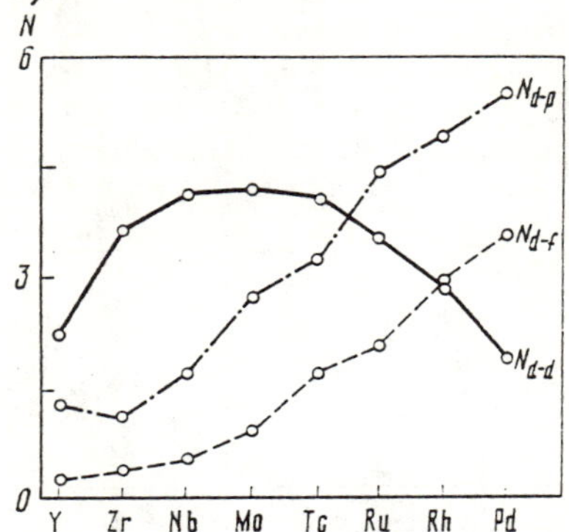

Figure 20. Forces of Oscillators of Basic Electron Transitions in 4d Metals.

Everything we have said concerning the value of the matrix element and the structure of $d-p$ transitions is also correct for $d-f$ interband transitions. Since the p and d bands are completely free, the total contribution of $d-p$ and $d-f$ transitions to the f–sum rule increases with increasing number of d electrons in the metal (Figure 20). This relation is near linear, with some deviation in metals at the beginning of the period caused by the rapid increase in $d-p$ and $d-f$ transition energy with transition from Y to Zr and from Zr to Nb (Figure 18). The curves of the contributions of $d-p$ and $d-f$ transitions ($N_{d\text{-}p}$ and $N_{d\text{-}f}$) as a function of atomic number (Figure 20) are much less like straight lines. This is probably because, due to the strong hybridization of the p and f bands, there is strong mutual influence of these groups of transitions.

In analyzing the results of calculations, we find that the matrix transition elements from the lower area of the d band to the lower area of the $p-f$ band and from the upper portion of the d band to the upper portion of the $p-f$ band are significantly greater than for transitions from the lower portion of the d band to the upper portion of the the $p-f$ band and from the upper portion of the d band to the lower portion of the $p-f$ band. Speaking in the language of chemical bonds, strong transitions occur either between bonding orbitals which have maximum electron density at the cell boundary, or between antibonding orbitals, for which the electron density at the boundary of the cell is near zero. This factor, which is related to the degree of localization of electron states, means that in metals at the beginning and middle of the period, where only the lower portion of the d band is occupied, $d-p$ transitions with the lowest energy dominate. In the metals at the end of the period, where the d band is almost completely filled, the beginning of $d-p$ transitions, corresponding to excitation of electrons from the upper portion of the d band into the lower portion of the p band, is quite weakened. Strongest here are $d-p$ transitions with high energies, and the $d-f$ transitions.

Our analysis shows that the basic characteristics of the electron transitions — their energy and contribution to the f–sum rule — are determined basically by the ordinal number of the element, with and mutual placement of d, p and f bands. The crystalline structure, which influences the fine energy structure of the bands, should not have a significant influence on these characteristics. Actually, calculations which we performed for technetium with bcc and fcc lattice yielded results quite close to each other (cf. Figure 19).

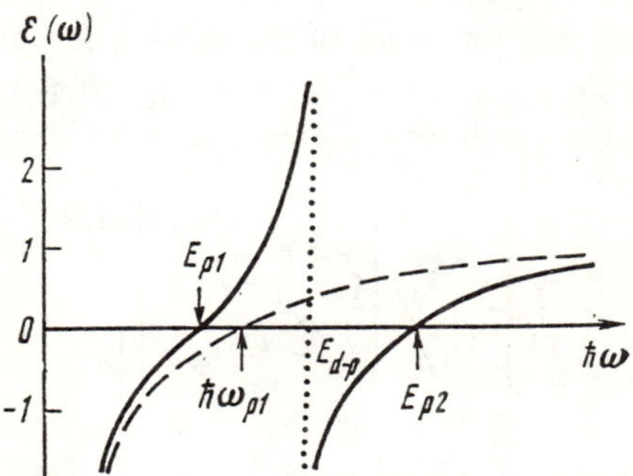

Figure 21. Development of Plasma Resonances for Model Dielectric Constant (3.2).

Now, with the picture of the basic electron transitions clarified, let us go on to show how this picture explains the electron energy–loss spectrum observed experimentally and generated by calculations. To simplify the analysis as much as possible, we shall use the model curve of $\sigma(\omega)$ consisting of two δ–shaped peaks: the first at energy $\omega \ll \omega_p$ corresponds to $d-d$ transitions, while the second with $\omega = \omega_{d-p}$ corresponds to $d-p$ transitions. Using Kramers–Kronig relation (1.2), it is not difficult to write the corresponding model dielectric constant:

$$\varepsilon_{\text{model}}(\omega) = 1 - \frac{N_{d-d}\Omega_p^2}{\omega^2} - \frac{N_{d-p}\Omega_p^2}{\omega^2 - \omega_{d-p}^2}, \tag{3.2}$$

containing the plasma frequency corresponding to a density of 1 electron per atom — $\Omega_p=(4\pi e^2/m\Omega_0)^{1/2}$ (Ω_0 being the atomic volume). The dielectric constant of (3.2) is shown in Figure 21. Previously, a very similar model was suggested on the basis of analysis of experimental data in [77]. It was shown in this work that the basic singularities of the electron energy-loss spectrum in d-transition metals at the beginning and middle of the period can be qualitatively correctly described by assuming that the parameters of the model satisfy the relations

$$\Omega_p N_{d-d}^{1/2} \ll \omega_{d-p} \ll \omega_p, \tag{3.3}$$

$$N_{d-d} + N_{d-p} \approx N_v. \tag{3.4}$$

Actually, if in (3.2) we ignore the term describing $d-p$ transitions, the condition $\epsilon(\omega)=0$ defines just one plasma mode with frequency $\omega_{p1}=\Omega_p N_{d-d}^{1/2}$ (the dielectric constant thus obtained is shown in Figure 21 by the dash line). Considering $d-p$ transitions, the frequency of this resonance decreases to

$$\omega_{p1} \approx \Omega_p \left[N_{d-d} \Big/ \left[1 + \frac{N_{d-p}\Omega_p^2}{\omega_{d-p}^2} \right] \right]^{1/2}$$

and a second resonance arises with frequency $\omega_{p2} \approx \Omega_p (N_{d-d}+N_{d-p})^{1/2} \approx \omega_p$. Thus was the experimentally observed double maximum in the electron energy-loss spectrum explained in [77], as well as the fact that the energy of the second resonance corresponded with good accuracy with $\hbar\omega_p$.

Analysis of Table 5 shows, however, that in real $4d$ metals relations (3.3) and (3.4) do not occur. They are replaced by the relations

$$\Omega_p N_{d-d}^{1/2} \sim \omega_{d-p} \lesssim \omega_p, \tag{3.5}$$

$$N_{d-d} + N_{d-p} \lesssim N_v. \tag{3.6}$$

The plasma oscillation frequencies are expressed in a somewhat more complex manner

$$\omega_{p1}^2 \approx \omega_{d-p}^2 \frac{N_{d-d}\Omega_p^2}{\omega_{d-p}^2 + (N_{d-d} + N_{d-p})\,\Omega_p^2}, \tag{3.7}$$

$$\omega_{p2}^2 \approx \Omega_p^2 (N_{d-d} + N_{d-p}) + (\omega_{d-p}^2 - \omega_{p1}^2). \tag{3.8}$$

According to (3.7), resonant frequency ω_{p1} is always somewhat less than ω_{d-p}, which also agrees with the numerical calculations (Table 5). Here, $\hbar\omega_{p1}$ falls in the gap between the the areas of intensive $d-d$ and $d-p$ transitions. In this energy area $\epsilon_2(\omega) \ll 1$, due to which this resonance in all $4d$ metals is a well defined collective oscillation of electron density.

All of this is clearly seen in Figure 21. It is also important that the introduction of $d-p$ transitions, in addition to changing the full force of oscillators with N_{d-d} to $N_{d-d}+N_{d-p}$, also shifts the frequency of the classical plasmon by $(\omega_{d-p}^2 - \omega_{p1}^2)$. Without this shift in real metals energy $\hbar\omega_{p2}$ would be 2–3 eV lower than $\hbar\omega_p$. Considering this shift, the frequency of the second resonance is almost precisely the same as ω_p in all metals at the beginning and middle of the $4d$ series. We should note that the resonance with frequency ω_{p2} is significantly more powerful than the first resonance. Simple calculations show that the contribution of low-frequency resonance to the sum rule (1.9) is

$$\frac{\omega_{p1}^2 \left(\omega_{d-p}^2 - \omega_{p1}^2\right)}{\left(\omega_{p2}^2 - \omega_{p1}^2\right) \Omega_p^2},$$

i.e., represents no more than 10% of N_v. The same thing is indicated by the data of Table 5. Thus we see that in $4d$ transition metals the position of the maxima in the electron energy-loss spectra is greatly determined by the nature of interband transitions. In particular, the presence of strong $d-p$ transitions unavoidably leads to the development of a low-frequency peak in the energy area corresponding to the gap between the $d-d$ and $d-p$ transitions. The accuracy of the position of the second peak is also largely determined not simply by the mean electron density in the system, i.e., the value of N_v, but rather by the interband transition spectrum. In this sense, the good agreement of the value of ω_{p2} with ω_p, repeatedly seen in experimental works, is to some extent accidental.

The attenuation of the main resonance at frequency ω_{p2} increases greatly from the beginning to the end of the period. In metals at the beginning of the period the strongest of the $d-p$ transitions are transitions at minimum energy. In these metals $\hbar\omega_{p2}$ falls in the area of higher energies where $d-p$ transitions are greatly attenuated. Since here $\epsilon_2(\omega_{p2}) \ll 1$, the main resonance is a rather well defined collective oscillation. As the atomic number increases, the intensive $d-p$ transitions are shifted into the area of higher energies and occupy an ever increasing energy interval. Here, energy $\hbar\omega_p$ falls in the area of rather intensive

$d-p$ transitions. Under their influence the main resonance is split into a group of closely located resonances and greatly attenuated. This effect is significant even in metals in the middle of the period, while in metals at the end of the $4d$ series the main resonance actually disappears. In these metals (Ru, Rh, Pd), the model dielectric constant of (3.2) in the area $\omega \sim \omega_p$ cannot be used, since at these frequencies $\epsilon_2(\omega) \sim 1$, while $\epsilon_1(\omega)$ does not even change its sign. Here the electron energy-loss spectrum has no universal nature, but is determined by the structure of the $p-f$ band. The difference in behavior of the main resonance in metals at the beginning, middle and end of the $4d$ series is clearly seen in Figures 11, 8 and 6, showing the functions $\epsilon_1(\omega)$ and $\epsilon_2(\omega)$ for Y, Mo and Pd.

The maximum in the electron energy-loss spectrum with energy E_{p3} mentioned above is genetically related to the $d-f$ transitions. It occurs at energies significantly greater than $\hbar\omega_p$, where in all $4d$ metals $\epsilon_2(\omega) \ll 1$, while $\epsilon_1(\omega) \approx 1$ (Figures 8 and 11). In this energy area $\mathrm{Im}[-1/\epsilon(\omega)]=\epsilon_2(\omega)/(\epsilon_1^2(\omega)+\epsilon_2^2(\omega))$, i.e., the maxima of the electron energy-loss spectrum coincide with the maxima of $\epsilon_2(\omega)$ resulting from the presence of interband transitions. (For more detail on electron energy-loss spectra of $4d$ transition metals, cf.[84, 85].)

Our discussions presented here are fully applicable to $3d$-transition metals as well. The calculations we performed for V and Cr show that these metals are largely analogues of Nb and Mo, respectively. The most significant difference between them is related to the fact that, although the energy difference E_p-E_d in $3d$ metals is almost the same as in $4d$ metals, the energy difference E_f-E_d in $3d$ metals is almost 13 eV greater. As a result, the strength of $d-f$ transition oscillators in $3d$ metals is significantly less, and the role of $d-f$ transitions is somewhat weaker. More detailed information on these metals can be found in Table 6 and Figures 9 and 10.

High-frequency collective density perturbations in the noble metals Cu and Ag do not differ in principle from the corresponding perturbations in the transition metals at the end of the period. The only significant difference appears in the low-frequency area. It is related to the fact that in the noble metals the d band is completely filled, so that interband transitions of d electrons begin only at energy $E_F=E_{d,\,\mathrm{top}}$ ($E_{d,\,\mathrm{top}}$ is the energy of the top edge of the d band). In copper this threshold is 2.1 eV, in silver 3.9 eV. As was shown by Erenreich and Phillip[55], the difference in the position of the d band relative to the Fermi level leads to a significant difference in the electron energy-loss spectrum in these metals. In the low-frequency area, the function $\epsilon(\omega)$ for the noble metals can be written as follows:

$$\epsilon(\omega)=11-\tilde{\omega}_p^2/\omega^2+\epsilon_{\mathrm{inter}}(\omega), \tag{3.9}$$

Table 6
Basic Singularities of Electron Energy-Loss Functions in 3*d* Transition Metals V and Cr, Noble Metals Cu and Ag and IIB Metals Zn and Cd

Metal	Calculated		Experimental	
	E_i eV	Type of Singularity	E_i eV	Type of Singularity
V	9.8	Peak	9.5	Bend
	12.0	Maximum	12.0	Bend
	21.5	Peak	21.0	Peak
Cr	6.6	Maximum	6.5	Bend
	10.3	Peak	10.8	Bend
	13.1	Maximum		
	21.0	Maximum		
	25.8	Maximum	25.1	Maximum
Cu	4.5	Maximum	4.2	Maximum
	7.7	Bulge	7.8	Maximum
	12	Peak	11.9	Maximum
			14.8	Bulge
	19	Maximum	20.0	Maximum
Ag	3.7	Sharp Maximum	3.78	Sharp Peak
	8.3	Sharp Maximum	7.5	Maximum
	17.0	Sharp Maximum	18.0	Maximum
Zn	10.4	Maximum	10.0	Step
	12.0	Maximum		
	14.2	Bend	14.0	Maximum
	20.0	Peak	20.1	Maximum
Cd	8.98	Sharp Peak	9.07	Sharp Peak
	11.0	Maximum	11.0	Bulge
	18.0	Weak Peak	18.1	Weak Peak

where $\tilde{\omega}_p$ is the plasma frequency, corresponding to oscillations of s electrons, while $\epsilon_{\text{inter}}(\omega)$ is the contribution to the dielectric constant of interband transitions from the d band to unoccupied states near the Fermi level. Near the threshold $\text{Re}[\epsilon_{\text{inter}}(\omega)]$ there is a sharp increase in $\text{Im}[\epsilon_{\text{inter}}(\omega)]$ by 4–6, which follows from the Kramers-Kronig relations. The difference in the behavior of these metals results from two factors. First of all, the value of $\hbar\tilde{\omega}_p$ in silver is 7.741 eV, in copper 9.687 eV, i.e., for Ag it is less. Secondly, the interband transition threshold in Ag (~3.9 eV) is almost twice as great as the threshold in Cu (~2.1 eV). Therefore, in silver $\epsilon_1(\omega)$ passes through zero at a frequency slightly less than the threshold frequency, while in Cu the increase in $\mathrm{Re}[\epsilon_{\mathrm{inter}}(\omega)]$ is insufficient for this. Therefore, $L(\omega)$ in copper in the low frequency area is not a maximum but rather a step, coinciding with the threshold of interband transitions. Our calculations describe these singularities rather well (cf. Table 6 and Figures 3 and 4).

Zinc and cadmium, following copper and silver in the periodic table, retain to some extent plasmon spectral features. Experimental and theoretical studies of these metals show that near the Fermi surface of the conductivity band, Zn and Cd are rather well described by a model of nearly free electrons with a density of two electrons per atom. On this basis, Zn and Cd are frequently considered simply bivalent metals, the properties of which depend little on the dynamics of the filled *d* bands and which satisfy the necessary conditions for successful application of the pseudopotential formalism quite well[20].

A good test of the adequacy of the concept of these metals as simple ones is found in experiments on excitation of plasma oscillations in the metals. If this concept corresponded to the truth, the electron energy-loss spectra would contain a strong resonant line close to $\hbar\omega_p^{(s)}$, where $\omega_p^{(s)}$ is the plasma frequency of *s* electrons with a density of two electrons per atom. It amounts to 13.5 eV in Zn and 11.3 eV in Cd.

However, experimental studies of the electron energy-loss spectra in Zn and Cd show that correct description of the spectrum requires accurate consideration of the filled *d* bands[22].

In cadmium there is a strong plasma peak at $\hbar\omega$=9.07 eV. The width of this peak $\Delta E_{1/2}$~0.9 eV is quite precisely described within the framework of the model of nearly-free electrons, the position is sharply shifted relative to $\hbar\omega_p^{(s)}$, which cannot be explained within the framework of the model. Nevertheless, the form of this resonance is characteristic for simple metals (Figure 16). As concerns the plasma resonance in Zn, externally it differs from a single peak (Figure 15). In the area $\hbar\omega$~9–10 eV there is a sharp increase in $L(\omega)$, which has a threshold nature. With energy $\hbar\omega$=14.3 eV, only slightly exceeding $\hbar\omega_p^{(s)}$, there is a broad but still well expressed resonant peak.

The reason for this behavior of $L(\omega)$ can be understood only by considering the contribution of the *d* bands. In zinc, the narrow 3*d* band lies within the conductivity band, slightly above its bottom[21], which is located, according to the estimate of Juras et al., [83], 11 eV below the Fermi level. However, at energies only slightly exceeding the area occupied by 3*d* states, the electron structure of this metal takes on a form characteristic for a model of nearly free electrons. In cadmium the narrow 4d band falls somewhat below the bottom of the conductivity band[21]. However, both metals are characterized by the fact that the threshold for excitation of transitions involving d electrons is close to $\hbar\omega_p^{(s)}$, i.e., at frequencies $\sim\hbar\omega_p^{(s)}$ both d and s electrons have strong polarizability, which leads to interesting results in the $L(\omega)$ spectra of these metals.

Since in cadmium the 4d band is approximately 2 eV lower than the 3d band in zinc (more precisely, $E_F-E_{3d}\approx 11$ eV in Zn, $E_F-E_{4d}\approx 13$ eV in Cd), this difference has approximately the same influence on the electron energy-loss spectrum as in the noble metals. In cadmium, where the threshold of interband transitions involving the d band is above $\hbar\omega_p^{(s)}$, there is a clear plasma peak with energy $\hbar\omega\sim 9$ eV, i.e., the effect resulting from polarization of d electrons is limited only by the shift in plasmon energy. In zinc, where $\hbar\omega_p^{(s)}$ falls in the area of intensive interband transitions, there is a threshold increase in the electron energy-loss function $L(\omega)$, somewhat remeniscent of the behavior in Cu. The remaining singularities of the electron energy-loss spectra of these metals are also rather well reproduced by our calculations (cf. Table 6 and Figures 15, 16). Detailed analysis of plasma oscillations in Zn and Cd was undertaken by us in [81].

CHAPTER 4
MICROSCOPIC THEORY OF DIELECTRIC CONSTANT OF CRYSTALS

In this chapter, we would like to discuss the degree of approximation achieved by our equations for calculation of the optical characteristics of metals in somewhat more detail. As is well known (for example, cf. review[86]), when there is no spatial homogeneity, the dielectric constant becomes a function of both coordinates individually, and in a momentum space — a function of the two momenta **q** and **q**′. In a crystal, due to its periodic structure, these momenta are not arbitrary, but differ by the inverse lattice vector. Correspondingly, the dielectric constant depends on the vectors **q**+**K**, **q**+**K**′, where the vector **q** is limited to the first Brillouine zone, while **K** and **K**′ are inverse lattice vectors. Thus, the dielectric constant of a crystal (as well as such quantities as polarizability and susceptibility) are matrices in the space of vectors **K**.

Let us define the dielectric constant of a crystal as a function relating the change in external electrostatic potential $\delta V_{\rm ext}$ and full electrostatic potential $\delta V_{\rm tot}$

$$\delta V_{\rm ext}(x) = \int dx' \varepsilon(x, x')\, \delta V_{\rm tot}(x'), \tag{4.1}$$

where we use the four-dimensional representation $x=(\mathbf{r},t)$. Going over to a momentum representation $(d^4q = d\mathbf{q}d\omega/(2\pi)^4)$,

$$\varepsilon(x, x') = \varepsilon(\mathbf{r}, \mathbf{r}', t - t') = \sum_{\mathbf{K},\mathbf{K}'} \int d^4q \varepsilon(\mathbf{q} + \mathbf{K}, \mathbf{q} + \mathbf{K}', \omega)\, e^{i\omega(t-t')} \times$$
$$\times \exp[-i(\mathbf{q} + \mathbf{K})\mathbf{r} + i(\mathbf{q} + \mathbf{K}')\mathbf{r}'], \tag{4.2}$$

we obtain

$$\delta V_{\text{ext}}(\mathbf{q}+\mathbf{K},\omega)=\sum_{\mathbf{K}'}\varepsilon(\mathbf{q}+\mathbf{K},\mathbf{q}+\mathbf{K}',\omega)\,\delta V_{\text{tot}}(\mathbf{q}+\mathbf{K}',\omega). \tag{4.3}$$

Let us introduce the inverse dielectric constant matrix $\epsilon^{-1}(\mathbf{q}+\mathbf{K},\mathbf{q}+\mathbf{K}',\omega)$:

$$\sum_{\mathbf{K}''}\varepsilon(\mathbf{q}+\mathbf{K},\mathbf{q}+\mathbf{K}'',\omega)\,\varepsilon^{-1}(\mathbf{q}+\mathbf{K}'',\mathbf{q}+\mathbf{K}',\omega)=\delta_{\mathbf{K}\mathbf{K}'}. \tag{4.4}$$

Using this matrix, we can write relations which are inverse to (4.1):

$$\delta V_{\text{tot}}(x)=\int dx'\,\varepsilon^{-1}(x,x')\,\delta V_{\text{ext}}(x'), \tag{4.5}$$

$$\delta V_{\text{tot}}(\mathbf{q}+\mathbf{K},\omega)=\sum_{\mathbf{K}'}\varepsilon^{-1}(\mathbf{q}+\mathbf{K},\mathbf{q}+\mathbf{K}',\omega)\,\delta V_{\text{ext}}(\mathbf{q}+\mathbf{K}',\omega). \tag{4.6}$$

For a crystal we can also introduce the dielectric constant which depends only on one momentum and relates the components of the full and external potentials to the same momentum

$$\delta V_{\text{ext}}(\mathbf{q},\omega)=\varepsilon(\mathbf{q},\omega)\,\delta V_{\text{tot}}(\mathbf{q},\omega). \tag{4.7}$$

Actually,

$$\delta V_{\text{tot}}(\mathbf{q}+\mathbf{K},\omega)=\sum_{\mathbf{K}'}\varepsilon^{-1}(\mathbf{q}+\mathbf{K},\mathbf{q}+\mathbf{K}',\omega)\,\delta V_{\text{ext}}(\mathbf{q}+\mathbf{K}',\omega),$$

where the right portion contains δV_{ext} which, in contrast to δV_{tot}, can be selected as desired. Let us select it so that only the component $\delta V_{\text{ext}}(\mathbf{q},\omega)$ is other than zero with $\mathbf{K}'=0$. Then, studying the response of the system corresponding to the same momentum $\mathbf{q}$, we have ($\mathbf{K}=0$)

$$\delta V_{\text{tot}}(\mathbf{q},\omega)=\varepsilon^{-1}(\mathbf{q}+0,\mathbf{q}+0,\omega)\,\delta V_{\text{ext}}(\mathbf{q},\omega).$$

From which it is clear that the role of the ordinary dielectric constant in the crystal is played by the quantity

$$\varepsilon(\mathbf{q},\omega)=\frac{1}{\varepsilon^{-1}(\mathbf{q}+0,\mathbf{q}+0,\omega)}. \tag{4.8}$$

With small $\mathbf{q}$, this quantity is the ordinary macroscopic dielectric constant, determining the relationship between quantities averaged over physically infinitesimal volumes.

4.1. Local Field Effects and Volume–Correlation Corrections

The matrix $\epsilon^{-1}(\mathbf{q}+\mathbf{K}, \mathbf{q}+\mathbf{K}', \omega)$ can be represented by the electron susceptibility matrix $\chi(\mathbf{q}+\mathbf{K}, \mathbf{q}+\mathbf{K}', \omega)$

$$\varepsilon^{-1}(\mathbf{q}+\mathbf{K}, \mathbf{q}+\mathbf{K}'', \omega) = \delta_{\mathbf{K}\mathbf{K}'} + \frac{4\pi e^2}{|\mathbf{q}+\mathbf{K}|^2}\chi(\mathbf{q}+\mathbf{K}, \mathbf{q}+\mathbf{K}', \omega). \tag{4.9}$$

In turn, matrix $\chi(\mathbf{q}+\mathbf{K}, \mathbf{q}+\mathbf{K}', \omega)$ satisfies the following equation:

$$\begin{aligned}\chi(\mathbf{q}+\mathbf{K}, \mathbf{q}+\mathbf{K}', \omega) &= \pi(\mathbf{q}+\mathbf{K}, \mathbf{q}+\mathbf{K}', \omega) + \\ &+ \sum_{\mathbf{K}''}\pi(\mathbf{q}+\mathbf{K}, \mathbf{q}+\mathbf{K}'', \omega)\frac{4\pi e^2}{|\mathbf{q}+\mathbf{K}''|^2}\chi(\mathbf{q}+\mathbf{K}'', \mathbf{q}+\mathbf{K}', \omega).\end{aligned} \tag{4.10}$$

Here $\pi(\mathbf{q}+\mathbf{K}, \mathbf{q}+\mathbf{K}', \omega)$ is the electron polarizability matrix, the set of diagrams, irreducible relative to the direct Coulomb interaction. It is easy to see that the macroscopic dielectric constant $\epsilon_M(\omega)$ can be written as follows:

$$\varepsilon_M(\omega) = 1 - \lim_{\mathbf{q}\to 0}\frac{4\pi e^2}{q^2}\tilde{\tilde{\chi}}(\mathbf{q}+\mathbf{O}, \mathbf{q}+\mathbf{O}, \omega). \tag{4.11}$$

Here the matrix $\tilde{\tilde{\chi}}(\mathbf{q}+\mathbf{K}, \mathbf{q}+\mathbf{K}', \omega)$ satisfies an equation quite similar to (4.10)

$$\begin{aligned}\tilde{\tilde{\chi}}(\mathbf{q}+\mathbf{K}, \mathbf{q}+\mathbf{K}', \omega) &= \pi(\mathbf{q}+\mathbf{K}, \mathbf{q}+\mathbf{K}', \omega) + \\ &+ \sum_{\mathbf{K}''\neq 0}\pi(\mathbf{q}+\mathbf{K}, \mathbf{q}+\mathbf{K}'', \omega)\frac{4\pi e^2}{|\mathbf{q}+\mathbf{K}''|^2}\tilde{\tilde{\chi}}(\mathbf{q}+\mathbf{K}'', \mathbf{q}+\mathbf{K}', \omega).\end{aligned} \tag{4.12}$$

The difference between equations (4.12) and (4.10) is that in the right portion of equation (4.12) with summation in the second component the term with $\mathbf{K}''=0$ is missing. The dielectric constant matrix $\epsilon(\mathbf{q}+\mathbf{K}, \mathbf{q}+\mathbf{K}', \omega)$ using the symbols introduced above becomes:

$$\varepsilon(\mathbf{q}+\mathbf{K}, \mathbf{q}+\mathbf{K}', \omega) = \delta_{\mathbf{K}\mathbf{K}'} - \frac{4\pi e^2}{|\mathbf{q}+\mathbf{K}|^2}\pi(\mathbf{q}+\mathbf{K}, \mathbf{q}+\mathbf{K}', \omega). \tag{4.13}$$

Even if we consider that we know the electron polarizability matrix $\pi(\mathbf{q}+\mathbf{K}, \mathbf{q}+\mathbf{K}', \omega)$, to determine the macroscopic dielectric constant $\epsilon_M(\omega)$ we must solve equation (4.12), which defines the function $\tilde{\tilde{\chi}}(\mathbf{q}+\mathbf{O}', \mathbf{q}+\mathbf{O}, \omega)$. Solution of this equation amounts to consideration of rapidly oscillating Coulomb fields in the crystal. Considering these fields is usually referred to as considering local–field effects.

In the simple single-frequency approximation for the matrix $\pi(\mathbf{q}+\mathbf{K}, \mathbf{q}+\mathbf{K}', \omega)$ it is easy to produce the following expression[86, 87]:

$$\pi(\mathbf{q}+\mathbf{K}, \mathbf{q}+\mathbf{K}', \omega) = 2 \sum_{\mathbf{k},\lambda,\lambda'} \frac{f_{\mathbf{k}\lambda} - f_{\mathbf{k}+\mathbf{q},\lambda'}}{\hbar\omega - E_{\mathbf{k}+\mathbf{q},\lambda'} + E_{\mathbf{k}\lambda} + i\delta} \times \\ \times \langle \mathbf{k}\lambda | e^{-i(\mathbf{q}+\mathbf{K})\mathbf{r}} | \mathbf{k}+\mathbf{q}, \lambda' \rangle \langle \mathbf{k}+\mathbf{q}, \lambda' | e^{i(\mathbf{q}+\mathbf{K}')\mathbf{r}} | \mathbf{k}\lambda \rangle. \tag{4.14}$$

It is easy to see that the dielectric constant $\epsilon(\omega)$ which we studied in the previous chapters is nothing but

$$\varepsilon(\omega) = 1 - \lim_{\mathbf{q}\to 0} \frac{4\pi e^2}{q^2} \pi(\mathbf{q}+\mathbf{O}, \mathbf{q}+\mathbf{O}, \omega), \tag{4.15}$$

where $\pi(\mathbf{q}+\mathbf{O}, \mathbf{q}+\mathbf{O}, \omega)$ is fixed by expression (4.10). This means that, first of all, in our calculations of dielectric constant we have ignored the difference between the quantities $\tilde{\tilde{\chi}}(\mathbf{q}+\mathbf{O}, \mathbf{q}+\mathbf{O}, \omega)$ and $\pi(\mathbf{q}+\mathbf{O}, \mathbf{q}+\mathbf{O}, \omega)$, i.e., we have ignored local-field effects. The second assumption made above is selection of the polarization operator as (4.14), i.e., completely ignoring the difference in the electron polarizability matrix $\pi(\mathbf{q}+\mathbf{K}, \mathbf{q}+\mathbf{K}', \omega)$ from the polarizability of a system of noninteracting particles. This corresponds to the so-called random phase approximation for the dielectric constant[88].

No orderly approach to calculation of the precise electron polarizability $\pi(\mathbf{q}+\mathbf{K}, \mathbf{q}+\mathbf{K}', \omega)$ has yet been undertaken. This problem has been studied with some accuracy only for the case of a homogeneous electron gas[88]. The polarizability of this system can be represented in the following form:

$$\pi(\mathbf{q}, \omega) = \frac{\pi_0(\mathbf{q}, \omega)}{1 + \frac{4\pi e^2}{q^2} G(\mathbf{q}, \omega) \pi_0(\mathbf{q}, \omega)}, \tag{4.16}$$

where $\pi_0(\mathbf{q},\omega)$ is the polarizability of a noninteracting electron gas, while $G(\mathbf{q},\omega)$ is the so-called exchange-correlation correction function, considering the contribution of exchange-correlation interactions to polarizability. As was shown in [87], within a generally self-consistent field for the matrix $\pi(\mathbf{q}+\mathbf{K}, \mathbf{q}+\mathbf{K}', \omega)$ we can obtain an expression which at the limit of the homogeneous system leads to the formula for $\pi(\mathbf{q},\omega)$, similar to (4.16)

$$\pi(\mathbf{q}+\mathbf{K}, \mathbf{q}+\mathbf{K}', \omega) = \sum_{\mathbf{K}''} \pi_0(\mathbf{q}+\mathbf{K}, \mathbf{q}+\mathbf{K}'', \omega) \Lambda(\mathbf{q}+\mathbf{K}'', \mathbf{q}+\mathbf{K}', \omega), \tag{4.17}$$

where $\pi_0(\mathbf{q+K}, \mathbf{q+K'}, \omega)$ is the polarizability of the system of noninteracting particles (4.14), while $\Lambda(\mathbf{q+K}, \mathbf{q+K'}, \omega)$ is a peak matrix defined by the equation

$$\Lambda^{-1}(\mathbf{q}+\mathbf{K}, \mathbf{q}+\mathbf{K}', \omega) = \delta_{\mathbf{KK}'} + \frac{4\pi e^2}{|\mathbf{q}+\mathbf{K}|^2} \times \\ \times \sum_{\mathbf{K}''} G(\mathbf{q}+\mathbf{K}, \mathbf{q}+\mathbf{K}'', \omega)\, \pi_0(\mathbf{q}+\mathbf{K}'', \mathbf{q}+\mathbf{K}', \omega). \tag{4.18}$$

For the matrix $G(\mathbf{q+K}, \mathbf{q+K'}, \omega)$, work[87] obtained a rather complex expression through the filling number, the spectrum of excited electrons and a form factor such as $\langle \mathbf{k}\lambda | e^{i(\mathbf{q+K})\mathbf{r}} | \mathbf{k+q}, \lambda' \rangle$. Unfortunately, the equations obtained in [87] are quite cumbersome and unsuitable for concrete numerical calculations.

To understand more specifically how the macroscopic dielectric constant varies with local-field effects and exchange-correlation corrections, let us write it in the form

$$\epsilon_M(\omega) = 1(1 + V_0(\chi)_{00}), \tag{4.19}$$

where $V_0 = \dfrac{4\pi e^2}{q^2} = \left.\dfrac{4\pi e^2}{|\mathbf{q+K}|^2}\right|_{\mathbf{K}=0}$ is the Coulomb interaction $\chi_{00} = \chi(\mathbf{q+G}, \mathbf{q+G'}, \omega)|_{\mathbf{G}=\mathbf{G}'=0}$. These symbols are introduced for convenience. The susceptibility matrix can be defined as (cf. [86])

$$\hat{\chi}^{-1} = \hat{\pi}_0^{-1} - (\hat{V} - \hat{I}_{xc}), \tag{4.20}$$

where $\hat{I}_{xc}$ is the exchange-correlation interaction operator, $\hat{V}$ is the Coulomb interaction operator. Considering that where $\mathbf{q}\to 0$, $I_{xc}(\mathbf{q}+G, \mathbf{q}+G', \omega) \sim 1$, while $\pi_0(\mathbf{q+G}, \mathbf{q+G'}, \omega) \sim |\mathbf{q+G}| \cdot |\mathbf{q+G'}|$, and using standard matrix inversion methods[89], we can show that

$$(\chi)_{00} = \frac{1}{(\hat{\chi}_0^{-1})_{00} - \sum\limits_{\mathbf{G},\mathbf{G}'\neq 0} (\hat{\pi}_0^{-1})_{0\mathbf{G}} \left[\left(\widetilde{\hat{\pi}_0^{-1} - (\hat{V} - \hat{I}_{xc})}\right)^{-1}\right]_{\mathbf{GG}'} (\hat{\pi}_0^{-1})_{\mathbf{G}'0}}. \tag{4.21}$$

Here $\hat{\chi}_0$ is the susceptibility of the system to ignoring exchange and correlation, while the wavy overline represents the matrix obtained from the initial matrix by crossing out the row and the column with $\mathbf{G}=0$. Elements $(\hat{\pi}_0^{-1})$ can be expressed explicitly through the elements of the operator $(\tilde{\tilde{\pi}}_0)^{-1}$, which has no singularities as $\mathbf{q}\to 0$, and elements of the operator $\hat{\pi}_0$:

$$(\hat{\pi}_0^{-1})_{\mathbf{GG}'} = (\tilde{\tilde{\pi}}_0)^{-1}_{\mathbf{GG}'} + \frac{1}{\beta} A^T_{\mathbf{G}} A_{\mathbf{G}'}, \tag{4.22a}$$

$$(\hat{\pi}_0^{-1})_{0\mathbf{G}} = -\frac{1}{\beta} A_{\mathbf{G}}, \tag{4.22b}$$

$$(\hat{\pi}_0^{-1})_{\mathbf{G}'0} = -\frac{1}{\beta} A^T_{G'}, \tag{4.22c}$$

$$\beta = (\hat{\pi}_0)_{00} - \sum_{\mathbf{G},\mathbf{G}'\neq 0} (\hat{\pi}_0)_{0\mathbf{G}} (\tilde{\tilde{\pi}}_0)^{-1}_{\mathbf{GG}'} (\hat{\pi}_0)_{\mathbf{G}'0}, \tag{4.22d}$$

$$A_{\mathbf{G}} = \sum_{\mathbf{G}_1\neq 0} (\tilde{\tilde{\pi}}_0)^{-1}_{\mathbf{GG}_1} (\hat{\pi}_0)_{\mathbf{G}_1 0}, \tag{4.22e}$$

T represents the operation of transposition.

It is clear from equation (4.21) that, ignoring exchange and correlation, we obtain an expression for the macroscopic dielectric constant in the random phase approximation (i.e., $(\chi)_{00}=(\chi_0)_{00}$). However, even in this case replacement of the matrix $\tilde{\tilde{\chi}}(\mathbf{q}+0, \mathbf{q}+0, \omega)$ by $\pi_0(\mathbf{q}+0, \mathbf{q}+0, \omega)$ generally leads to a different result due to local-field effects, i.e., due to rapidly oscillating Coulomb fields.

In many calculations of dielectric constant of semiconductors, attempts have been made within the framework of the single-frequency approach described above to consider local-field effects, i.e., by the use of expression (4.14) for π_0 to calculate the function $\tilde{\tilde{\chi}}(\mathbf{q}+0, \mathbf{q}+0, \omega)$ using equation (4.12). This is equivalent to ignoring exchange and correlation in equation (4.20), i.e., the random phase approximation (cf. review[90]). In our opinion, this approach cannot be considered justified, since it is not clear how local-field effects and exchange-correlation corrections are interrelated. We must therefore consider both simultaneously.

Many interesting results can be obtained by the use of the Vannier representation of the wave functions of electrons, in which we can write the following expression for the macroscopic dielectric constant $\epsilon_M(\mathbf{q}, \omega)$ [86, 90]:

$$\varepsilon_M(\mathbf{q}, \omega) = 1 - \frac{4\pi e^2}{q^2} \sum_{\substack{\lambda\lambda' \\ nn'}} \left(A^n_{\lambda\lambda'}(\mathbf{q})\right)^* \tilde{\tilde{\chi}}^{nn'}_{\lambda\lambda'}(\mathbf{q}, \omega)\, A^{n'}_{\lambda\lambda'}(\mathbf{q}). \tag{4.23}$$

The form factors $A^n_{\lambda\lambda'}(\mathbf{q})$ contained in (4.23) are defined by the equation

$$A^n_{\lambda\lambda'}(\mathbf{q}) = \int d\mathbf{r} a^*_\lambda(\mathbf{r}) e^{-i\mathbf{q}\mathbf{r}} a_{\lambda'}(\mathbf{r} - \mathbf{R}_n), \tag{4.24}$$

where $a_\lambda(\mathbf{r}\text{-}\mathbf{R}_n)$ is the Vannier function for an electron in band λ, centered on the atom at the node with coordinate $\mathbf{R}_n$. The matrix $\tilde{\tilde{\chi}}^{nn'}_{\lambda\lambda'}(\mathbf{q},\omega)$ in (4.23) is defined by the equation

$$[\tilde{\tilde{\chi}}^{-1}(\mathbf{q},\omega)]^{nn'}_{\lambda\lambda'} = [\hat{\pi}_0^{-1}(\mathbf{q},\omega)]^{nn'}_{\lambda\lambda'} + G^{nn'}_{\lambda\lambda'}(\mathbf{q},\omega) - V^{nn'}_{\lambda\lambda'}(\mathbf{q}), \tag{4.25}$$

where the matrix $[\hat{\pi}_0(\mathbf{q},\omega)]^{nn'}_{\lambda\lambda'}$ is the Vannier component of polarizability (4.14)

$$[\hat{\pi}_0(\mathbf{q},\omega)]^{nn'}_{\lambda\lambda'} = 2\sum_{\mathbf{k}} \frac{f_{\mathbf{k}\lambda} - f_{\mathbf{k}+\mathbf{q},\lambda'}}{\hbar\omega - E_{\mathbf{k}+\mathbf{q},\lambda'} + E_{\mathbf{k}\lambda}} \exp\left[i(\mathbf{k}+\mathbf{q})(\mathbf{R}_n - \mathbf{R}_{n'})\right], \tag{4.26}$$

while the quantities $G^{nn'}_{\lambda\lambda'}(\mathbf{q},\omega)$ and $V^{nn'}_{\lambda\lambda'}(\mathbf{q})$ describe the contributions from exchange-correlation effects and Coulomb short-wave fields

$$G^{nn'}_{\lambda\lambda'}(\mathbf{q},\omega) = \sum_{\mathbf{K},\mathbf{K}'} \left(A^n_{\lambda\lambda'}(\mathbf{q}+\mathbf{K})\right)^* \frac{4\pi e^2}{|\mathbf{q}+\mathbf{K}|^2} G(\mathbf{q}+\mathbf{K}, \mathbf{q}+\mathbf{K}', \omega) A^{n'}_{\lambda\lambda'}(\mathbf{q}+\mathbf{K}') \tag{4.27}$$

and

$$V^{nn'}_{\lambda\lambda'}(\mathbf{q}) = \sum_{\mathbf{K}\neq 0} \left(A^n_{\lambda\lambda'}(\mathbf{q}+\mathbf{K})\right)^* \frac{4\pi e^2}{|\mathbf{q}+\mathbf{K}|^2} A^{n'}_{\lambda\lambda'}(\mathbf{q}+\mathbf{K}). \tag{4.28}$$

We can see from equations (4.25)-(4.28) that exchange-correlation effects and local-field effects are included in the expression for the macroscopic dielectric constant so that they partially compensate each other. As equation (4.23) shows, we can also introduce the concept of the intraband contribution ($\lambda=\lambda'$) and interband contribution ($\lambda\neq\lambda'$) for the macroscopic dielectric constant. Furthermore, at the limit $\mathbf{q}\to 0$, we can obtain a precise expression for the intraband contribution to the dielectric constant $\epsilon_M(\omega)$. Considering that the quantities $G^{nn'}_{\lambda\lambda'}(\mathbf{q},\omega)$ and $V^{nn'}_{\lambda\lambda'}(\mathbf{q})$ approach a finite limit at $\mathbf{q}\to 0$, while $[\hat{\pi}_0^{-1}(\mathbf{q},\omega)]^{nn'}_{\lambda\lambda}$ diverges as ω^2/q^2, we can, by sequential iteration of equation (4.25), show that

$$\lim_{q\to 0} [\tilde{\tilde{\chi}}(\mathbf{q}, \omega)]^{nn'}_{\lambda\lambda} = \lim_{q\to 0} [\hat{\pi}_0(\mathbf{q}, \omega)]^{nn'}_{\lambda\lambda}. \tag{4.29}$$

This relationship means that at the limit $\mathbf{q}\to 0$ the intraband contribution to the macroscopic dielectric constant $\epsilon_M(\omega)$ coincides with the similar contribution to function $\epsilon(\omega)$, defined by equation (1.15), i.e., corrections for exchange–correlation interaction and local fields where $\mathbf{q}=0$ are completely absent in the intraband contribution to the macroscopic dielectric constant. Consequently, for $\epsilon_M(\omega)$ we can write the expression

$$\varepsilon_M(\omega) = 1 - \tilde{\omega}_p^2/\omega^2 + \varepsilon_M^{\text{inter}}(\omega), \tag{4.30}$$

where $\tilde{\omega}_p^2$ is determined by equation (1.17). The function $\epsilon_M^{\text{inter}}(\omega)$ represents the contribution of interband transitions which, of course, depends on local–field effects and exchange–correlation effects even where $\mathbf{q}=0$ and therefore differs from the corresponding contribution made by equation (1.15), which was used in our calculations. As our calculations and their comparison with the experimental data show, actually in most metals this contribution is rather small. A similar conclusion of the small nature of the contribution of local–field effects to optical characteristics of many semiconductors was reached by the authors of [91, 92]. In their opinion, the previously stated assumption of a significant role for such effects in semiconductors [93, 94] is related to insufficiently accurate calculation of the current operator matrix element in these works. Clearly, however, the problem requires more careful study.

4.2. Density Functional Method and Lifetime of Electron Levels

In considering exchange–correlation effects within the framework of the self–consistent field theory, there is an extremely important and still unmentioned task, the construction of an effective potential defining the electron spectrum. As we know, all modern computations of the band spectrum of electrons, including those which we have used to calculate the optical properties of metals, are based on the so–called density functional method.

The idea of this method is to reduce the problem of the interaction of the electron gas of a metal to the problem of a system of independent fermions, moving in a static external effective field [35, 36].

The theorem can be strictly proven that the energy of the base state of this system of interacting electrons in an external field acting only on the charge density (as a field of ions in a crystal or external electrostatic potential) is a single–valued functional of electron density

$$E_0 = E\{n(\mathbf{r})\}. \tag{4.31}$$

Naturally, the distribution of electron density in a crystal in the base state is determined in this case from the condition of the minimum of functional (4.31).

Functional (4.31) can be written in more detail as follows:

$$E\{n(\mathbf{r})\} = T\{n(\mathbf{r})\} + \int n(\mathbf{r})\, v_{\text{ext}}(\mathbf{r})\, d\mathbf{r} + \\ + \frac{e^2}{2}\int\int \frac{n(\mathbf{r})\, n(\mathbf{r}') - n(\mathbf{r})\, n^+(\mathbf{r}')}{|\mathbf{r} - \mathbf{r}'|}\, d\mathbf{r} d\mathbf{r}' + \widetilde{E}_{xc}\{n(\mathbf{r})\}. \tag{4.32}$$

Here $T\{n(\mathbf{r})\}$ is the kinetic energy of a heterogeneous system of interacting electrons. The second component describes the interaction with the external field, the third — the Coulomb (Hartree) interaction of the electrons with each other and with ions, the charge distribution of which is described by density $n^+(\mathbf{r})$, while the fourth component describes the exchange-correlation energy.

Unfortunately, we know of no precise expression for the kinetic and exchange-correlation energy. Kohn and Sham[35] suggested the following approach for the problem of seeking the energy of the base state of a system of interacting electrons, allowing us to avoid the problem of the form of the function $T\{n(\mathbf{r})\}$ and reducing the problem to solution of a single-frequency Schroedinger equation in a self-consistent external field. They rewrote expression (4.32) as

$$E\{n(\mathbf{r})\} = T_s\{n(\mathbf{r})\} + \int v_{\text{ext}}(\mathbf{r})\, n(\mathbf{r})\, d\mathbf{r} + \tag{4.32}$$

$$+ \frac{e^2}{2}\int\int \frac{n(\mathbf{r})\, n(\mathbf{r}') - n(\mathbf{r})\, n^+(\mathbf{r}')}{|\mathbf{r} - \mathbf{r}'|}\, d\mathbf{r} d\mathbf{r}' + T\{n(\mathbf{r})\} - \\ - T_s\{n(\mathbf{r})\} + \widetilde{E}_{xc}\{n(\mathbf{r})\}, \tag{4.33}$$

adding and subtracting $T_s\{n(\mathbf{r})\}$ — the kinetic energy of a system of noninteracting electrons at the same density. The last three components can be combined into one

$$E_{xc}\{n(\mathbf{r})\} = T\{n(\mathbf{r})\} - T_s\{n(\mathbf{r})\} + \widetilde{E}_{xc}\{n(\mathbf{r})\}, \tag{4.34}$$

so that $E_{xc}\{n(\mathbf{r})\}$ represents the contribution to total energy related to correlations among electrons both in potential and in kinetic energy. Now the expression for $E\{n(\mathbf{r})\}$ can be finally rewritten as

$$E\{n(\mathbf{r})\} = T_s\{n(\mathbf{r})\} + \int v_{\text{ext}}(\mathbf{r})\, n(\mathbf{r}) d\mathbf{r} + \\ + \frac{e^2}{2}\int\int \frac{n(\mathbf{r})\, n(\mathbf{r}') - n(\mathbf{r})\, n^+(\mathbf{r}')}{|\mathbf{r}-\mathbf{r}'|} d\mathbf{r} d\mathbf{r}' + E_{xc}\{n(\mathbf{r})\}. \tag{4.35}$$

Further, Kohn and Sham showed that, by formally introducing definition of electron density as the sum with respect to single-frequency states,

$$n(\mathbf{r}) = \sum_i \left|\psi_i(\mathbf{r})\right|^2, \tag{4.36}$$

using minimum condition (4.35) it is possible to obtain the following single-frequency Schroedinger equation, defining the wave functions $\psi_i(\mathbf{r})$ and energy levels ϵ_i:

$$\left[-\frac{\hbar^2\nabla^2}{2m} + v_{\text{ext}}(\mathbf{r}) + e^2\int \frac{n(\mathbf{r}') - n^+(\mathbf{r}')}{|\mathbf{r}-\mathbf{r}'|} d\mathbf{r}' + V_{xc}(\mathbf{r})\right] \psi_i(\mathbf{r}) = \varepsilon_i \psi_i(\mathbf{r}), \tag{4.37}$$

where $V_{xc}(\mathbf{r})$ is the single-frequency exchange-correlation potential, defined according to the relation

$$V_{xc}(\mathbf{r}) = \frac{\delta E_{xc}\{n(\mathbf{r})\}}{\delta n(\mathbf{r})}. \tag{4.38}$$

Equation (4.37) is the basis of modern band calculations as a model of the electron-ion system (assuming v_{ext} to be equal to the ion potential).

The specific form of the exchange-correlation energy and, therefore, the exchange-correlation potential, remain unknown in this approach. However, there is a method which gives a reasonable approximation of these quantities. At present, we know expressions for $E_{xc}(n)$ in the case of a homogeneous electron gas in various approximations. Assuming that in a real, heterogeneous electron gas the electron density changes with distance fairly slowly, we can introduce the so-called local approximation, i.e., consider that the density of exchange-correlation energy $\epsilon_{xc}(\mathbf{r})$ depends only on the electron density at the same point *r*, i.e.,

$$E_{xc}\{n(\mathbf{r})\} = \int \varepsilon_{xc}^h(n(\mathbf{r}))\, n(\mathbf{r})\, d\mathbf{r}, \tag{4.39}$$

where $\epsilon_{xc}^h(n)$ is the exchange-correlation energy density of a homogeneous electron gas of density *n*. It then becomes possible to use known expressions for a homogeneous electron gas. This is the approach generally used in the calculation of the electron band structure. In the simplest case in calculating only the exchange interaction of electrons, $V_{xc}(\mathbf{r})$ is proportional to the density to the 1/3 power.

At the present time, more complex expressions are usually employed for V_{xc}, also considering correlation corrections. This question is discussed in some detail, for example, in the review of Rajagopal[95].

If we fix the exchange–correlation potential and thus produce a reliable method for solving the Schroedinger equation in a crystal then, starting with any reasonable priming potential (or priming electron density), we can construct a self–consistent iterative procedure to compute both the electron density and wave functions $\psi_i(\mathbf{r})$ and the eigenvalues of energy ϵ_i.

Since in our calculations we used a self–consistent pontential calculated considering exchange and correlation, we cannot say that we have completely ignored these effects. They are partially included in the wave functions $\Psi_{\mathbf{k}\lambda}(\mathbf{r})$ and energy $E_{\mathbf{k}\lambda}$ which we defined. It is more correct to say that we have partially included the exchange and correlation effects in the electron mass operator and did not consider them at all in the peak portion of the irreducible polarization operator.

As we can see from the above, the single–frequency spectrum defined by equation (4.37) is included in the density functional method by a rather formal means, only in order to compute the base state energy. However, the density functional method, in addition to the base state energy, also allows us to find the static response functions of the electron system[35, 95]. We shall now briefly discuss the basic ideas of this conclusion. First of all, we assign a slight change to the external potential δV_{ext}. The electron density in system $n(\mathbf{r})$ can then be represented as

$$n(\mathbf{r})=n_0(\mathbf{r})+\delta n_{\text{ind}}(\mathbf{r}). \tag{4.40}$$

The effective potential $V_{\text{eff}}(\mathbf{r})$ included in expression (4.37) can also be written as an expansion with respect to the small quantity δn_{ind} or, which amounts to the same thing, with respect to δV_{ext}

$$V_{\text{eff}}(\mathbf{r}) = V_{\text{eff}}^{(0)}(\mathbf{r}) + \delta V_{\text{eff}}(\mathbf{r}). \tag{4.41}$$

The change in effective potential δV_{eff} in the lower order with respect to δn_{ind} is given by the following expression

$$\delta V_{\text{eff}}(\mathbf{r}) = \delta V_{\text{ext}}(\mathbf{r}) + e^2 \int d\mathbf{r}' \frac{\delta n_{\text{ind}}(\mathbf{r}')}{|\mathbf{r}-\mathbf{r}'|} + \int d\mathbf{r}' I_{xc}(\mathbf{r}, \mathbf{r}')\, \delta n_{\text{ind}}(\mathbf{r}'), \tag{4.42}$$

where

$$I_{xc}(\mathbf{r}, \mathbf{r}') = \frac{\delta^2 E_{xc}\{n(\mathbf{r})\}}{\delta n(\mathbf{r})\, \delta n(\mathbf{r}')}. \tag{4.43}$$

Solving equation (4.37) in the linear approximation with respect to $\delta V_{\text{eff}}(\mathbf{r})$, we immediately find that

$$\delta n_{\text{ind}}(\mathbf{r}) = \int dr' \pi_{\text{eff}}(\mathbf{r}, \mathbf{r}')\, \delta V_{\text{eff}}(\mathbf{r}'), \tag{4.44}$$

where $\pi_{\text{eff}}(\mathbf{r}, \mathbf{r}')$ is the polarizability of the system of noninterrupting electrons described by equation (4.37) with $V_{\text{eff}}(\mathbf{r}) = V^0_{\text{eff}}(\mathbf{r})$:

$$\pi_{\text{eff}}(\mathbf{r}, \mathbf{r}') = 2\,\text{Im} \int_{-\infty}^{\infty} \frac{d\omega}{2\pi} G(\mathbf{r}, \mathbf{r}', \omega)\, G(\mathbf{r}', \mathbf{r}, \omega). \tag{4.45}$$

The Green function $G(\mathbf{r}, \mathbf{r}', \omega)$ is

$$G(\mathbf{r}, \mathbf{r}', \omega) = \sum_i \frac{\psi_i^*(\mathbf{r})\, \psi_i(\mathbf{r})}{\omega - \varepsilon_i + i\delta}, \tag{4.46}$$

where $\psi_i(\mathbf{r})$ and ϵ_i are the eigenfunctions and eigenvalues of equation (4.37). It is easy to see that the Fourier component of polarizability $\pi_{\text{eff}}(\mathbf{q}+\mathbf{K}, \mathbf{q}+\mathbf{K}')$ can be reduced to equation (4.14), in which we must posit the value of ω equal to zero.

Considering further that the susceptibility of the system $\chi(\mathbf{r}, \mathbf{r}')$ is determined by the equation

$$\delta n_{\text{ind}}(\mathbf{r}) = \int d\mathbf{r}' \chi(\mathbf{r}, \mathbf{r}')\, \delta V_{\text{ext}}(\mathbf{r}') \tag{4.47}$$

and keeping also in mind the relationship (4.42) between the quantities $\delta V_{\text{eff}}(\mathbf{r})$ and $\delta V_{\text{ext}}(\mathbf{r})$, it is easy to produce the final equation for the static susceptibility

$$\chi(\mathbf{r}, \mathbf{r}') = \pi_{\text{eff}}(\mathbf{r}, \mathbf{r}') + \int\int d\mathbf{r}'' d\mathbf{r}'''\, \pi_{\text{eff}}(\mathbf{r}, \mathbf{r}'')\, K(\mathbf{r}'', \mathbf{r}''')\, \chi(\mathbf{r}''', \mathbf{r}'). \tag{4.48}$$

Here, the kernel $K(\mathbf{r}, \mathbf{r}')$ is expressed simply through the Coulomb interaction and $I_{xc}(\mathbf{r}, \mathbf{r}')$

$$K(\mathbf{r}, \mathbf{r}') = \frac{e^2}{|\mathbf{r} - \mathbf{r}'|} + I_{xc}(\mathbf{r}, \mathbf{r}'). \tag{4.49}$$

It is easy to see that in the Fourier representation, equation (4.48) coincides with the static version of equation (4.10) considering equations (4.17) and (4.18). The role of the peak matrix $\Lambda(\mathbf{q}+\mathbf{K}, \mathbf{q}+\mathbf{K}')$ is played here by the quantity

$$\Lambda^{-1}(\mathbf{r}, \mathbf{r}') = \delta(\mathbf{r} - \mathbf{r}') + \int I_{xc}(\mathbf{r}, \mathbf{r}'')\, \pi_{\text{eff}}(\mathbf{r}'', \mathbf{r}')\, d\mathbf{r}''. \tag{4.50}$$

In computing optical properties, we are interested by the response not to a static field, but rather to the field of an electromagnetic wave which varies over time. A simple extension of the density functional method to include time-dependent fields was suggested in the works of Singhal and Callaway[96], and Zangwill and Soven[97]. The essence of this suggestion is extremely simple and amounts to replacement in equation (4.48) of the static susceptiblity $\chi(\mathbf{r}, \mathbf{r}')$ and static effective polarizability by their dynamic analogues, i.e., by the quantites $\chi(\mathbf{r}, \mathbf{r}', \omega)$ and $\pi_{\text{eff}}(\mathbf{r}, \mathbf{r}', \omega)$. The Fourier component of $\pi_{\text{eff}}(\mathbf{r}, \mathbf{r}', \omega)$ in this case is determined by equation (4.14). The solution of equation (4.48) for the case of a crystal is just as difficult to obtain as the solution of equations (4.10), (4.17) and (4.18), derived within the framework of the ordinary multifrequency theory of perturbations. The situation is relieved to some extent only for simple metals within the framework of perturbations theory for the weak pseudopotential, where we can limit ourselves to summation with respect to the vectors of the inverse array by their relatively small number. The corresponding calculations for Al were undertaken by Singhal and Callaway[96]. As these calculations showed, local-field effects in this case are actually small and the quantity $\epsilon_M(\omega)$ corresponds with good accuracy to the diagonal matrix element $\epsilon(0, 0, \omega)$.

Calculations of the dynamic susceptibility of atoms[97, 98], molecules[99, 100] and small metallic clusters[101] were also performed using the method just described, reproducing the experimental data of[102] quite well.

From the standpoint of the precise theory, equation (4.48) for dynamic susceptibility with a kernel which is independent of frequency ω cannot correctly describe the behavior of function χ at all frequencies. It can be shown[88] that in the case of a homogeneous electron gas the function $\chi(\mathbf{q}, \omega)$ obtained within the framework of this approximation does not satisfy the precise sum rules.

An orderly description of the dynamic response requires an extension of the density functional method which would yield a frequency-dependent kernel $K(\mathbf{r}, \mathbf{r}', \omega)$. Unfortunately, no such approach presently exists. Therefore, in our analysis in Chapter 1 we were forced to introduce a correction to the intraband contribution, considering the processes of electron-phonon scattering, leading to the classical Drude–Siner equation for $\epsilon^{\text{intra}}(\omega)$.

It is clear from our analysis that in our computations, if we consider them from the standpoint of pure theory, a rather large number of approximate assumptions was used, and still the optical characteristics of the metals which we computed agree qualitatively and quantitatively with the experimental data within reasonable limits. The contributions of local field effects and space-correlation effects which we did not consider are rather small in most metals.

CHAPTER 5
OPTICAL PROPERTIES OF TRANSITION METALS IN THE INFRARED AREA OF THE SPECTRUM

The optical conductivity of metals at low frequencies ($\hbar\omega \lesssim 0.5$ eV) is explained basically by intrazonal electron transitions. This area of the spectrum is generally used to determine the parameters of the Drude conductivity — the intraband plasma frequency $\tilde{\omega}_p$ and the electron collision frequency γ. As concerns direct intraband transitions, their significance is less clear. It is well known that in univalent metals there is a threshold of interband absorption corresponding to a photon energy of 1–2 eV. The assumption of existence of such a threshold or of a smooth frequency variation $\epsilon_2^{inter}(\omega)$ in the infrared (IR) area of the spectrum is frequently used for transition metals as well. The question of how well this assumption agrees with the complex electron structure of the transition metals has practically never been discussed.

Experimental studies have yielded somewhat contradictory information on this point. Thus, in [103], a maximum of $\epsilon_2^{inter}(\omega)$ was found at $\hbar\omega \approx 0.3$ eV. Measurements undertaken in [69] found no maximum or singularity of $\epsilon_2^{inter}(\omega)$ in this frequency range. It was noted in a later work[104] that even where $\hbar\omega \sim 0.1$ eV, the function $\sigma(\omega)$ differs significantly from a Drude function. Contradictory data were also obtained for V, Ta and Rh[65, 71, 104-106]. In ferromagnetic iron, a strong anomaly of optical conductivity was observed at $\hbar\omega \sim 0.2$–0.3 eV[107, 108]. However, for molybdenum all others report no singularities of $\epsilon_2^{inter}(\omega)$ at $\hbar\omega \lesssim 0.5$ eV[104, 71]. Optical conductivity anomalies were observed also in other transition metals. More detailed information on this subject can be found in dissertation[105].

In analyzing experimental results, it is natural to question whether observed anomalies in $\sigma(\omega)$ are singularity in the volumetric optical properties of the transition metals or are related to surface electron states, or contamination of the surface layer of the metal. Such doubts were expressed in [69, 107]. Simple estimates help to simplify the situation. The field of an IR light wave ($\hbar\omega\sim0.1$ eV) penetrates into a transition metal to a $\delta\sim400$–800 Å, i.e., a layer containing 150–300 atoms. A characteristic value for an anomaly of $\epsilon_2^{inter}(\omega)$ is $\Delta\epsilon_2\sim1\cdot10^2$, corresponding to a transition with an oscillator force $\Delta f\approx\Delta\epsilon_2\cdot Z(\omega/\omega_p)^2\approx1\cdot10^{-2}$ (Z is the number of valent electrons per atom, ω_p is the classical plasma frequency, determined by all valent electrons). If this effect is caused not by all atoms in layer δ, but rather only by the surface electron states located on a single atom, Δf for this surface atom should be 150–300 times greater, i.e., on the order of 1.5–3.0. Such a powerful transition is possible only between two practically flat zones separated by $\hbar\omega\sim0.1$ eV, one of which is completely filled, while the other is entirely free. This type of electron situation would be rather exotic for the surface of a metal and could hardly be the cause of the anomalies observed in many transition metals. More frequently optical transitions with energy $\hbar\omega\sim0.1$ eV are caused by optically active oscillations of a lattice, but the oscillator force in this case is too small, on the order of $m/M\sim10^{-4}$ (where m and M are the masses of an electron and an ion), to explain the observed anomalies. Similar estimates indicate that if this effect is caused by contamination of the surface layer, the contamination must extend to a depth comparable with δ. Naturally, $\sigma(\omega)$ in other areas of the spectrum should also change in this case, which would hardly remain unnoticed. Keeping these considerations in mind, we shall interpret the observed IR singularities as purely volumetric optical properties of the transition metal.

Valuable information on the optical properties of a metal can be obtained from numerical calculations of $\epsilon^{inter}(\omega)$. Unfortunately, until recently such calculations were scarce and only a few of them included the IR area. In [11] a peak of $\epsilon_2^{inter}(\omega)$ was found in Nb at $\hbar\omega\approx0.2$ eV, in Mo it was found that $\epsilon_2^{inter}(\omega)$ smoothly approaches zero at $\hbar\omega\to0$. Calculations undertaken in [109] showed that in V and paramagnetic Cr there is no threshold of interband absorption. Similar calculations[106] found in Rh and Ir a threshold of interband absorption corresponding to the value of spin–orbital splitting. We note that all of these calculations were performed in the approximation of constant matrix elements, which does not allow a judgement to be made on the absolute value of $\epsilon^{inter}(\omega)$ and, as will be shown below, can lead to a basically incorrect function $\epsilon^{inter}(\omega)$. More careful analysis of the question was undertaken on the basis of both purely theoretical and numerical analysis in works of the authors [110, 111], which will serve as the basis for our further presentation.

5.1. Electron Structure and Singularities of Low-Frequency Optical Conductivity of Metals

The facts outlined above, in spite of their incompleteness and occasional contradictions, draw a picture of far from trivial behavior of the optical properties of the transition metals in the IR area. Here the central question in our opinion is: Are there intensive interband transitions with low energies? In this section, we shall attempt to determine which singularities of electron structure can lead to anomalies in $\epsilon^{\text{inter}}(\omega)$ in the low-frequency area. We shall limit ourselves to analysis of direct interband transitions, not touching upon processes involving defects or lattice oscillations.

The imaginary portion of the dielectric constant interband tensor in the chaotic phase approximation, ignoring local field effects, is

$$\varepsilon_2^{\text{inter}}(\omega)_{\alpha\beta} = \frac{2\pi^2 e^2}{m^2\omega^2\Omega} \sum_{\mathbf{k},\lambda\neq\lambda'} (\mathrm{p}^{\mathbf{k}}_{\lambda\lambda'})^*_\alpha (\mathrm{p}^{\mathbf{k}}_{\lambda\lambda'})_\beta f_{\mathbf{k}\lambda} (1 - f_{\mathbf{k}\lambda'})\, \delta(\hbar\omega - E_{\mathbf{k}\lambda'} + E_{\mathbf{k}\lambda}), \tag{5.1}$$

where $\left[\mathbf{p}^{\mathbf{k}}_{\lambda\lambda'}\right]_\beta = \left\langle \mathbf{k}\lambda \left| \frac{\hbar\nabla\beta}{i} \right| \mathbf{k}\lambda' \right\rangle$, integration with respect to $\mathbf{k}$ is conducted within limits of the first Brillouine zone, while $|\mathbf{k}\lambda\rangle$ represents single-electron states in band λ with energies $E_{\mathbf{k}\lambda}$ and filling numbers $f_{\mathbf{k}\lambda}$ (in nonmetallic metals bands differing only in spin direction are considered here to be different).

In the low-frequency area of interest to us, contributions to $\epsilon_2^{\text{inter}}(\omega)$, according to (5.1), are made only by electron states in the layer of $\hbar\omega$ near the Fermi level, i.e.,

$$E_{\mathbf{k}\lambda} \approx E_F \approx E_{\mathbf{k}\lambda'}, \tag{5.2}$$

where the approximate equality occurs with an accuracy of about $\hbar\omega$. With such small $\hbar\omega$, the probability of meeting (5.2) is rather small, so that only relatively small areas in the Brillouine zone take part in the formation of $\epsilon_2^{\text{inter}}(\omega)$.

Before going over to an analysis of the optical properties of specific metals, it is useful to clarify some general points of optical interband transitions where $\hbar\omega \to 0$. In this limiting case, approximate equality in (5.2) is converted to precise equality and we thus deal with degenerate bands λ and λ', intersecting the Fermi surface. Studies can be performed without difficulty, using the well-known results of group theory, where for simplicity we ignore the initial influence of spin-orbital interaction.

Nonrelativistic approximation. It is not difficult to note that the degeneration of the spectrum at isolated points of the Brillouine zone as $\hbar\omega \to 0$ is not of interest, since the probability that the degenerate level coincides precisely with E_F is negligible. It remains to study degeneration along the line or at the boundaries of the Brillouine zone, which can be only double if spin is ignored.

Let us study bands λ and λ', degenerate along a line in the Brillouine zone. The only important area for us is the vicinity of point $\mathbf{k}_0$, where the degenerate band intersects the Fermi level. Depending on the symmetry of the crystal and the position of the degeneration line in the Brillouine zone, two cases are possible: 1) on the degeneration line $(\mathbf{p}^{\mathbf{k}}_{\lambda\lambda'})_{\perp} \neq 0$, so that in the vicinity of the line degeneration is relieved linearly with respect to $\mathbf{k}$; 2) on the degeneration line $(\mathbf{p}^{\mathbf{k}}_{\lambda\lambda'})_{\perp} = 0$ and degeneration is relieved quadratically with respect to $\mathbf{k}$ (the characters $\perp(||)$ here and below represent projection in a direction perpendicular (parallel) to the degeneration line or surface).

In the first case, in the vicinity of point $\mathbf{k}_0$ the dispersion law is

$$E_{\mathbf{k},\lambda\lambda'} = E_F + \hbar(\mathbf{k} - \mathbf{k}_0)_{\parallel} \mathbf{v}_{\parallel} \pm \hbar(\mathbf{k} - \mathbf{k}_0)_{\perp} \mathbf{v}_{\perp}(\mathbf{n}_{\perp}), \tag{5.3}$$

where $n = \mathbf{k}/|\mathbf{k}|$. The contribution from the vicinity of point $\mathbf{k}_0$ to the dielectric constant is

$$\varepsilon_2^{\text{inter}}(\omega \to 0) = \frac{2}{3}\frac{\hbar}{v_{\parallel} m a_B}\frac{1}{2\pi}\int dn_{\perp}\left[\frac{|\mathbf{p}^{\mathbf{k}_0}_{\lambda\lambda'}|}{2mv_{\perp}(\mathbf{n}_{\perp})}\right]^2. \tag{5.4}$$

Here for simplicity we limit ourselves only to the case of cubic crystals, where $\epsilon_2^{\text{inter}}(\omega) = Sp\{\epsilon_2^{\text{inter}}(\omega)_{\alpha\beta}\}/3$, and introduce the Bohr radius a_B for convenience. We also consider double degeneration of bands with respect to the spin direction, which obtains in nonmagnetic crystals. The nonzero value of $\epsilon_2^{\text{inter}}(\omega \to 0)$ is caused by contraction of the factors ω^2 in the numerator and denominator of equation (5.1) (ω^2 in the numerator arises from the phase volume, making a contribution to $\epsilon_2^{\text{inter}}$). Using the Kramers–Kronig dispersion relation, it is not difficult to find that $\epsilon_2^{\text{inter}}(\omega \to 0) \sim \ln(\omega/\omega_0)$, where ω_0 is the cutoff frequency. Similar degeneration occurs, in particular, in bcc and fcc metals on the third order axis Λ (representation of Λ_3), and in bcc metals also on the third-order F axis (representation of F_3) (Figure 22). Furthermore, random degeneration is possible, when the degeneration line lies in the plane of symmetry of the Brillouine zone or passes through the general points of the Brillouine zone (the latter is possible only in crystals with inversion center, as in bcc and fcc metals).

In the second case, when degeneration is relieved quadratically, the dispersion rule in the vicinity of $\mathbf{k}_0$

$$E_{\mathbf{k},\lambda\lambda'} = E_F + \hbar(\mathbf{k}-\mathbf{k}_0)_{\parallel} v_{\parallel} + \hbar^2(\mathbf{k}-\mathbf{k}_0)^2_{\perp}/2m_{\lambda,\lambda'}(\mathbf{n}_{\perp}), \qquad (5.5)$$

while

$$|\mathbf{p}^{\mathbf{k}}_{\lambda\lambda'}| = m\hbar|\mathbf{k}-\mathbf{k}_0|_{\perp}/\mu(\mathbf{n}_{\perp}).$$

Here it is more convenient for us to replace the more common effective mass tensor with the direction-dependent effective masses:

$$m_{\lambda}(n_{\perp}),\ m_{\lambda'}(n_{\perp}),\ \mu(n_{\perp}) \text{ and } \mu_{-}(n_{\perp}) = 2\{1/m_{\lambda}(n_{\perp}) - 1/m_{\lambda'}(n_{\perp})\}^{-1}.$$

The contribution to $\epsilon_2^{\text{inter}}$ for cubic crystals here is:

$$\varepsilon_2^{\text{inter}}(\omega \to 0) = \frac{1}{3}\frac{\hbar}{v_{\parallel} m a_B}\frac{1}{2\pi}\int dn_{\perp}\left[\frac{\mu_{-}(n_{\perp})}{\mu(n_{\perp})}\right]^2. \qquad (5.6)$$

In this case the nonzero value of $\epsilon_2^{\text{inter}}(\omega \to 0)$ results from the fact that the phase volume $\sim\omega$, $|\mathbf{p}_{\lambda\lambda'}|^2 \sim \omega$, which together gives the factor ω^2 in the numerator of equation (5.3), which contracts with ω^2 in the denominator. This degeneration occurs, for example, in bcc and fcc metals on fourth-order axis Δ (representation of Δ_5) (Figure 22). We can judge how common these cases are from Figure 23, which shows the band structure of Nb (bcc lattice) and Pd (fcc lattice), as well as Table 7, where we show the values of $\epsilon_2^{\text{inter}}(\omega \to 0)$ for a number of metals.

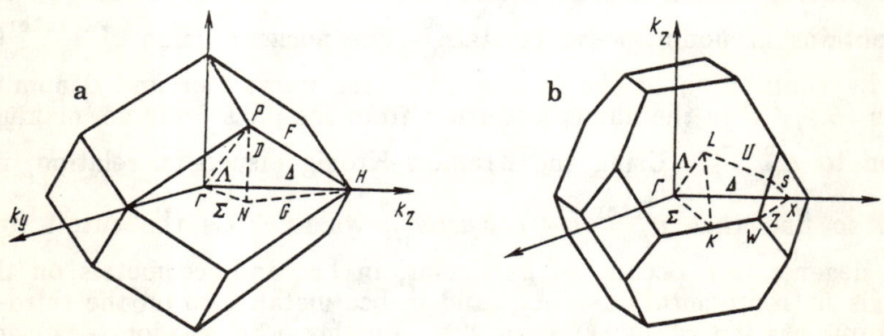

Figure 22. Brillouine Zone of bcc (a) and fcc (b) Lattices. Symmetrical lines and points are shown.

Degeneration of two bands at all points on the face of the Brillouine zone can occur only due to symmetry with respect to time inversion, if this face is perpendicular to the helical second order axis[112]. bcc and fcc crystals do not have such a symmetry element as a helical axis, and therefore degeneration in the plane does not occur in them. We know, however, that in metals with hcp structure this degeneration exists at the Brillouine zone boundary perpendicular to the sixth-order axis. It is not difficult to show that at this boundary $\left[\mathbf{p}^{\mathbf{k}_0}_{\lambda\lambda'}\right]_Z \neq 0$, and therefore degeneration is relieved linearly with respect to **k**. For infrared optics, only the vicinity of the line of intersection of the Fermi surface with the hexagonal Brillouine zone boundary (line L) is of interest. In the vicinity of this line

$$E_{\mathbf{k},\lambda\lambda'} = E_F + \hbar(\mathbf{k}-\mathbf{k}_0)_\perp \mathbf{v}_{\mathbf{k}_0\perp} \pm \hbar(\mathbf{k}-\mathbf{k}_0)_Z v_{\mathbf{k}_0 Z}, \tag{5.7}$$

where point $\mathbf{k}_0$ is assumed to lie on line L, while the character $\perp$ represents projection of the vector on the plane of the Brillouine zone boundary. The corresponding contribution to $\epsilon_2^{\text{inter}}$ is anisotropic:

$$\varepsilon_2^{\text{inter}}(\omega\to 0)_{ZZ} = \frac{1}{\pi}\oint_{(L)} \frac{dl\mathbf{k}_0\hbar^2}{\hbar\omega\cdot ma_B} \frac{|\mathbf{p}^{\mathbf{k}_0}_{\lambda\lambda'}|^2}{m^2 v_{\mathbf{k}_0 Z}\cdot|\mathbf{v}_{\mathbf{k}_0\perp}|}, \tag{5.8}$$

whereas $\epsilon_2^{\text{inter}}(\omega\to 0)_{Z\alpha}$=const, while $\epsilon_2^{\text{inter}}(\omega\to 0)_{\alpha\alpha}\sim\omega(\alpha\equiv x,\ y)$. This unusual behavior of $\epsilon_{2\,ZZ}^{\text{inter}}$ results from the fact that the phase volume here $\sim\omega$, due to which one factor ω in the denominator of (5.1) is contracted, and therefore $\epsilon_{2\,ZZ}^{\text{inter}}\sim 1/\omega$, while $\sigma_{ZZ}^{\text{inter}}(\omega\to 0)$=const. The Kramers–Kronig relation can be used to show that in this case $\epsilon_1^{\text{inter}}(\omega\to 0)=\epsilon_0>0$. Data on the Fermi surfaces of hcp metals[113, 114] show that a similar contribution to $\epsilon_2^{\text{inter}}$ should obtain for almost any metal with hcp lattice.

Based on our analysis it can be concluded that in the nonrelativistic approximation the frequency variation of $\epsilon_2^{\text{inter}}$ ω at low frequencies is always monotonic. To verify that this is not the case, let us now analyze a situation in which the energy of a level degenerate at the point of high symmetry of the Brillouine zone differs from E_F by $\sim\hbar\omega$. In bcc and fcc metals at points of high sym-

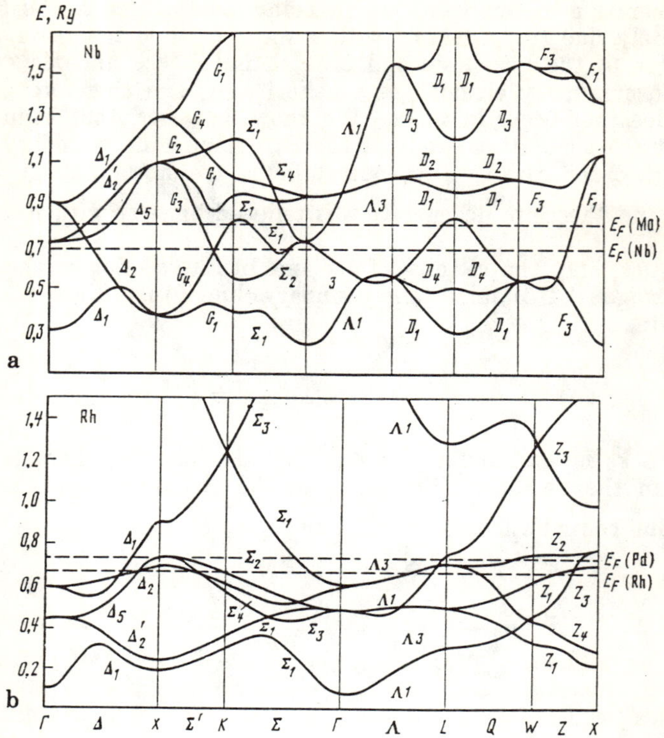

Figure 23. Band Structure of Niobium (bcc lattice) (a) and Rhodium (fcc lattice) (b). Positions of E_F approximately corresponding to Mo and Pd indicated by dash lines.

metry, having degenerate levels λ and λ', $\mathbf{p}^{\mathbf{k}_0}_{\lambda\lambda'}=0$. The dispersion rule near such a point:

$$E_{\mathbf{k},\lambda\lambda'} = E_{\mathbf{k}_0,\lambda\lambda'} + \hbar\,(\mathbf{k} - \mathbf{k}_0)^2/2m_{\lambda\lambda'}\,(\mathbf{n}) \tag{5.9}$$

and $|\mathbf{p}^{\mathbf{k}}_{\lambda\lambda'}|=\hbar|\mathbf{k}\text{-}\mathbf{k}_0|\mu(\mathbf{n})$. Two essentially different cases can arise here, schematically shown in Figure 24: a) the signs of m_λ and $m_\lambda{}'$ coincide, b) the signs of m_λ and $m_\lambda{}'$ are opposite. In case a), as is clear from Figure 24*a*, interband absorption begins at energy $\hbar\omega_1$ and ends at energy $\hbar\omega_2$. Clearly, this nonmonotic contribution can lead to a maximum on the curve of $\epsilon_2^{\text{inter}}(\omega)$. In case b) (Figure 24*b*), absorption is threshold in nature — it begins with $\hbar\omega_1$ and continues right up to high energies. We should note that the symmetry of Brillouine zone

Table 7

Behavior of ϵ_2^{inter} ω at Very Low Frequencies ($\hbar\omega \to 0$)

Metal	k_0, $4\pi/a$	Degeneration Type	$\epsilon_2^{inter}(\omega \to 0)$	$\epsilon_2^{inter}(\omega \to 0)_{full}$
V	(0.072; 0.072; 0.072)	Regular	4.7	38.6
	(0.131; 0.049; 0.049)	Random	33.9	
	(0.159; 0.063; 0)			
Nb	(0.056; 0.056; 0.056)	Regular	5.0	36.7
	(0.136; 0.033; 0.033)	Random	31.7	
	(0.156; 0.043; 0)			
Mo	(0.255; 0; 0)	Regular	6.4	6.4
Rh	(0.365; 0; 0)	Regular	67.3	76.7
	(0.185; 0.185; 0.185)	Regular	9.4	
Pd	(0.403; 0; 0)	Regular	45.5	45.5

points has no influence on whether the contribution will be of type a) or b). Furthermore, situations are possible in which in one direction $m_\lambda(\mathbf{n})$ and $m_\lambda'(\mathbf{n})$ have identical sign, while in the other direction the signs are different. Therefore, the full contribution from the vicinity of a high symmetry point may yield quite varied functions $\epsilon_2^{inter}(\omega)$, including functions with a maximum.

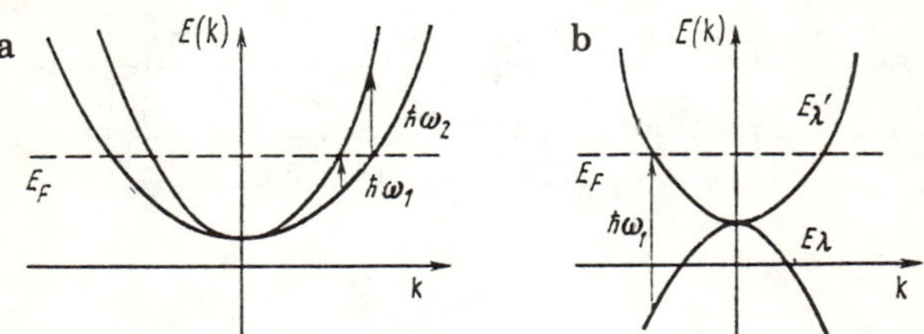

Figure 24. Schematic View of Band Structure Near Point of High Symmetry. a — $m_\lambda \gg m_\lambda' > 0$; b — $m_\lambda' > 0 > m_\lambda$.

The case we have studied of appearance of nonmonotonic variation of $\epsilon_2^{inter}(\omega)$ is not so rare as it might seem at first glance. For example, in bcc metals in the area of the *d* band at points of high symmetry there are six double or triple degenerate levels. Approximately the same degeneration occurs in fcc metals (Figure 23). Since the situation of interest to us arises when E_F lies near one of the degenerate levels in the interval $\hbar\omega \sim 0.3$ eV, and in sum these intervals represent about 1/3 of the width of the d band, the probability of appearance of contributions of type a) or b) is rather significant.

Influence of spin–orbital interaction. Consideration of relativistic effects greatly changes the behavior of $\epsilon_2^{\text{inter}}(\omega)$ at low frequencies. The basic role in this case is played by relief of degeneration of the electron spectrum, occurring under the influence of the spin–orbital interaction. For example, in bcc and fcc metals, all forms of degeneration along a line in the Brillouine zone disappear, while in hcp metals, degeneration continues on hexagonal Brillouine zone boundaries (degeneration along lines AL, connecting the centers of a boundary with the centers of the lines, remains). A typical value of splitting in 3d metals is about 0.15 eV, in 4d metals 0.3 eV, in 5d metals about 1.0 eV. These figures are quite approximate. At the beginning of the d series, splitting is significantly less, while at the end it is several times greater. In particular, in Y $\Delta E_{\text{so}} \approx 0.15$ eV, whereas in Pd $\Delta E_{\text{so}} \approx 0.6$ eV. It is important to note that under the influence of the spin–orbital interaction a new energy scale of ΔE_{so} appears in the electron structure of the metals, which is absent at the nonrelativistic limit. Its appearance cannot but reflect on the behavior of $\epsilon_2^{\text{inter}}(\omega)$ in the IR area of the spectrum. Let us analyze these effects in greater detail, limiting ourselves for simplicity to nonmagnetic crystals with inversion center.

In considering spin–orbital interaction, the dispersion rule in the vicinity of the $\mathbf{k}$ point where nonrelativistic bands λ and λ' were degenerate, is described by the equation

$$\text{Det}\left\| \begin{array}{c|c|c|c} E_{\mathbf{k}\lambda\uparrow} - E & 0 & \xi_{\lambda\uparrow,\lambda'\uparrow} & \xi_{\lambda\uparrow,\lambda'\downarrow} \\ \hline 0 & E_{\mathbf{k}\lambda\downarrow} - E & \xi^*_{\lambda\uparrow,\lambda'\downarrow} & -\xi_{\lambda\uparrow,\lambda'\uparrow} \\ \hline \xi_{\lambda\uparrow,\lambda'\uparrow} & \xi_{\lambda\uparrow,\lambda'\downarrow} & E_{\mathbf{k}\lambda'\uparrow} - E & 0 \\ \hline \xi^*_{\lambda\uparrow,\lambda'\downarrow} & -\xi_{\lambda\uparrow,\lambda'\uparrow} & 0 & E_{\mathbf{k}\lambda'\downarrow} - E \end{array} \right\| = 0 \tag{5.10}$$

Here $\xi_{\lambda\sigma,\lambda'\sigma'} = \langle \mathbf{k}\lambda\sigma | \hat{H}_{\text{so}} | \mathbf{k}\lambda'\sigma' \rangle$, while $E_{\mathbf{k}\lambda\uparrow} = E_{\mathbf{k}\lambda\downarrow}$ and $E_{\mathbf{k}\lambda\uparrow} = E_{\mathbf{k}\lambda'\downarrow}$. The characters $\uparrow$ and $\downarrow$ are included only to show which bands bind the spin–orbital interaction. Performing simple but somewhat cumbersome computations we find that the solution of equation (5.10) consists of bands α and β with dispersion:

$$E_{\alpha,\beta}(\mathbf{k}) = E_+(\mathbf{k}) \mp \sqrt{E_-^2(\mathbf{k}) + \xi_{\mathbf{k}}^2}, \tag{5.11}$$

where the symbols $\xi_{\mathbf{k}}^2=\xi_{\lambda\uparrow\lambda'\uparrow}^2+|\xi_{\lambda\uparrow,\lambda'\downarrow}|^2$ and $E_{\pm}(k)={}^1/_2(E_{\mathbf{k}\lambda'}\pm E_{\mathbf{k}\lambda})$ are used. Furthermore, there are two bands $\tilde{\alpha}$ and $\tilde{\beta}$ which are degenerate with α and β, respectively. The matrix momentum elements $\langle\tilde{\alpha}|\mathbf{p}|\beta\rangle=\langle\alpha|\mathbf{p}|\tilde{\beta}\rangle=0$. Optical transitions include the matrix elements

$$|\mathbf{p}_{AB}|^2={}^1/_2\,\{|\langle\alpha|\mathbf{p}|\beta\rangle|^2+|\langle\tilde{\alpha}|\mathbf{p}|\beta\rangle|^2\}=$$
$$=\frac{4\,|\mathbf{p}_{\lambda\lambda'}^{k}|^2\,E_-(\mathbf{k})+\xi_{\mathbf{k}}^2\,(\mathbf{p}_{\lambda\lambda}^{\mathbf{k}}-\mathbf{p}_{\lambda'\lambda'}^{\mathbf{k}})^2}{(E_\beta(\mathbf{k})-E_\alpha(\mathbf{k}))^2},\tag{5.12}$$

which convert directly to $|\mathbf{p}_{\lambda\lambda'}^{\mathbf{k}}|^2$ as $\xi\to 0$.

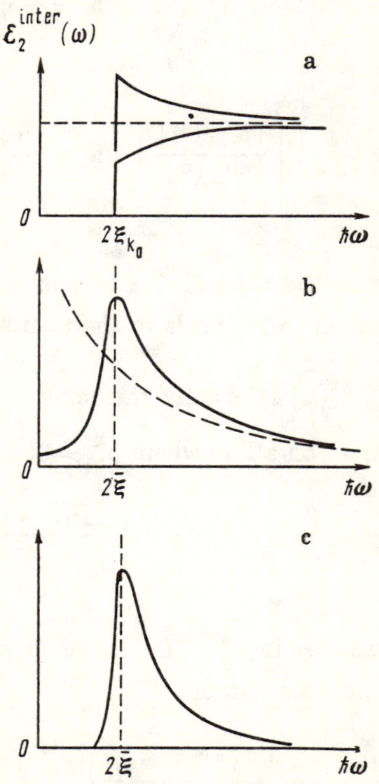

Figure 25. Behavior of $\epsilon_2^{\text{inter}}(\omega)$ at Low Frequencies Considering Spin–Orbital Interaction. a — bcc and fcc metals; b — hcp metals; c — ferromagnetic metals with cubic lattice. In Figures a and b, the dash line shows nonrelativistic functions.

Equations (5.10)–(5.12) can be used to obtain relativistic analogues of equations (5.4), (5.6) and (5.8). If the nonrelativistic bands λ, λ' are degenerate along a line in **k** space, while $p^{\mathbf{k}}_{\lambda\lambda'} \neq 0$, with small ω instead of (5.4) we have

$$\varepsilon_2^{\text{inter}}(\omega) = \frac{2}{3}\frac{\hbar}{v_{\parallel} m a_B}\frac{1}{2\pi}\int dn_{\perp}\left[\frac{|\mathbf{p}_{\lambda\lambda'}(\mathbf{n})|}{2mv_{\perp}(\mathbf{n}_{\perp})}\right]^2 \times$$
$$\times \frac{(\hbar^2\omega^2 - 4\xi^2) + \xi^2}{\hbar^2\omega^2}\theta(\hbar\omega - 2\xi). \tag{5.13}$$

The qualitative behavior of $\epsilon_2^{\text{inter}}(\omega)$ (5.13) is shown in Figure 25*a*. It is characteristic that there is a threshold of interband absorption where $\hbar\omega = 2\xi$, above which $\epsilon_2^{\text{inter}}(\omega)$ suddenly takes on the value $\frac{2}{3}\frac{\hbar}{v_{\parallel} m a_B}\cdot\frac{1}{4}$. Depending on the sign

$$\left\{\frac{1}{2\pi}\int\left[\frac{|\mathbf{p}_{\lambda\lambda'}(\mathbf{n})|}{2mv_{\perp}(\mathbf{n}_{\perp})}\right]^2 d\mathbf{n}_{\perp} - \frac{1}{4}\right\}$$

this value will be greater or less than $\epsilon_2^{\text{inter}}(\omega \to 0)$ in the nonrelativistic approximation. Where $\hbar\omega \gg \xi$ equation (5.13) leads to the nonrelativistic result (5.4).

The behavior of $\epsilon_2^{\text{inter}}(\omega)$ is exactly the same in the case of degeneration of nonrelativistic bands λ, λ' on a line where $\mathbf{p}^{\mathbf{k}}_{\lambda\lambda'} = 0$:

$$\varepsilon_2^{\text{inter}}(\omega) = \frac{1}{3}\frac{\hbar}{v_{\parallel} m a_B}\frac{1}{2\pi}\int d\mathbf{n}_{\perp}\left[\frac{\mu_{-}(\mathbf{n}_{\perp})}{\mu(\mathbf{n}_{\perp})}\right]^2 \cdot \frac{(\hbar^2\omega^2 - 4\xi^2) + \xi^2}{\hbar^2\omega^2}\theta(\hbar\omega - 2\xi). \tag{5.14}$$

More significant changes in $\epsilon_2^{\text{inter}}(\omega)$ arise in crystals with hcp lattice. Instead of equation (5.8) here we obtain

$$\varepsilon_2^{\text{inter}}(\omega)_{ZZ} = \frac{1}{\pi}\oint_{(L)}\frac{dl_{\mathbf{k}_0}\hbar^2\theta(\hbar\omega - 2\xi_{\mathbf{k}_0})}{ma_B\sqrt{\hbar^2\omega^2 - 4\xi^2_{\mathbf{k}_0}}}\left\{\frac{|\mathbf{p}^{\mathbf{k}_0}_{\lambda\lambda'}|^2\left(\hbar^2\omega^2 - 4\xi^2_{\mathbf{k}_0}\right)}{m^2 v_{\mathbf{k}_0 Z}|v_{\mathbf{k}_0\perp}|} + \right.$$
$$\left. + 4\xi^2_{\mathbf{k}_0}\frac{|\mathbf{v}_{\mathbf{k}_0\perp}|}{v_{\mathbf{k}_0 Z}}\right\}\frac{1}{\hbar^2\omega^2}. \tag{5.15}$$

In deriving it we assume that the radius of curvature of line L is much greater than $\xi_{\mathbf{k}}/\hbar|\mathbf{v}_{\mathbf{k}_{0\perp}}|$. If the parameter of spin–orbital splitting $\xi_{\mathbf{k}_0}$ is constant along line L (the line of intersection of the Fermi surface with the hexagonal boundary of the Brillouine zone), then expression (5.15) has where $\hbar\omega=2\xi$ a radical singularity coinciding with the threshold of interband absorption. However, in actuality $\xi_{\mathbf{k}_0}$ changes rather significantly as a function of the position of $\mathbf{k}_0$ on the face of the Brillouine zone and, in particular, vanishes on line AL. This leads to smoothing of the radical singularity. Furthermore, depending on the specific form of the Fermi surface the value of the threshold changes sharply and, if the Fermi surface intersects line AL, the threshold disappears, and $\epsilon_2^{\text{inter}}(\omega\to 0)_{ZZ}$ is not equal to zero. The qualitative course of the curve of $\epsilon_2^{\text{inter}}(\omega)_{ZZ}$ for hcp metals is shown in Figure 25*b*.

One interesting singularity of interband absorption appears when a spin–orbital interaction occurs in ferromagnetic metals. In the nonrelativistic case each band corresponds to a spin direction with or against the direction of the magnetic moment. Optical transitions between bands with different directions of spin are strictly forbidden, since the momentum operator is diagonal with respect to the spin projection. Consideration of spin–orbital interaction leads to hybridization of the bands $\uparrow$ and $\downarrow$. At those points in the Brillouine zone where the nonrelativistic bands $\uparrow$ and $\downarrow$ intersected, the hybridized bands contain 50% $\uparrow$ and 50% $\downarrow$ states. Optical transitions between hybridized bands in these points of $\mathbf{k}$ space are now, of course, permitted. Where bands with opposite spin direction are separated by a large energy interval, mixing of $\uparrow$ states with a $\downarrow$ band and $\downarrow$ states with an $\uparrow$ band is weak. Optical transmissions remain practically forbidden. Of greatest interest for infrared optics are Brillouine zone areas lying near the line of intersection of sheets of the Fermi surface with different spin directions. Here the dispersion rule is defined by the equation

$$\mathrm{Def}\begin{Vmatrix} E_{\mathbf{k}\lambda\uparrow}-E & \xi_{\mathbf{k}} \\ \xi_{\mathbf{k}}^{*} & E_{\mathbf{k}\lambda'\downarrow}-E \end{Vmatrix}=0, \tag{5.16}$$

which has solutions similar to (5.11). The matrix transition elements are given by equation (5.12) in which, however, only the term with $|\mathbf{p}^{\mathbf{k}}_{\lambda\uparrow,\lambda\uparrow}-\mathbf{p}^{\mathbf{k}}_{\lambda'\downarrow,\lambda'\downarrow}|^2$ remains, since $\mathbf{p}^{\mathbf{k}}_{\lambda\uparrow,\lambda'\downarrow}=0$. The contribution of this group of transitions to $\epsilon_2^{\text{inter}}(\omega)$ is

$$\Delta\varepsilon_2^{\text{inter}}(\omega) = \frac{1}{3\pi} \oint\limits_{(L)} \frac{dl_{\mathbf{k}_0} \hbar^2 \theta\left(\hbar\omega - 2\,|\,\xi_{\mathbf{k}_0}|\right)}{a_B m \sqrt{\hbar^2\omega^2 - 4\,|\,\xi_{\mathbf{k}_0}|^2}} \cdot \frac{|\,\xi_{\mathbf{k}_0}|^2}{\hbar^2\omega^2} \frac{v_-^2(\mathbf{k}_0)}{|\,[\mathbf{v}_+(\mathbf{k}_0)\cdot\mathbf{v}_-(\mathbf{k}_0)]\,|}, \tag{5.17}$$

where

$$E_{\lambda\uparrow}(\mathbf{k}) = E_F + \hbar(\mathbf{k} - \mathbf{k}_0)\,\mathbf{v}_{\lambda\uparrow}(\mathbf{k}_0),$$
$$\mathbf{v}_\pm(\mathbf{k}_0) = {}^1\!/_2\,[\mathbf{v}_{\lambda'\downarrow}(\mathbf{k}_0) \pm \mathbf{v}_{\lambda\uparrow}(\mathbf{k}_0)],$$

while integration with respect to $\mathbf{k}_0$ is performed along the line of intersection of nonrelativistic Fermi surface sheets with opposite spin directions. It is not difficult to note that this contribution is concentrated in the area of energies $\hbar\omega \sim 2|\xi|$, since it arises where $\hbar\omega = 2|\xi|_{\min}$, and decreases at greater energies as $(\hbar\omega)^{-3}$ (Figure 25*c*). Its power is characterized by the force of the oscillator

$$\Delta f = \frac{m\Omega}{2\pi^2 e^2} \int\limits_0^\infty \omega\Delta\varepsilon_2^{\text{inter}}(\omega)\,d\omega = \frac{1}{24\pi^2}\frac{\Omega m}{\hbar^2} \oint\limits_{(L)} \frac{dl_{\mathbf{k}_0}\,|\,\xi_{\mathbf{k}_0}|^2 v_-^2(\mathbf{k}_0)}{|\,[\mathbf{v}_+(\mathbf{k}_0)\cdot\mathbf{v}_-(\mathbf{k}_0)]\,|} \tag{5.18}$$

(Ω is the atomic volume). Equations (5.17) and (5.18) are written for ferromagnetics with cubic lattice. With sufficient length of line L the contribution from this group of transitions is quite strong and causes a significant anomaly of optical conductivity. In ferromagnetics with hcp lattice a similar contribition to $\epsilon_2^{\text{inter}}(\omega)_{zz}$ is difficult to resolve against the background of (5.15). However, the anomaly of $\epsilon_2^{\text{inter}}(\omega)_{xx}$ and $\epsilon_2^{\text{inter}}(\omega)_{yy}$ should be easily seen.

These cases do not of course exhaust all situations which are possible in infrared optics. Our purpose has been much more modest — to show that under certain conditions anomalies in the IR area truly can occur. Now after our theoretical analysis, let us turn to numerical computations of specific metals and see to what extent the predictions of the theory correspond to the experimental data.

5.2. Numerical Investigation of Optical Properties of Transition Metals in the IR Area

Calculations of the optical properties of transition metals at low frequencies ($\hbar\omega \lesssim 0.5$ eV) have certain specific features. Due to the small value of the energy interval studied (it represents ~5% of the width of the *d* band) in equations of the linear muffin-tin orbitals method (cf. (2.19)) it is best to assign

$E_\nu = E_F$. This increases the accuracy of band structure contributions in the vicinities of the Fermi level and, which is particularly important, decreases the error in calculation of matrix transition elements to 1–2%. Certain difficulties arise in performing integration in **k** space. They are related to the fact that at small values of $\hbar\omega$, a contribution to $\epsilon_2^{\text{inter}}$ is not made by the entire Brillouine zone, but only by individual parts of the band where condition (5.2) is met. Therefore, it is best at first to segregate these areas of **k** space, then perform integration only with respect to them. To achieve an accuracy of ~10%, we must utilize a rather fine grid, corresponding to about 1500, and sometimes even ~10,000 **k** points in 1/48 of the Brillouine zone.

Detailed nonrelativistic computations of $\epsilon_2^{\text{inter}}(\omega)$ where $\hbar\omega < 0.5$ eV and calculation of the Drude parameters $\tilde{\omega}_p$ and γ were undertaken by us for the following metals: V, Nb, Mo, Rh, Pd and ferromagnetic Fe. Although in cubic metals the quantity $\epsilon_2^{\text{inter}}(\omega \to 0)$ cannot be measured against the background of the strong intraband contribution, we, for methodological reasons, calculated it for all of the nonmagnetic metals analyzed. The results are presented above in Table 7, where we also present the coordinates of points $\mathbf{k}_0$, the vicinities of which yield contribution where $\hbar\omega \to 0$, and the nature of degeneration of the electron spectrum at these points. The major conclusion which follows from these results is that in transition metals absence of an interband absorption threshold (in the nonrelativistic approximation) is the rule rather than the exception.

Let us now turn to analysis of the computed frequency functions $\epsilon_2^{\text{inter}}(\omega)$ (Figure 26). The group VA elements V and Nb have similar band structures and Fermi surfaces. Glancing at the band structure (cf. Figure 23*a*), it seems that the maximum of $\epsilon_2^{\text{inter}}(\omega)$ results from transitions $\Sigma_2 \to \Sigma_1$, quite similar to the interband transitions of Figure 24*a*. In direction Λ the two bands merge, in direction Δ they lie above E_F. To imagine the behavior of the bands in intermediate directions, let us turn to cross sections of the Fermi surface produced by planes of high symmetry (Figure 27). Band Σ_2 corresponds to the hole surface in the second band near point Γ ("octahedron"), while band Σ_1 corresponds to the tubular hole surface in the third band in direction ΓH ("jungle gyms"). Calculations have shown that the major contribution is that of the area of approach of two points of random degeneration (cf. Table 7). This same area yields the maximum of $\epsilon_2^{\text{inter}}(\omega)$.

In Mo, analysis of the band structure (Figure 23*a*) and Fermi surface [113, 114] gives no basis to expect a nonmonotonic course of the curve $\epsilon_2^{\text{inter}}(\omega)$. Numerical computations (Figure 26*c*) confirm this conclusion.

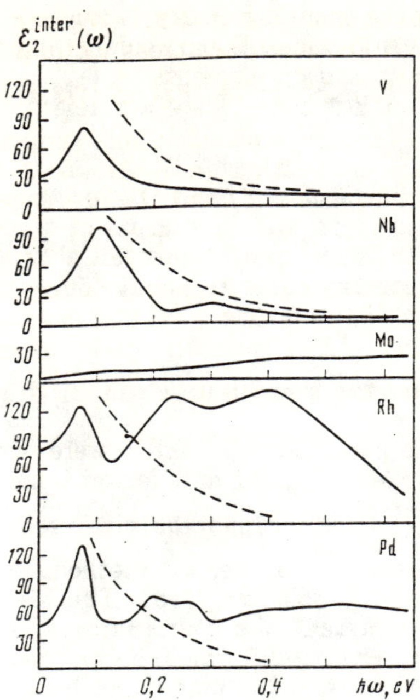

Figure 26. Calculated Variation of $\epsilon_2^{\text{inter}}(\omega)$ in Low-Frequency Area for the Metals V, Nb, Mo, Rh and Pd. Dash lines show variation $0.05 \cdot \epsilon_2^{\text{Drude}}(\omega)$.

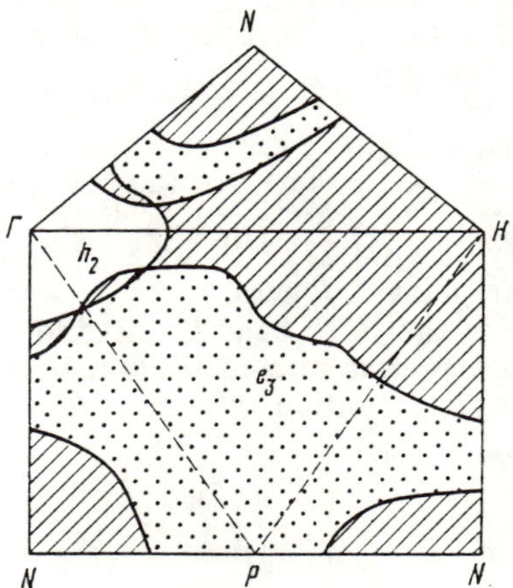

Figure 27. Cross Sections of Fermi Surface of Niobium Produced by Planes of High Symmetry.

The situation is most complex in Rh. We can conclude from the band structure (Figure 23*b*) and Fermi surface cross sections (Figure 28) that the major contribution to $\epsilon_2^{\text{inter}}(\omega)$ should be that of two **k**-space areas. The first is the vicinity of point *X*, where the embedded hole surfaces in the third and fourth bands ("ellipsoids") create a situation quite similar to that of Figure 24*a*. The major difference is that the zones X_2 and X_5 are degenerate, and therefore at point *X* the matrix element $|\mathbf{p}^{\mathbf{k}_0}_{\lambda\lambda'}|\neq 0$. The second area extends parallel to direction *LX*. In this direction due to random factors, a large electronic surface in the sixth band around point Γ and a tubular hole surface in the fifth band, extended along direction *LX*, approach each other quite closely. The contributions of these two areas cause both the low-frequency peak of $\epsilon_2^{\text{inter}}(\omega)$ at $\hbar\omega$=0.05 eV, and the two main maxima at 0.21 and 0.4 eV.

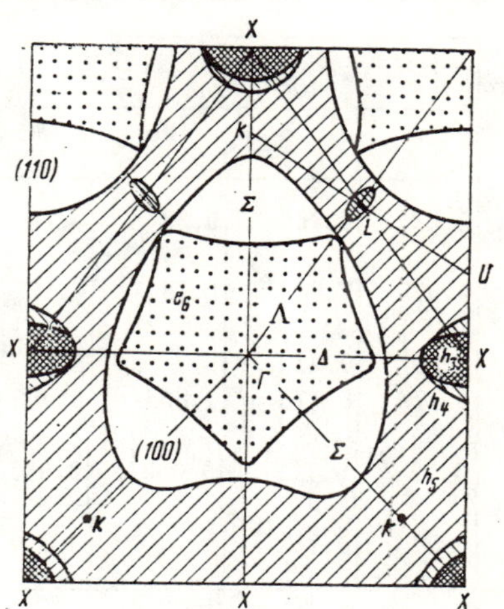

Figure 28. Cross Sections of Fermi Surface of Rhodium Produced by Planes of High Symmetry.

In Pd, analysis of the band structure (cf. Figure 23*b*) and Fermi surface[113, 114], indicates that the clear maxima on the curve of $\epsilon_2^{\text{inter}}(\omega)$ may arise only from the vicinity of point *X*. The computation we performed (Figure 26*d*) shows that, actually, the maxima at $\hbar\omega$=0.05 eV and 0.22 eV arise from this area in **k** space. The gentle maximum at $\hbar\omega\approx$0.5 eV is caused by transitions between bands 5 and 6, which occur near the boundary of the large Fermi surface sheet surrounding point Γ.

We know that in ferromagnetic Fe the band structure and Fermi surface for electrons with down spin (opposite the direction of the magnetic moment) are similar to the band structure and Fermi surface in Mo[113, 114]. The corresponding contribution to $\epsilon_2^{inter}(\omega)$, as in Mo, has a slight frequency variation and is comparatively small in absolute value — about 25% of the full contribution. The main role here is played by transitions of electrons from up spin 3→4.5 and 4→5 in the vicinity of point *H*. The electron structure near this point differs from that shown in Figure 24*a* only in the triple degeneration of the spectrum at a point of high symmetry (point *H*) instead of double degeneration. These transitions form two maxima of the curve of $\epsilon_2^{inter}(\omega)$ — the weak one at $\hbar\omega$=0.07 eV and the main one at $\hbar\omega$=0.2 eV.

Table 8

Experimental and Calculated[6] Parameters of Intraband Conductivity

	Metal					
	V	Nb	Mo	Rh	Pd	Fe
$\hbar\tilde{\omega}_{p\ theor}$, eV	8.07	9.76	9.23	9.76	9.99	6.85
$\hbar\tilde{\omega}_{p\ exp}$, eV	4.92[104]	7.91[103]	5.78[104]	4.16[106]	5.67[115]	3.94[107]
	5.29[109]	7.21[104]		5.19[117]	2.77[117]	
$\hbar\gamma_{theor}$, eV	0.176	0.173	0.056	0.055	0.063	0.496
$\hbar\gamma_{exp}$, eV	0.09[104]	0.171[103]	0.059[104]	0.033[106]	0.03[115]	0.524[107]
	0.04[109]	0.169[104]		0.016[117]	1.01[117]	

We see that in all of the metals studied except Mo, the frequency variation of $\epsilon_2^{inter}(\omega)$ in the IR area, computed in the nonrelativistic approximation, is complex in nature. A more frequent cause of the appearance of singularities such as a maximum or threshold is one variant of the situation schematically shown in Figure 24*a*. Let us now analyze how these calculation singularities relate to the anomalies of optical conductivity experimentally observed.

Comparison of the results of calculation with experimental data can usefully be started with the intraband contribution. Table 8 shows the values we have calculated for $\tilde{\omega}_p$ and γ, as well as the experimental data of various authors[1].

As we can see from the table, the theoretical values of $\tilde{\omega}_p$ *and* γ are always somewhat greater than the experimental values (by an average of 20–50%). The experimental data differ by 10–20%. However, we should note that the experimental Drude conductivity cannot always be interpreted as a contribution to conductivity met by intraband transitions. Actually, with this approach the distance $(\epsilon_2(\omega)-\epsilon_2^{\text{Drude}}(\omega))$ determined experimentally must be considered the contribution to dielectric constant by interband transitions. In practice, this quantity is essentially negative at some frequencies. This absurd result shows that the widely used procedure of finding $\tilde{\omega}_p$ and γ by means of Arganda diagrams is not always correct. A more complex question is also unclear — how accurate is the Drude equation in its description of the contribution to conductivity made by intraband transitions where $\hbar\omega \sim 0.1$ eV.

As was noted repeatedly above, the intraband contribution to conductivity in the IR area is the basic one and it is rather difficult to distinguish against its background anomalies related to interband transitions. We shall say somewhat arbitrarily that only singularities, the amplitude of which is no less than 5% of the Drude contribution, are accessible to experimental observations. To facilitate this differentiation of singularities in $\epsilon_2^{\text{inter}}(\omega)$ into observable and nonobservable ones, we have illustrated on Figure 26 with the dash line the curve of $0.05\cdot\epsilon_2^{\text{Drude}}(\omega)$. Singularities above this line are observable, those below it are not. As we can see from Figure 26, the observable contributions include the maxima of the calculated curves of $\epsilon_2^{\text{inter}}(\omega)$ in Ph at 0.21 and 0.4 eV, and also in Pd at 0.22 and 0.5 eV. They agree well with the maxima of the experimental curves of $\sigma(\omega)$: in Rh at 0.2 and 0.4 eV[106] and at 0.2 and 0.35 eV[117], in Pd at 0.19 eV[118] and at 0.5 eV[115]. Most low-frequency maxima of the calculated curves, however, fall in the nonobservable category. These include the maximum at $\hbar\omega$=0.07 eV in V, the maxima at 0.09 and 0.3 eV in Nb, the maxima at 0.13 and 0.05 eV, which we computed for Rh and Pd, respectively, as well as the maxima at 0.07 and 0.2 eV in ferromagnetic Fe. It is interesting to note that in experiments with similar frequencies a number of anomalies are observed, the values of which, however, are several times greater than on the calculated curves. Thus, in V and Nb there are strong singularities of $\sigma(\omega)$ at $\hbar\omega \sim 0.1$

---

[1]Our previous values of γ [110] did not consider the factor 0.185, which greatly increased their value.

eV[104, 105], and in Rh — a clearly observed conductivity maximum at $\hbar\omega=0.07$ eV[106]. In Mo neither calculations nor experiments[104, 71] indicate the existence of any singularities in $\epsilon_2^{\text{inter}}(\omega)$.

The reason for the significant disagreement between the calculated and experimental values of low-frequency anomalies may be the effects of the spin-orbital interaction, not considered in our computations. Their role is very significant at $\hbar\omega \leqslant 0.1$–$0.2$ eV. A more concrete idea of their magnitude can be obtained from an estimate made for ferromagnetic iron[111]. In Fe, the experimental curve of $\epsilon_2^{\text{inter}}(\omega)$[107 2] agrees with our nonrelativistic calculation for $\hbar\omega\sim0.75$ eV quite well, whereas at $\hbar\omega\sim0.2$–0.4 eV the calculated values of $\epsilon_2^{\text{inter}}$ are almost 4 times less than the experimental values (Figure 29). To verify that this disagreement does not relate to an incorrect value of spin moment, we performed calculations for $\mu=2.04$; 2.10 (cf. Figure 21) and 2.19 μ_B (the experimental value of $\mu=2.12$ μ_B[119]). The form of the curve of $\epsilon_2^{\text{inter}}(\omega)$ in the low-frequency area in this case changed slightly, but the disagreement with the experimental values remained virtually the same. We get the impression that our calculations do not include the group of interband transitions with energy 0.2–0.3 eV and oscillator force

$$\Delta f \approx \frac{m\Omega}{2\pi^2 e^2} \int \omega \left\{ \varepsilon_2^{\text{inter}}(\omega)_{\text{exp}} - \varepsilon_2^{\text{inter}}(\omega)_{\text{theor}} \right\} d\omega \approx 4{,}7 \cdot 10^{-2}.$$

We assumed that this difference is related to the interband transitions near the line of intersection of the Fermi surfaces with the up spin and the down spin, which are resolved as a result of the spin-orbital interaction. In this case the line of the intersection consists of six circles around the directions ΓH and 24 circles encompassing the tubular Fermi surfaces extended in the direction *HN*. The total length of the lines is about $45a_B^{-1}$, the value of spin-orbital splitting $\xi\approx0.07$ eV. Assuming for our estimates the factor $v_-^2/|[\mathbf{v}_+\times\mathbf{v}_-]|$ to be equal to 1, we obtain, according to (5.18), $\Delta f\sim4\cdot10^{-2}$.

---

[2]In [107] the function $\epsilon_2^{\text{inter}}(\omega)$ was rather correctly determined from $\epsilon_2(\omega)$.

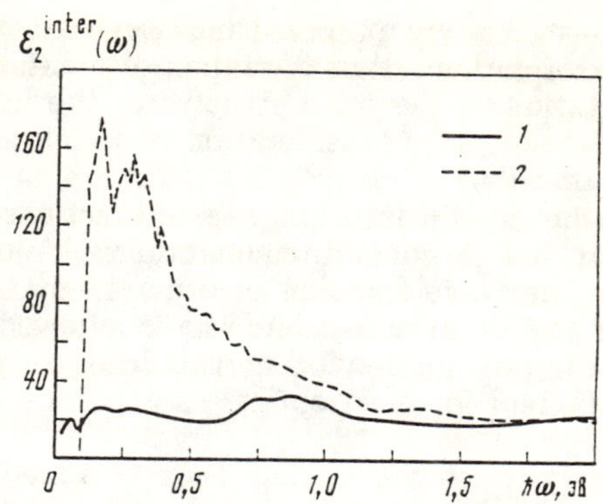

Figure 29. Calculated *(1)* and Experimental *(2)*[107] Curves of $\epsilon_2^{inter}(\omega)$ for Ferromagnetic Iron.

This estimate, of course, is quite rough, but the fact that the values of Δf obtained from the difference between the experimental and theoretical curves and from equation (5.18) are of the same order of magnitude shows how significant this effect is. In other cases the influence of the spin–orbital interaction unfortunately cannot be numerically estimated by any reasonable simple method. However, the fact that it can be quite significant is beyond doubt. This is indicated also by the results obtained in section 3.1. We hope that in the near future calculations of optical properties in the IR area considering the spin–orbital interaction will be performed. This will allow a more orderly analysis of the question of agreement between the theoretical predictions and experimental results in this spectral area.

CONCLUSIONS

Let us now list the basic results and conclusions obtained in this work.

1. Calculation of optical characteristics of metals in the single-frequency approximation using an energy spectrum and wave functions of electrons obtained from band calculations lead to results which agree fully with the experiments. In the energy area $\hbar\omega \leqslant 10$ eV, the error in calculation of $\epsilon_2(\omega)$ is not over 10–15%. Comparison of the optical characteristic of 15 simple, transition and noble metals computed by the LMTO method with experimental data shows that in the broader frequency area ($\hbar\omega$=0–35 eV) the agreement is good not only as to the position of the basic singularities in the optical characteristics, but also as to their absolute value.

2. When calculating optical characteristics, one must correctly consider the operator current matrix elements. Due to the approximate nature of band

computations, they must be computed using approximate Bloch functions. Therefore, the accuracy of computation of matrix elements is always lower than the accuracy of computation of the band structure. We have found expressions for corrections to matrix elements, consideration of which allows computation of $\langle \mathbf{k}\lambda | \mathbf{j} | \mathbf{k}\lambda' \rangle$ with the same degree of accuracy with which the spectrum $E_{\mathbf{k}_\lambda}$ is computed. However, the specific form of these corrections is cumbersome, and they are inconvenient to use in concrete calculations. A more convenient path to follow is to improve the wave function in order to make it approximate the precise eigenfunction of the Hamiltonian. We have followed this procedure within the LMTO method (l_{max}=3) to show that this leads to accurate computation of $\langle \mathbf{k}\lambda | \mathbf{j} | \mathbf{k}\lambda' \rangle$, quite sufficient for most optical calculations.

3. A precise analytic expression has been obtained for the macroscopic dielectric constant of a crystal considering both local field effect and exchange and correlation, which are converted at the limit to the equations of the single-frequency approximation. It has been shown that these effects are closely interconnected, and both the local field and exchange and correlation must be considered simultaneously. It has been shown in the approximation of the Vannier function that the intraband contribution to the macroscopic dielectric constant does not contain exchange-correlation corrections or local-field corrections, i.e., they are contained only in the interband contribution. Apparently, their role is not very significant for the calculation of the optical constants of the metals, and considering them leads to no qualitatively new effects.

4. In calculating optical characteristics of metals in the single-frequency approximation, we completely ignored electron-phonon and electron-electron scattering processes, resulting in finite lifetime of electron levels. Therefore, the optical constants which we have calculated frequently have more singularities than are experimentally observed, and all singularities on the calculation curves are sharper.

5. We calculated the characteristic loss spectra, i.e., the functions $L(\omega)=\mathrm{Im}[-1/\epsilon(\omega)]$ in a number of simple, transition and noble metals. It was shown that in the simple metals (Al) the position of the plasma peak and its width are calculated by the method developed in this work with very good accuracy. This is also true of the plasma peak in Cd, in which the shift of the peak into the area of lower energies due to polarization of the filled $3d$ shell is quite significant. The characteristic loss spectrum was studied in Zn, in which there are no plasma oscillations of electrons, since the energy of these oscillations falls in the area of intensive interband transitions from the filled d shell.

In $4d$-series transition metals there is well-defined plasma excitation at energy (~10 eV), significantly less than the energy of plasma oscillations in a homogeneous electron gas $\hbar\omega_p$. The broad peak in the energy area near $\hbar\omega_p$ observed experimentally, as our calculations have shown, consists of two or more peaks of similar energies, related to the presence in this frequency area (~20–30 eV) of interband d–p transitions. We studied plasma excitation in the $3d$ metals V and Cr, as well as the noble metals Cu and Ag.

6. The specifics of optical characteristics of the transition metals in the infrared ($\hbar\omega \leqslant 0.5$ eV) area are related to the problem of degeneration of the electron spectrum in the Brillouine zone. It was shown that in the transition metals the threshold of interband absorption may be absent right up to $\hbar\omega=0$. This occurs in the case of intersection of a doubly degenerate Fermi surface band, which is a rather frequent phenomenon in the transition metals. At the qualitative level, an analysis was performed for various cases of degeneration by methods of group theory in the nonrelativistic and relativistic cases. It was shown that the low-frequency peaks observed in transition metal spectra may be related to the closeness of the degeneration points to the Fermi surface. Furthermore, in ferromagnetic metals such anomalies may arise due to transitions between bands which in the nonrelativistic approximation are degenerate (with opposite spins), while consideration of the spin-orbital interaction relieves the degeneration of these bands and strongly hybridizes them. A numerical analysis was performed of singularities in the optical characteristics of five transition metals (V, Nb, Mo, Rh, Pd), as well as ferromagnetic iron. Some of the singularities found in the calculations correlate well with the experimental data. Other singularities in the calculation curves are apparently difficult to observe experimentally due to the small nature of their contribution to $\epsilon_2(\omega)$ in comparison to the dominant Drude contribution.

REFERENCES

1. Payns, D., Elementarnye vozbuzhdeniya v tverdykh telakh [Elementary Excitation in Solids], Moscow, Mir Press, 382 pp., 1965.

2. Payns, D., Nosier, F., Teoriya kvantovykh zhidkostey [Theory of Quantum Liquids], Moscow, Mir Press, 380 pp., 1967.

3. Smith, D. Y., Shiles, E., "Finite-Energy *f*-Sum Rules for Valence Electrons," Phys. Rev. B., Vol. 17, pp. 4689–4694, 1978.

4. Glotzel, D., McMahan, A. K., "Relativistic Effects, Phonon and Isostructural Transition in Cesium," Phys. Rev. B., Vol. 20, pp. 3210–3216, 1979.

5. Ehrenreich, H., Cohen, M. H., "Self-Consistent Field Approach to the Many-Particle Problem," Phys. Rev., Vol. 115, pp. 786–792, 1959.

6. Mazin, I. I., Savitskii, E. M., Uspenskii, Yu. A., "Electron-Phonon Effects in $4d$-metals: Calculation of Coupling Constant and Resistivity," J. Phys. F., Vol. 14, pp. 167–178, 1984.

7. Christensen, N. E., "Electronic Structure of Palladium and Its Relation to UV Spectroscopy," Phys. Rev. B., Vol. 14, pp. 3446–3458, 1976.

8. Christensen, N. E., Seraphin, B. O., "Relativistic Band Calculation and the Optical Properties of Gold," Phys. Rev. B., Vol. 4, pp. 3321–3332, 1971.

9. Christensen, N. E., "Band Structure of Rhodium and Its Relation to Photoemission Experiments," Phys. Stat. Sol. b., Vol. 55, pp. 117–126, 1973.

10. Christensen, N. E., "Band Structure of Silver and Optical Interband Transitions," Phys. Stat. Sol. b., Vol. 54, pp. 551–563, 1972.

11. Pickett, W. E., Allen, Ph. B., "Optical Properties of Nb and Mo Calculated from APW Band Structure," Phys. Rev. B., Vol. 11, pp. 3599–3608, 1975.

12. Ching, W. Y., Callaway, J., "Energy Bands, Optical Conductivity and Compton Profile of Sodium," Phys. Rev. B., Vol. 11, pp. 1324–1332, 1975.

13. Uspenskii, Yu. A., Maksimov, E. G., Rashkeev, S. N., Mazin, I. I., "Calculations of the Optical Properties of Metals by LMTO Method," Z. Phys. B., Vol. 53, pp. 263–270, 1983.

14. Kubo, Y., Wakoh, S., "Optical Properties of $3-d$ Transition Metals," J. Phys. Soc. Japan, Vol. 50, pp. 835–847, 1981.

15. Yamashita, J., Kubo, Y., Wakoh, S., "Optical Properties of Mo and Nb Calculated from APW Band Structure," J. Phys. Soc. Japan, Vol. 42, pp. 1906–1917, 1977.

16. Petroff, Irene, Viswanathan, C. R., "Calculation of Photoelectric Emission from Tungsten, Tantalum and Molybdenum," Phys. Rev. B., Vol. 4, pp. 799–808, 1971.

17. Nilsson, P. O., "Optical Properties of Metals and Alloys," Solid State Physics, Ed. H. Ehrenreich, F. Seitz, D. Turnball, N. Y.: Academic Press, Vol. 29, pp. 139–232, 1974.

18. Sturm, K., "Electron Energy Loss in Simple Metals and Semiconductors," Adv. Phys., Vol. 31, pp. 1–83, 1982.

19. Alward, J. F., Fong, C. Y., Guha Sridhar, C., "Theoretical Studies of the Optical and Electronic Properties of V, Nb and Ta," Phys. Rev. B., Vol. 18, pp. 5438–5445, 1978.

20. Kasowski, R. V., "Temperature-Dependent Optical Properties of Zn and Cd: A Theoretical Study," Phys. Rev., Vol. 187, pp. 885–895, 1969.

21. Moruzzi, V. L., Williams, A. R., Janak, J. F., "Calculated Electronic Properties of Metals," N. Y., Pergamon, 187 p., 1979.

22. Himples, E. S., Fastman, D. E., Koch, E. E., Williams, A. R., "Experimental $E(\mathbf{k})$ Dispersions for the $Zn—3-d$ States: Evidence for Itinerant Character," Phys. Rev. B., Vol. 22, pp. 4604–4612, 1980.

23. Lasser, R., Smith, N. V., Bendow, R. L., "Empirical Band Calculations of the Optical Properties of d-Band Metals. I, Cu, Ag and Au," Phys. Rev. B., Vol. 24, pp. 1895–1907, 1981.

24. Lasser, R., Smith, N. V., "Empirical Band Calculations of the Optical Properties of d–Band Metals. II. $Ag_{1-x}Cd_x$ Alloys," Phys. Rev. B., Vol. 24, pp. 1910–1920, 1981.

25. Lasser, R., Smith, N. V., "Empirical Band Calculations of the Optical Properties of d–Band Metals. III. Rh and Pd," Phys. Rev. B., Vol. 25, pp. 806–818, 1982.

26. Smith, N. V., Lasser, R., Chiang, S., "Empirical Band Calculations of the Optical Properties of d–Band Metals. IV. Optical Absorption, Photoemission and Magneto-Optical Properties of Ferromagnetic Nickel," Phys. Rev. B., Vol. 25, pp. 793–801, 1982.

27. Smith, N. V., "Photoemission Spectra and Band Structures of d–Band Metals. VII. Extensions of the Combined Interpolation Scheme," Phys. Rev. B., Vol. 19, pp. 5019–5030, 1979.

28. Bendow, R. L., Smith, N. V., "Photoemission Spectra and Band Structures of d–Band Metals. X. Relativistic Momentum Matrix Elements," Phys. Rev. B., Vol. 27, pp. 3144–3153, 1983.

29. Williams, A. R., Janak, J. F., Moruzzi, V. L., "One–Electron Analysis of Optical Data in Copper," Phys. Rev. Lett., Vol. 28, pp. 671–675, 1972.

30. Ray, P. N., Chowdhury, J., Chatterjee, S., "Electronic Energy Bands and Optical Properties of Pd," J. Phys. F., Vol. 12, pp. 2531–2540, 1982.

31. Laurent, P. G., Callaway, J., Wang, C. S., "Optical Conductivity of Iron and Nickel," Phys. Rev. B., Vol. 20, pp. 1134–1142, 1979.

32. Laurent, P. G., Callaway, J., Fry, J. L., Brener, N. E., "Band Structure, Fermi Surface, and Optical Conductivity of Paramagnetic Chromium," Phys. Rev. B., Vol. 23, pp. 4977–4986, 1981.

33. Janak, J. F., Williams, A. R., Moruzzi, V. L., "Self-Consistent Band Theory of the Fermi–Surface, Optical and Photoemission Properties of Copper," Phys. Rev. B., Vol. 11, pp. 1522–1530, 1975.

34. Chen, A. B., "Calculation of Momentum Matrix Elements Using the Green's Function Method," Phys. Rev. B., Vol. 14, pp. 2384–2390, 1976.

35. Kohn, W., Sham, L. J., "Exchange and Correlation Effects," Phys. Rev., Vol. 140, pp. A1133–A1140, 1965.

36. Hohenberg, P., Kohn, W., "Inhomogeneous Electron Gas," Phys. Rev., Vol. 136, pp. B864–B876, 1964.

37. Hedin, L., Lundquist, B. I., "Explicit Local Exchange–Correlation Potential," J. Phys. C., Vol. 4, pp. 2064–2076, 1974.

38. Andersen, O. K., Jepsen, O., "Advances in the Theory of One-Electron Energy States," Physica B., Vol. 91, pp. 317–329, 1977.

39. Andersen, O. K., "Linear Methods in Band Theory," Phys. Rev. B., Vol. 12, pp. 3060–3080, 1975.

40. Jepsen, O., Glotsel, D., Mackintosh, A. R., "Potential Band Structures and Fermi Surfaces in the Noble Metals," Phys. Rev. B., Vol. 23, pp. 2684–2701, 1981.

41. Jepsen, O., "Electronic Structure and Magnetic Breakdown in Titanium," Phys. Rev. B., Vol. 12, pp. 2988–3001, 1975.

42. Jepsen, O., Andersen, O. K., Mackintosh, A. R., "Electronic Structure of hcp Transition Metals," Phys. Rev. B., Vol. 12, pp. 3084–3103, 1975.

43. Blount, E. I., "Formalism of Band Theory," Solid State Physics, Edited by F. Seitz, O. Turnball, N. Y.: Acad. Press, Vol. 13, pp. 306–408, 1962.

44. Fano, V., Cooper, J. W., "Spectral Distribution of Atomic Oscillators Strengths," Rev. Mod Phys., Vol. 40, pp. 441–505, 1968.

45. Smrcka, L., "Soft X-Ray Transition Probability Calculation Based on the Muffin-Tin Potential," Czech. J. of Phys. B., Vol. 30, pp. 1012–1020, 1980.

46. Volkov, B. A., Gorbatsevich, A. A. Kapayev, Yu. V., Tugushchev, V. V., "Macroscopic Current States in Crystals," ZhETF, Vol. 81, pp. 726–740, 1981.

47. Varshalovich, D. A., Moskalev, A. N., Khersonskiy, V. K., Kvantovaya teoriya uglovogo momenta [Quantum Angular Moment Theory], Leningrad, Nauka Press, 440 pp., 1975.

48. Barlett, R. J., Lynch, D. W., Rosei, R., "Optical Properties of Single-Crystal Cadmium," Phys. Rev. B., Vol. 3, pp. 4074–4081, 1971.

49. De Sorbo, W., "Effect of Dissolved Phases on Some Superconductivity Properties of Miobium," Phys. Rev., Vol. 132, pp. 107–120, 1963.

50. Shiles, E., Sasaki, T., Inokuti, M., Smith, D. Y., "Self-Consistent and Sum-Rule Tests in the Kramers-Kronig Analysis of Optical Data: Application to Al," Phys. Rev. B., Vol. 22, pp. 1612–1626, 1980.

51. Bos, L. W., Lynch, D. W., "Low-Energy Absorption Peak in Aluminum and Al-Mn Alloys," Phys. Rev. Lett., Vol. 25, pp. 156-162, 1970.

52. Pells, P., Shiga, M., "The Optical Properties of Copper and Gold as a Function of Temperature," J. Phys. C., Vol. 2, pp. 1835-1843, 1969.

53. Johnson, P. B., Christy, R. W., "Optical Constants of Noble Metals," Phys. Rev. B., Vol. 6, pp. 4370-4378, 1972.

54. Nilsson, P. O., "Optical Properties of Copper," Phys. Kondens. Mater., Vol. 11, pp. 1-12, 1970.

55. Beaglehole, D., De Crescenti, M., Theye, M. L., Vuye, G., "Dielectric Constant of Gold, Copper and Gold-Copper Alloys Between 18 and 35 eV," Phys. Rev. B., Vol. 19, pp. 6306-6317, 1979.

56. Ehrenreich, H., Phipipp, H. K., "Optical Properties of Ag and Cu," Phys. Rev., Vol. 128, pp. 1622-1628, 1962.

57. Kubo, Y., Wakoh, S., Yamashita, J., "Energy Loss Spectra of Copper," J. Phys. Soc. Japan, Vol. 41, pp. 1556-1562, 1976.

58. Misell, D. L., Atkins, A. S., "Electron-Energy Loss in the Transition Metals," Phil. Mag., Vol. 27, pp. 95-104, 1973.

59. Daniels, J., "Determination of Optical Constants of Palladium and Silver from 2 to 90 eV from Energy Loss Measurements," Z. Phys. , Vol. 227, pp. 234-251, 1969.

60. Irani, G. B., Huen, T., Wooten, F., "Optical Properties of Ag and α-Phase Ag-Al Alloys," Phys. Rev. B., Vol. 3, pp. 2385-2394, 1971.

61. Lynch, M. S., Swan, J. B., "The Characteristic Loss Spectra of the Second and Third Series Transition Metals," Aust. J. Phys., Vol. 21, pp. 811-819, 1968.

62. Seignac, S., Robin, S., "Reflectivity of Rhodium and Vanadium," C. r. Acad. Sci. Paris B., Vol. 271, pp. 919-924, 1970.

63. Weaver, J. H., Lynch, D. W., Olson, C. G., "Optical Investigation of the Electronic Structure of Bulk Rh and Ir," Phys. Rev. B., Vol. 15, pp. 4115-4122, 1977.

64. Pierce, D. T., Spicer, W. E., "Optical Properties of Rhodium," Phys. Stat. Sol. b., Vol. 60, pp. 689-695, 1973.

65. Weaver, J. H., "Optical Properties of Rh, Pd, Ir, Pt," Phys. Rev. B., Vol. 11, pp. 1416-1424, 1975.

66. Vehse, R. C., Arakawa, E. T., Williams, M. W., "Optical and Photoemission Properties of Pd in the Vacuum Ultraviolet Spectral Region," Phys. Rev. B., Vol. 1, pp. 517–526, 1970.

67. Savtskiy, Ye. N., Rashkeyev, S. N., Uspenskiy, Yu. A., Mazin, I. I., "Calculation of the Coefficient of Reflection of Copper, Paladium and Rhodium from 'First Principles', ", DAN SSSR, Vol. 277, pp. 91–93, 1984.

68. Uspenskiy, Yu. A., Rashkeyev, S. N., Mazin, I. I., "Calculation of the Optical Properties of Metals Using LMTO Band Structure," Proc. 13 Ann. Int. Symp. on Electronic Structure of Metals and Alloys," Dresden, pp. 52–54, 1983.

69. Weaver, J. H., Lynch, D. W., Olson, C. G., "Optical Properties of Niobium from 0.1 to 36.4 eV," Phys. Rev. B., Vol. 7, pp. 4311–4319, 1973.

70. Vilesov, F. I., Zagrubskiy, A. A., Kirillova, M. M., "Photoemission Method of Determining Diffusion Reflection Spectra in the Vacuum Ultraviolet," Optika i spektroskopiya, Vol. 23, pp. 153–162, 1967.

71. Weaver, H. H., Lynch, D. W., Olson, C. G., "Optical Properties of V, Ta and Mo from 0.1 to 35 eV," Phys. Rev. B., Vol. 10, pp. 501–516, 1974.

72. Mayevskiy, V. N., Druzhinin, A. V., Kirillova, M. M., Nomerovannaya, L. V., "The Kramers-Kronig Method for Determining Optical Characteristics of Metals in the High-Energy Range. The Transition Metals Mo and Re," Sverdlovsk, 40 pp., 1981, Deposited at VINITI, No. 5151–81.

73. Szalkowski, F. J., Bertrand, P. A., Somorjai, G. A., "Characteristic-Energy-Loss Spectra of Vanadium and V_2O_3," Phys. Rev. B., Vol. 9, pp. 3369–3378, 1974.

74. Nestell, J. E., Christy, R. W., "Optical Conductivity of bcc Transition Metals: V, Nb, Ta, Cr, Mo, W.," Phys. Rev. B., Vol. 21, pp. 3173–3179, 1980.

75. Weaver, J. H., Lynch, D. W., "Absorptivity of Single-Crystal Yttrium at 4.2 k," Phys. Rev. B., Vol. 7, pp. 4737–4742, 1973.

76. Weaver, J. H., Olson, C. G., "Optical Properties of hcp Yttrium," Phys. Rev. B., Vol. 15, pp. 590–594, 1977.

77. Gorobchenko, V. L., Zharnikov, M. V., Maksimov, Ye. G., Moldosanov, K. A., "Study of Plasma Excitation in Technicium by Inelastic Scattering of Electrons," ZhETF, Vol. 86, pp. 597–608, 1984.

78. Gluskin, Ye. S., Druzhinin, A. V., Kirillova, M. M., et al., "Optical Properties of Single Crystals of Rhuthenium, Osmium and Their Compounds in the UV Area of the Spectrum," Optica i spektroskopiya, Vol. 55, pp. 891–897, 1983.

79. Zharnikov, M. V., Rashkeyev, S. N., "Theoretical Analysis of Plasma Excitation in Technicium," FTT, Vol. 26, pp. 3385–3388, 1984.

80. Weaver, J. H., Lynch, D. W., Rosei, R., "Optical Properties of Single-Crystal Zinc," Phys. Rev. B., Vol. 5, pp. 2829–2832, 1972.

81. Gorobchenko, V. D., Zharnikov, M. V., Maksimov, Ye. G., Rashkeyev, S. N., "Experimental and Theoretical Study of Plasma Oscillations of Electrons in Zinc and Cadmium," ZhETF, Vol. 88, pp. 677–691, 1985.

82. Maksimov, E. G., Mazin, I. I., Rashkeev, S. N., Uspenskii, Yu. A., "Microscopical Calculations of the Optical Properties of Metals" Preprint, FIAN N 189, Moscow, 35 pp., 1985.

83. Juras, G. E., Segall, B., Sommers, C. B., "Electronic Structure of Zinc," Sol. St. Comm., Vol. 10, pp. 427–430, 1972.

84. Maksimov, E. G., Mazin, I. I., Rashkeev, S. N., Uspenskii, Yu. A., "The First-Principle Calculations of the Optical Properties of Transition Metals. Electron Energy Loss Spectra. The Optical Properties in the Infrared Region," Preprint FIAN N 188, Moscow, 45 pp., 1985.

85. Mazin, I. I., Maksimov, Ye. G., Rashkeyev, S. N., Uspenskiy, Yu. A., "Microscopic Analysis of Energy-Loss Spectra of Fast Electrons in 4*d* Metals," ZhETF, Vol. 90, pp. 1092–1110, 1986.

86. Dolgov, O. V., Maksimov, Ye. G., "Local Field Effects and Disruption of Kramers-Kronig Relations for Dielectric Constants," UFN, Vol. 135, pp. 441–447, 1981.

87. Gorobchenko, V. D., Maksimov, Ye. G., "Dielectric Matrix of Band Electrons Considering Space-Correlation Effects," ZhETF, Vol. 81, pp. 1847–1859, 1981.

88. Gorobchenko, V. D., Maksimov, Ye. G., "Dielectric Constant of Interacting Electron Gas," UFN, Vol. 130, pp. 65–130, 1980.

89. Voyevodin, V. V., Kuznetsov, Yu. A., Matritsy i vychisleniya [Matrices and Computation], Moscow, Nauka Press, 318 pp., 1984.

90. Hanke, W., "Dielectric Theory of Elementary Excitations in Crystals," Adv. Phys., Vol. 27, pp. 287–343, 1978.

91. Isomaki, H. M., Boehm, J. von., "Optical Response of Uniaxial Semiconductors. II. Optical and Energy-Loss Spectra of ZrS_2 and $ZrSe_2$," Phys. Rev. B., Vol. 26, pp. 5807–5815, 1982.

92. Isomaki, H. M., Boehm, J. von, Stubb, T., "Optical Response of Uniaxial Semiconductors III. Optical and Energy-Loss Spectra of Te," Phys. Rev. B., Vol. 26, pp. 5815–5829, 1982.

93. Hanke, W., Sham, L. J., "Many Particle Effects in the Optical Spectrum of a Semiconductor," Phys. Rev. B., Vol. 21, pp. 4656–4673, 1980.

94. Nizzoli, F., "A Model Calculation of the Dielectric Function in Trigonal Se and Te with Local-Field Corrections Included," J. Phys. C., Vol. 11, pp. 673–683, 1978.

95. Rajagopal, A. K., "Theory of Inhomnogenous Electron Gas: Spin-Density Functional Formalism," Adv. Chem. Phys., Vol. 41, pp. 59–191, 1980.

96. Singhal, S. P., Callaway, J., "Exchange Correction to the Dielectric Function in the Local Exchange Approximation," Phys. Rev. B., Vol. 14, pp. 2347–2352, 1976.

97. Zangwill, A., Soven, P., "Density-Functional Approach to Local-Field Effects in Finite Systems: Photoabsorption in Rare Gases," Phys. Rev. A., Vol. 21, pp. 1561–1572, 1980.

98. Zangwill, A., Soven, P., "Resonant Photoemission in Barium and Cerium," Phys. Rev. Lett., Vol. 45, pp. 204–207, 1980.

99. Levine, Z. H., Soven, P., "Electron Correlation Effects in Photoemission from $1\pi_u$ Level in Acetylene," Phys. Rev. Lett., Vol. 50, pp. 2074–2078, 1983.

100. Levine, Z. H., Soven, P., "Time-Dependent Local-Density Theory of Dielectric Effects in Small Molecules," Phys. Rev. A., Vol. 29, pp. 625–635, 1984.

101. Ekardt, W., "Size-Dependent Photoabsorption and Photoemission of Small Metal Particles," Phys. Rev. B., Vol. 31, pp. 6360–6370, 1985.

102. Zangwill, A., Liberman, D. A., "On the Interplay Between Dielectric Screening and Core Hole Relaxation in Soft X-Ray Absorption," J. Phys. B., Vol. 17, pp. L253–L257, 1984.

103. Golovashkin, A. I., Leksina, I. Ye., Motulevich, G. P., Shubin, A. A., "Optical Properties of Nb," ZhETF, Vol. 56, pp. 51–58, 1969.

104. Kirillova. M. M., Nomerovannaya, L. V., "Optical Properties of d–Transition Metals with bcc Lattice," FTT, Vol. 20, pp. 984–988, 1978.

105. Kirillova, M. M., "Optical Spectroscopy of Transition d–Metals," Dissertation for Doctor of Physical-Mathematical Sciences, Sverdlovsk, Institute of Metal Physics, Urals Science Center, USSR Academy of Sciences, 342 pp., 1984.

106. Topolskiy, V. G., Shirokovskiy, V. P., Kirillova, M. M., Nomerovannaya, L. V., "Optical Conductivity of Rhodium and Iridium," Fizika metallov i metallovedeniye, Vol. 52, pp. 745–751, 1981.

107. Shirokovskiy, V. P., Kirillova, M. M., Shilkova, N. A., "Anomalies of Optical Absorption in Iron," ZhETF, Vol. 82, pp. 784–791, 1982.

108. Stashchuk, V. S., Kravchuk, A. N., Yashinskaya, M. N., "Optical and Electronic Characteristics of Iron," Ukr. fiz. zhurn., Vol. 29, pp. 1720–1725, 1984.

109. Ganin, G. V., Kirillova, M. M., Nomerovannaya, L. V., Shirokovskiy, V. P., "Optical Absorption Spectra of V and Cr. Experiment and Theory," FMM, Vol. 43, pp. 907–916, 1977.

110. Rashkeyev, S. N., Uspenskiy, Yu. A., Mazin, I. I., "Specifics of Optical Properties of Transition Metals in the Infrared Range," ZhETF, Vol. 88, pp. 1687–1698, 1985.

111. Khalilov, S. V., Rashkeyev, S. N., Uspenskiy, Yu. A., "Spin Moment in the Formation of Optical Properties of Ferromagnetic Ion," FTT, Vol. 29, pp. 1467–1469, 1987.

112. Herring, C., "Time-Inversion Symmetry and Crystal Energy Bands," Phys. Rev., Vol. 52, pp. 361–384, 1937.

113. Cracknell, A., Young, K., Poverkhnost' Fermi [The Fermi Surface], Moscow, Atomizdat Press, 270 pp., 1978.

114. Landolt-Bornstein. "Metals: Phonon and Electron States, Fermi Surfaces," Berlin, Heidelberg: Springer-Verlag, New Series, Group III, Vol. 15, Subvol. b., 1983.

115. Duysebayeva, Zh., Korsunskiy, M. I., Motulevich, G. P., "Optical Properties of Palladium," Optika i spektroskopiya, Vol. 34, pp. 535–538, 1973.

116. Golovashkin, A. I., Mash, I. D., Motulevich, G. P., "Optical Properties of Vanadium Obtained by Atomization in a Vacuum," Kratkiye soobshcheniya po fizike, Moscow, FIAN SSSR, No. 9, pp. 51–59, 1970.

117. Bolotin, G. A., Chukina, T. P., "Optical Properties of Rh in the IR Range," Optika i spektroskopiya, Vol. 23, pp. 620–622, 1967.

118. Bolotin, G. A., Kirillova, M. M., Nomerovannaya, L. V., Noskov, M. M., "Optical Properties and Fermi Surface of Palladium," Fizika metallov i metallovedeniye, Vol. 23, pp. 463–470, 1967.

119. Danan, H., Herr, A., Meyer, J. P., "New Determination of the Saturation Magnetization of Nickel and Iron," J. Appl. Phys., Vol. 39, pp. 669–670, 1968.

DISPERSION OF COMPLEX DIELECTRIC CONSTANT AND ELECTRONIC CHARACTERISTICS OF THE COMPOUNDS Nb-Al AND Nb-Ge WITH A15 STRUCTURE

N. D. Kuz'michev, I. S. Levchenko and G. P. Motulevich

ABSTRACT

The dispersions of complex dielectric constant of the compounds Nb-Al and Nb-Ge with A15 structure, used for determination of electronic characteristics and their variations with temperature, are measured in the 0.177-3.1 eV spectral interval at 295 and 670 K. The squares of the plasma frequencies ω_p^2 of conduction electrons are obtained. In both compounds $\omega_p^2 \approx 19$ eV^2, which is 3.2 times less than for niobium. In this spectral interval, Nb-Al has four interband transition zones at 0.2, 0.35, 1.45 and 3.1 eV, while Nb-Ge has five such bands: 0.21, 0.32, 0.48, 0.95 and 2.0 eV. As the temperature increases, ω_p^2 of conduction electrons increases somewhat more than for usual metals (in both compounds), and the decrease in the analogous characteristic in the long-wave band for Nb-Ge is unusually great, preserving the sum of the changes, and there is also a significant decrease in the width of the long-wave band. These anomalies can be explained by thermal transfer of electrons from the base state of the long-wave band to the conduction band. Figures 21, Tables 7, References 34.

INTRODUCTION

The interaction of the electromagnetic field of a light wave with the valent electrons of metals is described by the frequency-dependent complex dielectric constant $\epsilon(\omega)=\epsilon_1(\omega)-i\epsilon_2(\omega)$. It was demonstrated in [1, 2] that measurements of the dispersion of the dielectric constant over a broad spectral range allow valuable information to be produced on the electronic characteristics of metals, relating both to conduction electrons and to electrons involved in interband transitions. In this work, the dispersion of the complex dielectric constant of the compounds Nb–Al and Nb–Ge with A15 structure is measured in the 0.177–3.1 eV spectral interval at 295 and 670 K, the electronic characteristics and their variation with temperature are determined. Both compounds are interesting both due their superconducting properties, and due to anomalies in the temperature variation of a number of static characteristics, which are determined by the peculiarities of the energy spectrum of the valent electrons. Previous studies are discussed in Chapter 3.

CHAPTER 1
RELATIONSHIP OF OPTICAL PROPERTIES OF A METAL WITH ITS ELECTRONIC CHARACTERISTICS

The concept of the pseudopotential, developed in solid-state theory by a number of authors (for example, cf. [3]) was used to establish the relationship of the complex dielectric constant with the basic electronic characteristics of the transition metals.

The pseudopotential considers the shielding of the potential of an ion by electrons and averaging of fast oscillation movements of an electron within the ion core. The pseudopotential is much less than the initial potential of an ion and is a significantly smoother function than the true potential. The concept of pseudopotential has allowed analysis of the optical properties of metals on the basis of the concept of "almost free" electrons. However, the properties of these "almost free" electrons differ from the properties of free electrons. Coherent scattering of electrons by lattice planes, even with a weak pseudopotential, has a significant influence on the optical properties of a metal. Consideration of the influence of the periodic lattice potential on the basic integral characteristics and on the optical properties of metals has yielded a number of equations relating the optical dielectric constant to the basic microcharacteristics of a metal[1, 2].

Let us now study the qualitative picture of the interaction of an electromagnetic field with a metal. The optical properties of the metals are determined by the valent electrons, the concentration of which $N_{val} \sim 10^{23} cm^{-3}$. A metal is a strongly degenerate system. The states of the electrons are described by the distribution function which, when there is no external field present, is the Fermi function. Where $T \equiv 0$, all energy states with energies $E \ll E_F$ (E_F is the Fermi energy) are occupied. For metals, $E_F \sim 10$ eV. Where $T \neq 0$, the step on the distribution function is not as clear. Spreading of the step is on the order of $k_B T \ll E_F$, where k_B is Boltzman's constant. The electrons occupy a certain closed volume in momentum space.

Without a lattice, the surface delimiting the volume of occupied states (the Fermi surface) is a sphere. A crystalline lattice distorts the Fermi surface. In momentum space, the lattice is characterized by the Bragg planes, perpendicular to the vectors **g** of the reciprocal lattice, and the Fourier components V_g[1] of the pseudopotential. $|V_g|{\sim}1$ eV, i.e., $|V_g|/E_F \ll 1$.

Figure 1 below illustrates the area of intersection of the Fermi surface with one Bragg plane. The lattice has almost no influence on the states of electrons far from the Bragg plane. Corrections to the electron energy are of second order with respect to $|V_g|/E_F$, the speed of the electrons is practically unchanged, the Fermi surface remains a portion of a sphere. The states of the electrons near the Bragg plane are distorted. Corrections appear to the first order energies with respect to $|V_g|/E_F$, the energy levels are split, two energy bands arise, the speeds of the electrons decrease (the component perpendicular to the Bragg plane decreases), the Fermi surface becomes a surface of rotation with an axis perpendicular to the Bragg plane. On the Bragg plane itself, the energy gap $\Delta E=2|V_g|$. The Fermi surface is broken. The area of occupied electron states is limited not only by the Fermi surface, where $E=E_F$, but also by the circles on the Bragg planes, where $E \neq E_F$. The total area of the Fermi surface S_F decreases. Only those Bragg planes which intersect the Fermi surface have a significant influence. Therefore, the crystalline structure is significant for polyvalent metals and has little significance for univalent ones.

The field of a light wave changes the equilibrium distribution function. The addition to the equilibrium distribution function determines the optical properties of the metal. The energy of a light quantum $\hbar\omega \ll E_F$, and therefore the optical properties are determined by electrons located near the surface delimiting the volume of occupied states, i.e., near the Fermi surface and corresponding to sections of the Bragg planes.

As was noted earlier, the response of electrons to the electromagnetic field of a light wave is described by the complex dielectric constant[2] $\epsilon(\omega)=\epsilon_1(\omega)-i\epsilon_2(\omega)$. The contribution to ϵ of electrons located near the Fermi surface, called conduction electrons, and of electrons near the rings on the Bragg planes, which we shall call "Bragg" electrons, will be represented by ϵ_{1e}, ϵ_{2e}, ϵ_{1g} and ϵ_{2g}. We can consider the contribution of these groups of electrons to be additive, i.e.,

---

[1]In subscripts below, the vector **g** will be represented by g.

[2]Some works use the optical conductivity σ_{opt}, which is related to ϵ_2 by the equation $\epsilon_2=4\pi\sigma_{\mathrm{opt}}/\omega$.

$$\varepsilon_1(\omega) = 1 + \varepsilon_{1e}(\omega) + \varepsilon_{1g}(\omega), \quad \varepsilon_2(\omega) = \varepsilon_{2e}(\omega) + \varepsilon_{2g}(\omega). \tag{1}$$

The one represents the contribution of the vacuum to ϵ_1.

The energy spectrum of conduction electrons has no gaps. An arbitrarily small quantum is sufficient to shift them to the excited state. A static field leads to static currents described by the static conductivity σ_{st}. A light field leads to variable currents described by $\epsilon(\omega)$. The frequency variation $\epsilon_{1e}(\omega)$ and $\epsilon_{2e}(\omega)$ is monotonic (cf. Figure 2 below). Both $|\epsilon_{1e}(\omega)|$ and $\epsilon_{2e}(\omega)$ decrease with increasing frequency (under the influence of electron mass). The conduction electrons determine the optical properties of metals in the long-wave (infrared) area of the spectrum.

The energy spectrum of the "Bragg" electrons has a gap $\Delta E=2|V_g|$. In the base state, the energy of an electron $E<E_F$, in the excited state $E>E_F$. Therefore, these electrons do not react to a static field and do not participate in transfer phenomena. However, if a light quantum $\hbar\omega>\Delta E$, they participate in interband transitions (internal photoeffect). The frequency variations $\epsilon_{1g}(\omega)$ and $\epsilon_{2g}(\omega)$ are illustrated by Figure 2. They are not monotonic. In the area $\hbar\omega\sim\Delta E$, $\epsilon_{2g}(\omega)$ has a maximum, while $\epsilon_{1g}(\omega)$ has an area of anomalous dispersion. The jump $\Delta\epsilon_{1g}$ is close to the quantity max ϵ_{2g}. In other words, in the frequency area $\hbar\omega\approx\Delta E$, there is a band of absorption of light. Due to the strong interaction of electrons with each other and with the lattice potential, the absorption bands are usually quite broad $\Delta\omega/\omega\sim 0.1$–$0.5$. Generally, there are several absorption bands, i.e.,

$$\varepsilon_{1g}(\omega) = \sum \varepsilon_{1gj}(\omega), \quad \varepsilon_{2g}(\omega) = \sum \varepsilon_{2gj}(\omega), \tag{2}$$

(j is the band number). The summation is performed with respect to all interband transition zones. Usually, the interband transition zones are in the short-wave area of the spectrum (visible and near-infrared areas). However, in polyvalent metals and alloys with complex lattices, they may be located in the long-wave (infrared) area as well.

To generate information on the electron characteristics of a metal we require equations relating the imaginary and real parts of $\epsilon(\omega)$ to these characteristics.

1.1. Conduction Electrons

The relationship of ϵ_{1e} and ϵ_{2e} with the characteristics of the conduction electrons is obtained by solving an equation system including a kinetic equation for the electron distribution function and the equations of Maxwell. The solution depends on the nature of the skin effect. For polyvalent metals at high (room) temperatures, corrections for the nonlocal nature of the relation between current and field can be ignored. Limiting ourselves to this case, we obtain for cubic crystals[1, 2]

$$\varepsilon_e(\omega) = \varepsilon_{1e}(\omega) - i\varepsilon_{2e}(\omega) = -\frac{\omega_p^2}{\omega(\omega - i\nu)}, \tag{3}$$

$$\omega_p^2 = 4\pi e^2 \frac{N}{m}, \tag{3a}$$

$$\nu = \nu_{ep} + \nu_{ee} + \nu_{ed}, \tag{3b}$$

where ω_p^2 is the square of the plasma frequency of the conduction electrons, e is the charge of an electron, m is the mass of a free electron, N is the effective concentration of conduction electrons[3], v is the relaxation frequency (v=-$1/\tau$, τ is the relaxation time), v_{ep}, v_{ed} and v_{ee} are the relaxation frequencies related with processes of interaction of electrons with phonons, defects and electrons, respectively[4]. Equation (3b) considers that the absorption of light by electrons in the metal is possible only upon simultaneous interaction of an electron with a phonon, impurity or electron. Usually the term v_{ee} can be ignored, since $v_{ee} \ll v_{ep}$.

Further, in the case of the normal skin effect for cubic crystals

$$\frac{N}{m} = \frac{2}{3}\frac{\langle v_F\rangle S_F}{(2\pi\hbar)^3}, \quad \frac{N_{\text{val}}}{m} = \frac{2}{3}\frac{v_F^0 S_F^0}{(2\pi\hbar)^3}, \tag{4}$$

where N_{val} is the concentration of valent electrons, $\langle v_F\rangle$ is the mean speed of an electron on the Fermi surface, S_F is the full area of the Fermi surface, v_F^0 and S_F^0 relate to the sphere of free electrons at a concentration equal to the valent concentration. It follows from (4) that

$$\frac{N}{N_{\text{val}}} = \frac{\langle v_F\rangle S_F}{v_F^0 S_F^0}. \tag{5}$$

---

[3]In optical experiments only the ratio N/m is determined.

[4]The one related to the vacuum is omitted in equations (3).

The crystalline lattice potential decreases the full area of the Fermi surface, the mean speed of electrons on the Fermi surface and the square of the plasma frequency of conduction electrons (or the effective concentration of conduction electrons).

The conditions of the normal skin effect [4]

$$l \ll \delta = c/\omega\kappa, \tag{6a}$$

$$v \ll c/n \tag{6b}$$

where l is the free path length on a metal, c is the speed of light, $n-i\kappa$ is the complex index of refraction, related to the dielectric constant by the equation

$$\epsilon = \epsilon_1 - i\epsilon_2 = (n-i\kappa)^2. \tag{7}$$

Expression (3) for $\epsilon_e(\omega)$ is based on the classical kinetic Boltzman equation. The quantum kinetic equations for electrons in metal located in the field of an electromagnetic wave was obtained by Gurzhi [5]. All the basic relations remain unchanged. A significant difference is obtained for the temperature variation of v_{ep}. Where $T \gg \theta$ (θ is the Debye temperature), both the classical and the quantum the theories yield a linear variation in relaxation frequency v as a function of temperature T, $v = v_\theta T/\theta$. At low temperatures, $v^{cl}_{ep} \sim T^5$, whereas v^{opt}_{ep} depends very little on T and at the limit $T \to 0$, $v^{opt}_{ep} \to 0.4 v_\theta$. The reason for this difference is the quantum nature of absorption of light by an electron. After absorbing a photon with energy $\hbar\omega \gg k_B\theta$, an electron can radiate the entire spectrum of phonons, leading to the finite value of v^{opt}_{ep} where $T=0$.

If the variation in the static resistance of the metal as a function of T has Kruneisen nature, indicating weak dependence of effective conduction electron concentration on T, additional measurements of the static conductivity σ_{st} and the relation $R_{res}/R(T)$ allows us to obtain v^{cl}_{ep} and the electron–phonon interaction constant λ_{ep}. We utilize the equations

$$v^{cl}_{ep} + v_{ed} = \frac{e^2}{m}\frac{N}{\sigma_{st}}, \quad \frac{v_{ed}}{v^{cl}_{ep} + v_{ed}} = \frac{R_{res}}{R_T}, \tag{8a}$$

$$\lambda_{ep} = \frac{\hbar}{2\pi}\frac{v^{cl}_{ep}}{k_B T}, \quad T \gg \theta. \tag{8b}$$

Here R_{res} and R_T represents the residual resistance and the resistance at temperature T. Equation (8b) was obtained in [6].

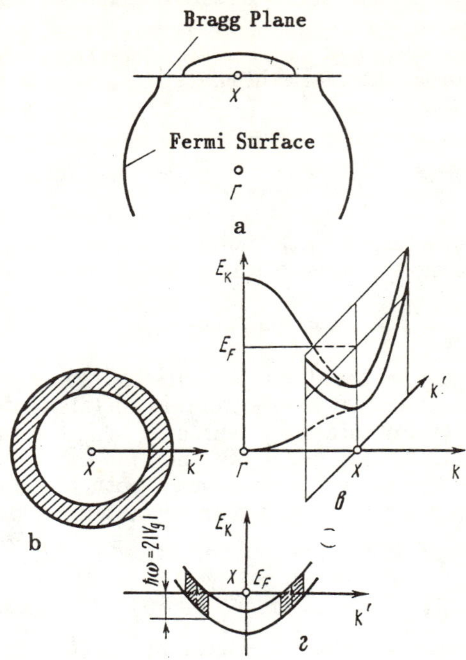

Figure 1. a, b — Fermi Surface Intersected by Bragg Plane; c — energy bands in the direction from Γ to X and in direction **K′** (**K′** is perpendicular to direction ΓX and lies in the Bragg plane); d — energy bands in direction **K′**. Possible lines of optical transitions between occupied and free levels are shaded.

1.2. Interband Transitions

In computing $\epsilon_g(\omega)$, we must consider the "vertical" interband transitions, i.e., transitions which preserve quasimomentum. In transition metals, the major contribution to $\epsilon(\omega)$ is made by transitions related to the "Bragg" splitting[5] of the energy spectrum of the electrons[1]. The band spectrum $E(\mathbf{k})$ splits for them. Two parallel sections (sheets) of isoenergy surfaces appear in $\mathbf{k}$ space, leading to

[5]"Bragg" interband transitions result from the transfer of electrons by an electromagnetic wave through energy gap Δ, resulting from splitting of the energy spectrum of the electrons near the Bragg plane.

a maximum for the combined state density and a related maximum on the function $\epsilon_{2g}(\omega)$. Figure 1 shows the band spectrum of an electron in the direction from Γ to X (Γ is the center of the Brillouine zone, X is the center of the ring formed by intersection of the Bragg plane by the Fermi surface) and in the direction $\mathbf{k}'$, where $\mathbf{k}'$ is the wave vector of an electron on the Bragg plane. For clarity, the function $E(\mathbf{k}')$ is shown on the separate Figure 1*d*. We can see from the figure that $E(\mathbf{k}')$ on the Bragg plane is split into two parallel branches. The distance between them is $2|V_g|$. The possible optical transitions between the occupied and free levels are shaded.

Figures 2*a* and *b* show the frequency variation of the real and imaginary parts $\epsilon_g(\omega)$ and $\epsilon_e(\omega)$, respectively.

Let us study the intersection of the Fermi surface by one Bragg plane. Computation of linear polarizability χ of a metal, related to interband transitions of "Bragg" electrons in the case of one Bragg plane is undertaken in [7].

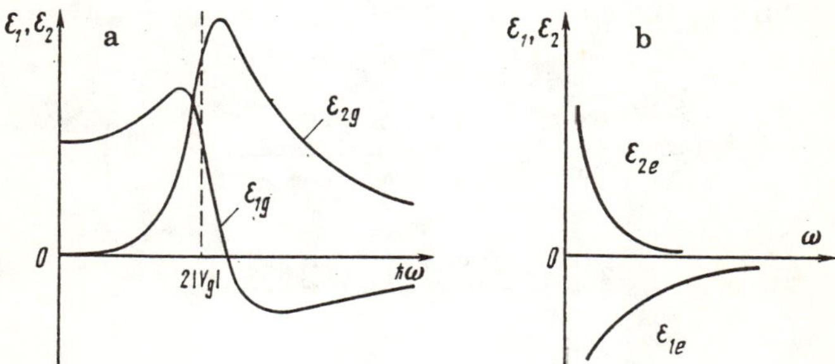

Figure 2. Frequency Variations. a — $\epsilon_{1g}(\omega)$ and $\epsilon_{2g}(\omega)$, related to one Bragg plane; b — $\epsilon_{1e}(\omega)$ and $\epsilon_{2e}(\omega)$, related to conduction electrons.

The following expression is obtained for the quantity χ:

$$\chi_g(\omega) = \frac{e^2}{4\pi^2}\frac{k_g}{\Delta_g}\int_0^\infty \frac{dx}{(1+x^2)^{3/2}}\left\{\frac{1}{\sqrt{1+x^2}-\omega'+i\nu'} + \frac{1}{\sqrt{1+x^2}+\omega'-i\nu'}\right\}. \tag{9}$$

Here $\omega'=\omega/\omega_g$, $v'=v/\omega_g$, k_g is the distance from the center of the Brillouine zone Γ to the Bragg plane, $\Delta_g=\hbar\omega_g=2|V_g|$. V_g is the Fourier component of the pseudopotential, corresponding to Bragg plane g, $v=1/\tau$, τ is the relaxation time. The dielectric constant, ignoring the dielectric constant of the vacuum, is related to χ_g by the expression

$$\epsilon_g = 4\pi\chi_g \tag{10}$$

Figure 2*a* shows the computed variation of Re $\epsilon_g(\omega)$ and Im $\epsilon_g(\omega)$ according to the data of [7]. We can see from the figure that the maximum in $\epsilon_{2g}(\omega)$ is displaced in the direction of increasing energy from the value of ω_g and has clear asymmetry $\epsilon_{2g}(\omega)$ with respect to a line passing through the maximum. Where $\omega<\omega_g$ the drop is quite rapid, more gentle where $\omega>\omega_g$. Where $\omega \gg \omega_g$, $\epsilon_{2g}(\omega) \sim 1/\omega^3$, i.e., the Drude variation.

This model considers interband transitions related only with the Bragg planes, and ignores transitions related to the lines and points of intersection of these planes. The contribution to $\epsilon_2(\omega)$ of these transitions is significantly less than the contribution from transitions related to the Bragg plane[8]. Therefore, in the structure of $\epsilon_2(\omega)$ of the metal in question, interband transitions related to the lines and points of intersection of Bragg planes against the background of the "Bragg" interband transitions should appear as a fine structure.

Let us return to the dielectric constant ϵ_g. Computing the integral (9), we obtain [9, 10]

$$\varepsilon_g(\omega) = \frac{\omega_p^2}{(\omega - i\nu)^2}\left(\frac{\omega_0}{\sqrt{\omega_0^2 - (\omega - i\nu)^2}} - 1\right), \tag{11}$$

$$\omega_p^2 = \left(\frac{e}{\hbar}\right)^2 k_g(\hbar\omega_0) = \frac{4\pi e^2}{m} N_g, \tag{12}$$

N_g is the effective number of electrons participating in the interband transition related to the Bragg plane with subscript *g*.

For cubic crystals the contribution to ω_p^2 of the equivalent Bragg planes is equal to

$$\omega_p^2 = \left(\frac{e}{\hbar}\right)^2 k_g(\hbar\omega_0)\frac{n_g}{3}, \tag{13}$$

where n_g is the number of equivalent Bragg planes.

The equations for Re $\epsilon_g(\omega)$ and Im $\epsilon_g(\omega)$ are presented in Chapter 4. They are equivalent to the corresponding expressions which were presented in [11, 11a]

The asymptotic expression for $\epsilon_g(\omega)$ as $\omega \to 0$, obtained in [11], is

$$\varepsilon_g(\omega)\,|_{\omega\to 0} = \varepsilon_{1g}(\omega)\,|_{\omega\to 0} = \frac{\omega_p^2}{\nu^2}\left(1 - \frac{\omega_0}{\sqrt{\omega_0^2 + \nu^2}}\right). \qquad (14)$$

It is equivalent to equation (47) of [12].

Let us compute the integral

$$\int_0^\infty \omega\,[-\operatorname{Im}\varepsilon_g(\omega)]\,d\omega.$$

To compute this integral we take the expression for ϵ_2 from [12], which, with dimensionless ω' and v', is as follows:

$$\varepsilon_2 = \left(\frac{e}{\hbar}\right)^2 \frac{n_g}{3\pi}\frac{k_g}{\omega_0}\frac{I_2(\omega)}{\omega'},$$

where

$$I_2(\omega) = \omega'\nu'\left\{\int_0^\infty \frac{dx}{(1+x^2)^{3/2}\left[(\sqrt{1+x^2}-\omega')^2+\nu'^2\right]} - \int_0^\infty \frac{dx}{(1+x^2)^{3/2}\left[(\sqrt{1+x^2}+\omega')^2+\nu'^2\right]}\right\}.$$

Then

$$\int_0^\infty \omega\varepsilon_2(\omega)\,d\omega = \frac{e^2}{3\pi\hbar^2} n_g k_g \int_0^\infty I_2(\omega)\,d\omega.$$

After computing the integral $\int_0^\infty I_2(\omega)\,d\omega$ we obtain

$$\int_0^\infty \omega\varepsilon_2(\omega)\,d\omega = \frac{\pi}{2}\frac{e^2}{3\hbar^2} n_g k_g(\hbar\omega_0), \qquad (15)$$

or, considering that $\frac{e^2}{3\hbar} n_g k_g (\hbar\omega_0) = \omega_p^2$

$$\int_0^\infty \omega\varepsilon_2(\omega)\, d\omega = \frac{\pi}{2}\,\omega_p^2. \qquad (16)$$

Expression (16) coincides with the sum rule. For all valent electrons, the sum rule is as follows:

$$\int_0^\infty \omega\varepsilon_2(\omega)\, d\omega = \frac{\pi}{2}\,\Omega_p^2. \qquad (17)$$

If $\epsilon_2(\omega)$ contains a contribution of only $(n-1)$ "Bragg" interband transitions and intraband transitions (conduction electrons), the sum rule becomes

$$\sum_{j=1}^{n} \omega_{pj}^2 = \Omega_p^2, \qquad (18)$$

where ω_{pj}^2 is the square of the plasma frequency of the corresponding transition. The oscillator strength of an optical transition, related to the equivalent Bragg planes with subscript j, is ω_{pj}^2/Ω_p^2.

Thus, optical measurements of the contribution to dielectric constant by "Bragg" interband transitions allow, for each system of physically equivalent Bragg planes, determination of the square of the plasma frequency ω_p^2 and the corresponding oscillator strength ω_p^2/Ω_p^2, the value of the dielectric gap $\Delta=\hbar\omega_0$ and the relaxation frequency v. (In the pseudopotential approximation $\Delta=2|V_g|$, where V_g is the Fourier component of the pseudopotential for a given system of physically equivalent Bragg planes.)

The model we have presented for dispersion of the complex dielectric constant of metals was developed for nontransition polyvalent metals. It agrees with the experimental data of [1]. For transition metals and compounds based upon them, the situation is more complex. Theoretical works have been published in which the optical properties of various metals have been computed, including certain transition metals. The final results are obtained by numerical methods.

The basis of all computations, as in our model, is the single-frequency approximation, although essentially the problem is a multifrequency one. As yet, no methods have been found for solution of the multifrequency problem. The best theoretical work on the transition metals, in our opinion, is that of I. I. Mazin, Ye. G. Maksimov, S. N. Rashkeyev and Yu. A. Uspenskiy, "Microscopic Computation of Dielectric Response Functions of Metals," (elsewhere in this collection). These authors obtained the functions $\epsilon_1(\omega)$ and $\epsilon_2(\omega)$ for a number of $3d$– and $4d$–transition metals, agreeing with experimental data.

Unfortunately, no similar computations are available for the compounds of interest to us.

In processing the experimental data, we considered that in all cases the main structure of the interband transition zones is related to the presence of "parallel" (or near-parallel) parts (sheets) of isoenergy surfaces, which corresponds to a Van-Hoff type Q_0 singularity in the combined state density. For the corresponding group of electrons, the form of the interband transition zone (i.e., the function $\epsilon(\omega)$) can be described with good approximation by equation (11). The parameters ω_p^2, ω_0 and v are taken from the experimental data. ω_p^2 determines the effective square of the plasma frequency (effective oscillator strength), $\Delta=\hbar\omega_0$ — the effective distance between two "parallel" sheets, v is the effective relaxation frequency, determined by the widths of the corresponding bands. The parallel sheets may be either at the boundary of the Brillouine zone, or within it.

CHAPTER 2
EXPERIMENTS

2.1. Specimens

In this work, we studied the compounds Nb–Ge and Nb–Al with A15 structure.

Figure 3 shows a unit cell of an A15 lattice (chemical formula A_3B). The atoms of *B* form a body-centered cubic lattice, on the faces of which are two atoms of *A*. The atoms of *A* form three orthogonal systems of linear chains. The unit cell of an A15 lattice contains eight atoms, six *A* and two *B*, i.e., it contains two molecules of A_3B.

Films of these compounds were produced by simultaneous evaporation of the components of the compound in a vacuum[13] from different evaporators. Refractory niobium was evaporated by an electron beam. A simple electron gun was used [14]. The lower-melting-point components Al or Ge were evaporated thermally from tantalum or tungsten boats. The vacuum was maintained by an ordinary oil vacuum pump with a nitrogen trap. At the start of evaporation, the pressure in the vacuum chamber was $(1.3\text{–}4)\cdot 10^{-4}$ Pa, in the process of evaporation it was $8\cdot 10^{-4}\text{–}2.6\cdot 10^{-3}$ Pa. The gas mixture of the components of the compound condensed onto heated substrates of sapphire or ruby. The substrate temperature is extremely significant for the production of A15 phases with good superconducting parameters. The optimal substrate temperature for each element was selected individually. Variations in substrate temperature by even 100 K in either direction result in a great decrease in superconducting transition temperature T_C.

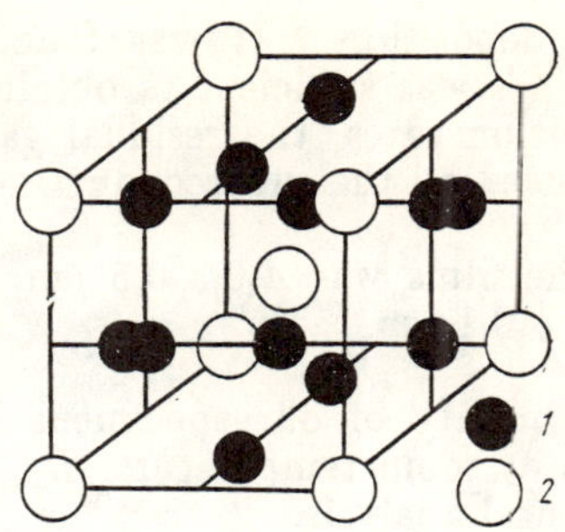

Figure 3. Unit Cell of A15 Lattice (Chemical Formula A_3B). 1 — Atoms of element A; 2 — atoms of element B.

Table 1
Technological Modes of Production of Films

Conditions	Nb-Al	Nb-Ge No. 1	Nb-Ge No. 2
Deposition Temperature, Torr	$(5\text{-}7)\cdot 10^{-6}$	$1.5\cdot 10^{-5}$	$2\cdot 10^{-5}$
Substrate Temperature, °C	800	900	950
Annealing Temperature, °C	---	1300	---
Annealing Time, s	---	15	---
Deposition Rate, Å/s	50	10	10
Film Thickness, μm	0.3	0.3	0.5

The substrate material selected was one that did not form compounds with the elements deposited on it at the operating temperatures. The influence of reactions between deposited elements and the substrate of sapphire and ruby was undetectible right up to ~1700°C.

The film condensed on the substrate was then high-temperature annealed. The annealing time and temperature were selected individually for each compound. The technological conditions of production of the Nb-Al and Nb-Ge films are shown in Table 1. In order to clean and degas the substrate surfaces they were heated before deposition to a temperature greater than the maximum annealing temperature. The substrates were heated by passing current through foil masks.

The geometry of the evaporator was as follows: the distance between evaporators was 2 cm, the distance from evaporators to substrate 8 cm. The rate of deposition of the films was 10–50Å/s (cf. Table 1). It was determined by the rate of evaporation of the refractory component. The deposition rate was higher for Nb-Al (in comparison to Nb-Ge), in order to decrease the influence of oxidation of aluminum during deposition. It was found that with a vacuum of 10^{-3}–10^{-4} Pa, a rate of ~50 Å/s was sufficient to obtain good superconducting parameters. At lower deposition rates, the residual gases, oxides and other contaminants significantly reduced all the superconducting properties of Nb–Al.

The thickness d of the films was d~0.3–0.5 μm, several times greater than the depth of the skin layer $\delta \lesssim 0.1$ μm.

We determined the quality of our specimens by measuring T_c, R_r/R_{res} (where R_r is the resistance at room temperature, R_{res} is the resistance near T_c) and performed x-ray structural analysis.

The critical temperature was measured by the resistive method. A four-contact potentiometric system was used. The temperature sensor used was an Allen–Bradley thermometer, calibrated by means of type TSG-1 germanium. The specimen and the thermometer were placed in a massive copper cup to eliminate any temperature gradient between specimen and thermometer. The accuracy of measurement of T_c was 0.1°.

The static characteristics of the specimens are presented in Table 2. The x-ray diffractograms, like the measurements of T_c and R_r/R_{res}, were performed on the specimens used in the optical studies after they were completed.

The x-ray diffractogram of the Nb–Al specimen was measured on a type DRON-3 installation by N. N. Lobanov. It contains the 12 lines belonging to the A15 lattice. Of these, nine lines, labeled (200), (210), (211), (222), (320), (321), (400), (420), (421), (332), and (432) are the base lines, while the two lines labeled (110) and (220) are difference lines. The presence of difference lines indicates the ordering of the A15 phase. In addition to the A15 lattice lines, the diffractogram also contains maxima not related to the A15 phase. The content of A15 phase is approximately 80%.

The x-ray diffractograms of Nb–Ge specimens No. 1 and No. 2 were measured on the DRON-2 installation by S. P. Merkulova[6]. The x-ray diffractogram of Nb–Ge specimen No. 1 is shown in Figure 4. We can see from the figure that the specimen has 17 lines belonging to phase A15, one of which is a difference line. This indicates the ordering of the A15 phase of specimen No. 1. Furthermore, we can see from the figure that the diffractogram has one maximum which can be related to the σ phase. Its intensity is so small in comparison to the A15 phase lines that we can conclude virtually 100% content of the A15 phase in specimen No. 1. We note that the diffractogram of Nb–Ge specimen No. 1 was measured at four points over the area of the mirror. The analysis showed that identical diffractograms were obtained, indicating the homogeneity of the specimen.

---

[6]The authors express their gratitude to N. N. Lobanov and S. P. Merkulova for measuring the x-ray diffractograms of the specimens studied.

Table 2
Static Characteristics of Specimens

Characteristic	Specimen		
	Nb-Al	Nb-Ge No. 1	Nb-Ge No. 2
T_c, K*	14.1-16.0	14.2-15.3	11.9-13.0
R_r/R_{res}	2.55	1.19	1.16
Lattice Constant a, Å**	5.19	5.187±0.002	5.16
Mirror Surface Area, mm^2	7×20	20×60	20×60

* Beginning and ending temperatures of transition from superconducting state to normal state are given as T_c.

** If no error is indicated, it is less than one in the last digit listed.

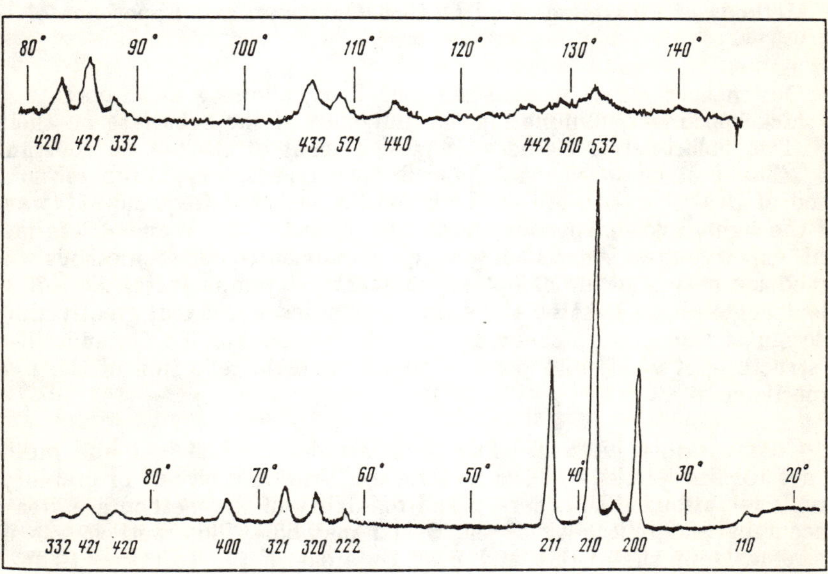

Figure 4. X-Ray Diffractogram of Nb–Ge Specimen No. 1. Numbers above curve correspond to doubled angle of diffraction, beneath curve — atomic plane index.

The x-ray diffractogram specimen No. 2 showed that the specimen has 18 lines belonging to the A15 lattice, of which 15 lines — (200), (210), (211), (222), (320), (321), (400), (420), (421), (332), (521), (440), (442), (610), (532) — are base lines, while three — (110), (220), (330) are difference lines. The presence of difference lines indicates that the A15 phase of specimen No. 2 is ordered. In addition to the A15 lines, the diffractogram contains seven σ-phase lines and two weak lines of the NbO phase. We can conclude from this diffractogram that the quantity of A15 phase is greater than the quantity of the other phases.

For all of the A15 compounds studied, i.e., Nb-Al, Nb-Ce No. 1, Nb-Ge No. 2, the x-ray diffractograms were used to compute the lattice constant, which is presented in Table 2. The specimen with excess Ge has an A15 lattice constant which is smaller than that of Nb-Ge specimen No. 1 which, possibly, is related to the presence of the σ phase. We determined the lattice constants using the extrapolation function $\varphi(\Theta)=\cos^2\Theta(1/\sin\Theta+1/\Theta)$, for which the proportionality $\Delta d/d \sim \varphi(\Theta)$ is preserved for both larger and smaller angles Θ. Here d is the distance between planes, Δd is the error (for example, cf. [15, 16]).

2.2. Methods of Measurement of Optical Constants and Experimental Installations

Our measurements studied reflected light, allowing us to operate with rather thick specimens, eliminating the influence of the substrate on their structure. Two polarimetric methods of measurement of the optical constants were used. The first method could be called "compensatory,", the second — the method of Beatti[17]. In both methods for the assigned frequency ω (wavelength λ) of the light, two independent quantities ϵ_1 and ϵ_2 (or n and κ) are measured. In the experimental installations which implemented these methods and were designed for measurement of optical constants at temperatures 290—700 K, the incident angle of the light on the specimen studied was fixed, greatly simplifying the design of the vacuum chamber. For the compounds Nb-Ge and Nb-Al with A15 structure, it was found possible to use a single reflection of the light from the specimen.

Compensatory method. The compensatory method is a null method. It does not require linearity of the receiver and amplifier circuit or stability of the source of radiation. The accuracy and reliability of the method are greater than the accuracy and reliability of the Beatti method. This method was used for measurements in the visible and near IR areas of the spectrum (0.4-2.5 μm). The installation developed for the method allows measurements to be conducted at 290-700 K.

The quantities directly measured are: 1) Δ — the phase shift between the R_p and R_s components of the electrical vector $\mathbf{R}$ of the wave reflected from the metal (R_p is in the incident plane, R_s — in the perpendicular plane); 2) angle 2ρ, where $\tan\rho=|R_p|/|R_s|$ (when the plane of polarization of the incident light is at an angle of 45° to the incident plane).

These quantities are quite sufficient for unambiguous determination of n and κ or the complex dielectric constant[7] $\epsilon=\epsilon_1-i\epsilon_2=(n-i\kappa)^2$. The quantities Δ and 2ρ are related to ϵ by the following equation (cf. [13]):

$$\frac{\sqrt{\varepsilon-\sin^2\varphi}}{\sin\varphi\,\mathrm{tg}\,\varphi}=\frac{\cos 2\rho-i\sin 2\rho\cdot\sin\Delta}{1-\sin 2\rho\cdot\cos\Delta}. \tag{19}$$

Here angle φ is the incident angle. In this area, we can limit ourselves to one reflection of the incident light from the mirror studied.

A schematic diagram of the polarization method with a single reflection for measurement of the dispersion of optical constants is shown in Figure 5. According to this method, Δ and 2ρ are determined as follows: a parallel monochromatic light beam first passes through polarizer *2*, which transmits oscillations of the electrical vector of the incident wave at an angle of 45° to the incident plane, and strikes the surface of the metal studied *3* at known angle φ. The reflected beam, after passing through Soleil compensator *4* and analyzer *5*, enters recording system *6*.

The compensator is installed so that its main directions coincide with the p and s directions. By changing the phase shift Δ_k between the p and s components introduced by the compensator, the total shift $\Delta_k+\Delta$ is adjusted either to 0 or to π (Δ is the phase shift introduced by mirror *3*), which corresponds to linear polarization. Linear polarization can be recorded by rotating the analyzer to achieve full extinction of the light. Actually, full extinction is not achieved due to the nonideal nature of the polarizer, the finite aperture of the beams, the finite spectral width of the incident light and a number of other factors. The measurements record the minimum of the signal leaving system *1–4*. After these operations are performed, the compensator reading and analyzer azimuth are recorded. The compensator reading is directly proportional to Δ. The polarizer is then rotated by 90°, measured from the initial position of the polarizer, then the light striking the metal is polarized at 45° to the incident plane. The analyzer is once more rotated to achieve "complete" extinction of the signal and the analyzer azimuth is recorded. The difference between the two azimuths is 2ρ.

A diagram of the installation is presented in Figure 6[18]. Light source S_1 is a halogen-filled incandescent lamp with a quartz envelope ("Narva"). It has greater light output than an ordinary incandescent lamp. S_2 is an ordinary incandescent lamp. It forms the reference signal for the synchronous detector. Lens $Л_1$ projects the image of source S_1 on the slit in monochromator M (for the 0.4–1 μm spectral area, a type UM-2 monochromator is used; for the 0.8–2.5 μm area — a monochromator manufactured by Carl Zeiss Jena). The light leaving the monochromator strikes rotating mirror 3 and passes through lens $Л_2$, used to

[7] $\kappa>0$ corresponds to attenuation of the plane wave in the depth of the metal.

obtain a parallel light beam. The parallel monochromatic light beam then passes through the system П-$З_{иссл}$-К-А, used to measure the optical constants of the specimen $З_{иссл}$ as was described above.

П and A are film polarizers, K is a Soleil compensator. The operating principle of the Soleil compensator was described in detail in [19, 20]. Diaphragm *D* changes the diameter of the light beam striking the specimen, matching the cross section of the beam to the aperture ratio of the installation. The dimensions of diaphragm *D* must be individually selected to eliminate light which is scattered by the other parts of the installation. Apertures O_1, O_2, O_3 are made of fused quartz. They are location perpendicular to the incident beam of light and do not displace the beam.

After passing through the optical system, the light strikes receiver *6*. (The receiver used for the 0.4–0.75 μm structural area was an FEU-38 (FEU-79) photomultiplier, for the 0.7-1.1 μm area — a type FEU-62 (FEU-28) unit and for the 1-2.5 μm area a type FSA-G2 instrument.) The light signal is converted to an electrical signal by detector *6* and is sent to selective microvolt meter *7* (type V6-9), which is tuned to the frequency of the modulator, 215-230 Hz. This narrow-band volt meter cuts off the noise frequency spectrum sharply. The output of the volt meter is connected to oscilloscope *8* and synchronous detector *9*. The oscilloscope is used for visual monitoring of the signal. The synchronous detector increases the accuracy of recording the position of "full" extinction, but increases the time required to perform measurements. A reference signal is generated by system S_2–*2*–*3*. During measurements of optical constants at room temperature, there is no need for the synchronous detector, since the noise level is significantly lower than that of the useful signal (volt meter *7* and oscilloscope *8* suffice). At temperatures of ~500 K and higher, the natural radiation of the hot parts of the installation creates a parasitic signal, which increases with increasing wave length. The use of the synchronous detector significantly improves the signal to noise ratio.

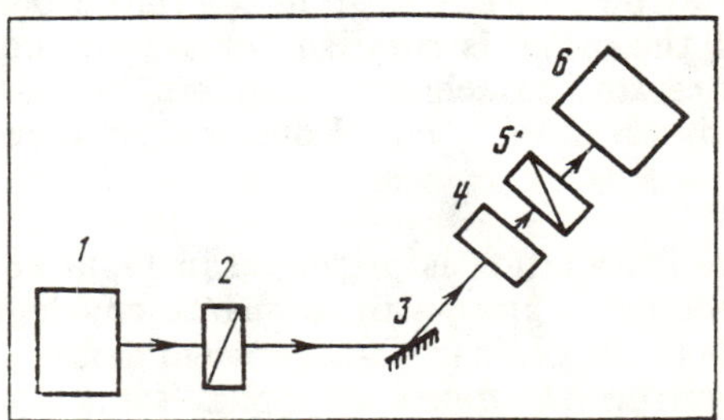

Figure 5. Schematic Diagram of "Null" Polarization Method Using Compensator. 1 — Source of parallel monochromatic light beam; 2 — polarizer; 3 — metal; 4 — compensator; 5 — analyzer; 6 — recording system.

Light sources S_1 and S_2 are powered by low-voltage stabilized dc power supplies. This eliminates the 100 Hz parasitic signal. The photomultipliers and photoconductive cells are also powered by stabilized power supplies. Primary attention is given to eliminating vibrations. For this purpose, all optical elements are mounted on a massive plate installed on a laboratory stage. A cushion is located between the plate and the stage. The modulator is mounted on a separate rail isolated from the plate by a soft cushion to decrease vibrations.

The main problem in conducting the experiments was that of protecting the specimens from oxidation at high temperatures. A vacuum chamber was conducted for this purpose, in which the specimen was placed in contact with the heater. The chamber was evacuated to $P \cong 8 \cdot 10^{-4}$ Pa unheated, $P \cong 1.3 \cdot 10^{-4}$ Pa at $T \sim 500$–700 K.

A general view of the vacuum chamber is shown in Figure 7, a general view of the slide with heater and specimen holder in Figure 8. For more convenient adjustment, the chamber was designed as a rider on rail *9*. This design allows it to be moved in three axes. The vacuum tube with bellows allows small movements to be made without moving the vacuum pump. All parts of the chamber, the surfaces of which are exposed to the vacuum, were made of 12Kh18N10T steel. Cap *1* contains adapters for the entry and exit windows *5*, the design of which allows the window material to be changed. They allow studies to be performed at three incident angles: 0°, 60° and 80°.

Let us now look at Figure 8, which shows clamp *1*, in which heater *4* and specimen *6* are held. The body of the heater is made of copper. It has two penetrating apertures, in which two ceramic tubes are inserted. Inside the ceramic tubes is tungsten coil *12*. The copper, thanks to its good heat conductivity, decreases the temperature gradients. Specimen *6* is mounted in a slot in the heater. The temperature of the hot specimen is measured by chromel–alumel thermocouple *5*. The junction of the thermocouple is attached to the middle of the upper portion of the specimen so that the thermocouple is not heated by the incident or reflected light. This allows the temperature of the film to be measured. The free ends of the thermocouple are kept at $T \approx 65$°C, the accuracy of temperature measurement is ±7°C. This design allows temperatures to be achieved of up to 700 K without decreasing the vacuum or forced cooling. Higher temperatures and long-term measurement require additional cooling of the vacuum seals (rubber inserts). If the temperature does not exceed 700 K, measurements can be performed continuously for 12–15 hours of operation.

Clamp *1* can be rotated around its axis, sliding on the plane of slide *3*. Stop *2* of the clamp can be fixed in three positions. Cylindrical depressions *7* and *8* in the body of the slide serve as sockets to receive stop *2*. The bottom plane of the clamp and the plane of the slide are fitted accurately to each other. After each setting of the specimen, the incident angle is verified with the GS-5 goniometer.

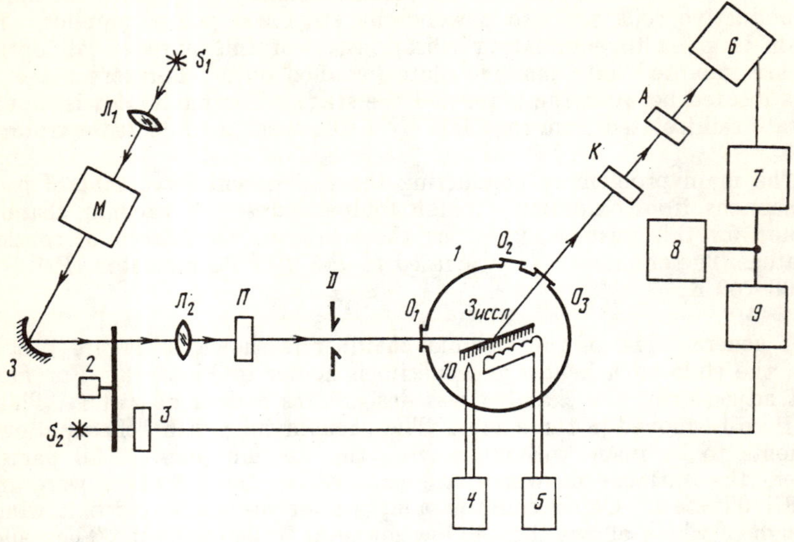

Figure 6. Diagram of Installation for Measurement of Optical Constants of Metal Compounds in Visible and Near–IR Spectral Areas. S_1, S_2 — Light sources; $Л_1$, $Л_2$ — lenses; M — monochromator; З — rotating mirror; П — polarizer; O_1, O_2, O_3 — entry and exit apertures of vacuum chamber; $З_{иссл}$ — metal being studied; K — Soleil compensator; A — analyzer; 1 — vacuum chamber; 2 — modulator; 3 — type FSA–G2 photoconductive cell; 4 — type Shch1516 dc volt meter; 5 — heater power supply; 6 — photoreceiver; 7 — V6—9 selective microvolt meter; 8 — S1–18 oscilloscope; 9 — KZ–2 synchronous detector; 10 — thermocouple; 11 — tungsten spiral.

The optical constants are measured in parallel beams of monochromatic light (cf. Figure 6). Optical system S_1–$Л_1$–M–З–$Л_2$ is used for this purpose. The parallel light beam is generated by moving lens $Л_2$, which resolves the image of the output slot of the monochromator at distances of ~100f (f is the focal length of lens $Л_2$). The required incident angle is set as follows: first it is made equal to 0°; diaphragm D is reduced and the adjusting screws are rotated until the reflected light beam once more strikes the diaphragm; the clamp is then rotated, set to the required incident angle, and fixed by the stop. Optical system П–$З_{иссл}$–K–A is then adjusted.

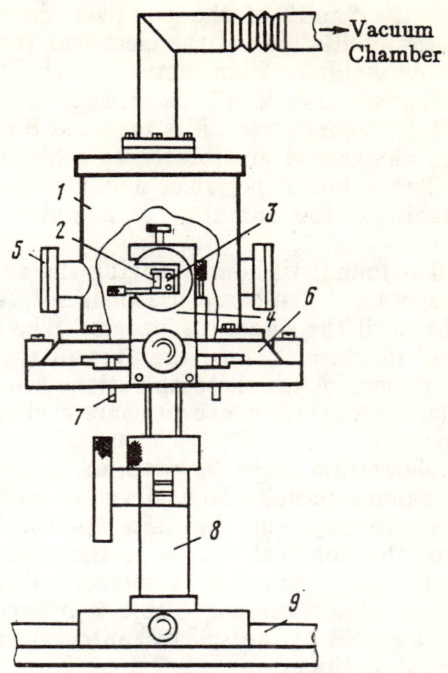

Figure 7. Vacuum Chamber. 1 — Cap; 2 — specimen; 3 — heater; 4 — clamp; 5 — window; 6 — slides; 7 — electrical leads; 8 — rider; 9 — rail; 10 — to diffusion pump.

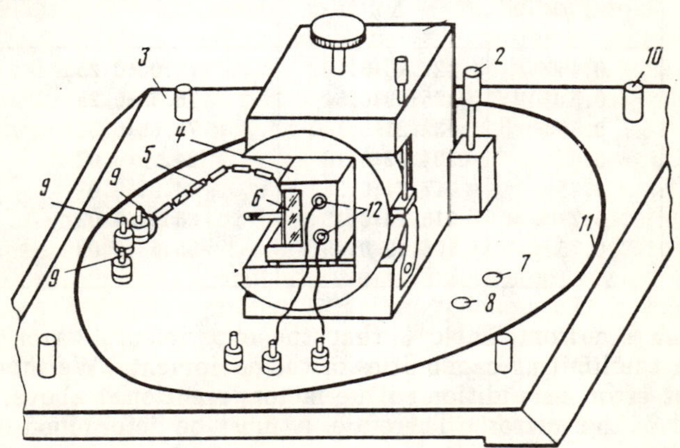

Figure 8. Slides with Specimen Holder. 1 — Clamp; 2 — stop; 3 — slides; 4 — copper heater; 5 — thermocouple; 6 — specimen; 7,8 — stop sockets; 9 — eletrical leads; 10 — pin; 11 — channel; 12 — ceramic tubes containing tungsten coil.

We first find the "null" of the polarizer, i.e., the polarizer azimuth at which it transmits the oscillations of the electrical vector parallel either to the s or the p direction. As we know from metal optics[3, 20, 21], in this case the electrical vector of the reflected wave **R** will be directed in the s or p direction (i.e., **R** is R_s or R_p). For this, compensator K is removed from the system Π–$З_{\text{ИССЛ}}$–K–A and polarizer Π and analyzer A are rotated to achieve the minimum signal leaving system Π–A. The value of polarizer azimuth is then recorded and used as the beginning of reading of the azimuth, i.e., as null.

After the null is found, without changing the adjustment, the compensator is returned to the system. The compensator is rotated around the axis parallel to the beam of light until the signal disappears. The minimum is found and the compensator is fixed in place. Thus, the axes of the compensator are adjusted for the s and p directions. After these operations are performed, the installation is ready for use. The compensator can measure a phase shift in the spectral area used with an accuracy of 0.07° (4′). The polarizers allow measurement of the doubled restored polarization angle 2ρ with an accuracy of 0.003° (10″). However, this accuracy cannot actually be achieved due to the insufficient sensitivity and noise of the recording system, the finite apertures and spectral width of the incident light beam, the nonideal nature of the polarizers, etc. Therefore, the measurement error of these quantities is determined by the spread of the results of this series of tests. For example, Table 3 presents the results of measurements of Δ and 2ρ for gold in the spectral interval of 0.45–0.75 μm. The mean dispersion is presented as the error.

Table 3
Results of Measurement of Δ and 2ρ of Gold

λ, μm	Δ, °	2ρ, °
0.45	127.76±0.10	75.00±0.25
0.5	125.69±0.34	75.33±0.25
0.55	132.31±0.15	78.88±0.05
0.6	124.72±0.18	84.25±0.07
0.65	117.71±0.26	84.95±0.07
0.7	115.16±0.13	85.92±0.02
0.75	106.79±0.49	85.68±0.09

We can see from Table 3 that the experimental error is several times greater than the limiting capabilities of the equipment. We should note that the measurement error, in addition to the factors mentioned above, also depends on the material of the mirror. Therefore, it must be determined experimentally in each case.

Method of Beatti. The method of Beatti[17] has been found suitable for measurement of optical constants of metals and alloys at high temperatures in the infrared range. The method of Beatti is not a null method and therefore the requirement for linearity of the amplifier system over a broad dynamic range and quality of optical elements (particularly polarizers) becomes important. This method measures the intensity of light passing through the polarizer-specimen-analyzer (PSA) system with known values of analyzer (polarizer) azimuth. The essence of the Beatti method is as follows. Suppose a parallel monochromatic beam of light strikes a PSA system (we shall represent the intensity of the light after passing through the polarizer as J_0)

$$J_0 = \frac{c}{8\pi} E^2,$$

where c is the speed of light, E is the amplitude of the electrical vector of the light. In what follows, we shall omit the coefficient $c/8\pi$. Suppose Ψ_P and Ψ_A are the azimuths of the polarizer and analyzer, respectively. $\Delta = \delta_p - \delta_s$, where δ_p and δ_s are the phase shifts of the p and s components of the wave after reflection from the specimen. The subscripts p and s indicate membership of the corresponding quantities in the groups of components of the electrical vector of the wave E which lie in the planes parallel and perpendicular to the incident plane, respectively[8]. The intensity of light passing through the PSA system can be recorded as:

$$J(\Psi_P, \Psi_A) = E^2 r_s(\cos^2\Psi_P \cos^2\Psi_A + \rho^2 \sin^2\Psi_P \sin^2\Psi_A + \\ + 1/2 \rho \sin 2\Psi_P \sin 2\Psi_A \cos\Delta),$$

where $\rho = r_p/r_s$ (r_p and r_s are the amplitude reflection factors of the p and s components of the wave). In the Beatti method $\Psi_A = 0^\circ, 45^\circ, 90^\circ, -45^\circ$, $\Psi_P = 45^\circ$ or $\Psi'_P = 0^\circ, 45^\circ, 90^\circ, -45^\circ$, $\Psi_A = 45^\circ$ [17]. There are other modifications of the Beatti method[22, 23].

A schematic diagram of an installation for measurement of optical constants in the IR area by the Beatti method is shown in Figure 9[24]. In this version, four light "intensities" are measured[9]:

$$J_1 = J(0, 45^\circ);\ J_2 = J(90^\circ, 45^\circ);\ J_3 = J(45^\circ, 45^\circ);\ J_4 = (-45^\circ, 45^\circ),$$

(the parentheses indicate the azimuths of the polarizer and analyzer). The analyzer azimuth remains fixed. This eliminates error related to polarization of the radiation by the monochromator, which is located at the output of the optical system. For transition metal-based alloys an analyzer azimuth of 45° is suffi-

---

[8]The incident plane is defined by a perpendicular to the surface of the specimen studied and the light beam.

[9]Actually, J_1, J_2, J_3 and J_4 are quantities proportional to the corresponding light intensities.

cient. In the 1–8 μm spectral area, an incident angle of φ=80° was used. Measurements of optical constants performed with an incident angle of 60° do not produce the desired accuracy, since phase shift Δ between the p and s components of the electrical vector **R** is small. (**R** is the electrical vector of the wave reflected from the metal specimen.)

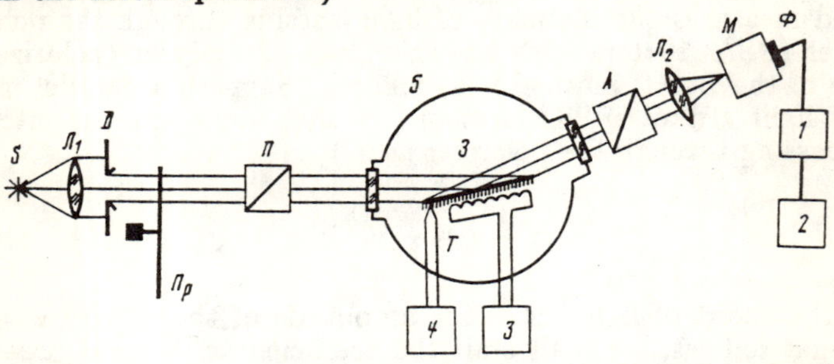

Figure 9. Schematic Diagram of Installation for Measurement of Optical Constants in IR Area of Spectrum. S — Source of IR radiation; $Л_1$, $Л_2$ — lenses (BaF_2); D — diaphragm; Π_p — mechanical interrupter; T — thermocouple; 3 — metal specimen studied; A — analyzer; M — monochromator; Φ — radiation receiver; 1 — V6-9 selective microvolt meter; 2 — S1-18 oscilloscope; 3 — heater power supply; 4 — Shch1516 dc volt meter; 5 — vacuum chamber.

The four measured values of light intensity are used to compute the two independent quantities ϵ_1 and ϵ_2 by the following equations:

$$\cos\Delta = \frac{J_3 - J_4}{4\sqrt{J_1 J_2}}\left(1 + \frac{J_1 + J_2}{J_3 + J_4}\right), \tag{20}$$

$$n_0 = \sin\varphi\tan\varphi \cdot \frac{J_2 - J_1}{J_1 + J_2 - 2\sqrt{J_1 J_2}\cos\Delta}, \tag{21}$$

$$\varkappa_0 = 2n_0 \frac{\sqrt{J_1 J_2 (1 - \cos^2\Delta)}}{J_2 - J_1}, \tag{22}$$

$$\varepsilon_1 = n_0^2 - \varkappa_0^2 + \sin^2\varphi, \tag{23}$$

$$\varepsilon_2 = 2n_0\varkappa_0, \tag{24}$$

where ϵ_1 and ϵ_2 are the real and imaginary parts of the complex dielectric constant $\epsilon=\epsilon_1-i\epsilon_2$. Equations (21) and (22) were obtained from the following expressions[10]:

$$n_0=\sin\varphi\tan\varphi\frac{\cos 2\rho}{1-\sin 2\rho\cos\Delta};\ \kappa_0\sin\varphi\tan\varphi\frac{\sin 2\rho\sin\Delta}{1-\sin 2\rho\cos\Delta}. \tag{25}$$

With the azimuth of the analyzer $\Psi_A=\pi/4$ $\tan\rho=\sqrt{J_1/J_2}=|R_p|/|R_s|$. Expression (20) was obtained by averaging of the values generated by the three equations[17]:

$$(\cos\Delta)_1 = \frac{2J_3 - J_1 - J_2}{2\sqrt{J_1J_2}},$$

$$(\cos\Delta)_2 = \frac{J_1 + J_2 - 2J_4}{2\sqrt{J_1J_2}},$$

$$(\cos\Delta)_3 = \frac{1}{2}\left(\sqrt{\frac{J_1}{J_2}} + \sqrt{\frac{J_2}{J_1}}\right)\frac{J_3 - J_4}{J_3 + J_4}.$$

Experience has shown that this averaging decreases the error of determination of Δ.

The experimental installation for measurement of optical constants by the Beatti method in the IR area was assembled on the basis of the installation described in the section on the compensator method. The vacuum and temperature portions of the installation were not modified. The optical elements, operating in the visible and near IR areas, were replaced by elements operating in the far IR area. The lenses and windows of the vacuum chamber were made of barium fluoride. The use of lenses instead of mirrors significantly decreases the size of the installation. The monochromator with NaCl prism is placed at the output of the optical system. This location yields a number of advantages. First of all, it is convenient to adjust the optical portion of the installation, and secondly, the monochromator cuts off a significant fraction of the thermal radiation from the heated portions of the installation, increasing the sensitivity of the radiation receivers by decreasing the background noise.

Calibration of the monochromator with its NaCl prism in the spectral interval λ=0.5–14 μm, performed earlier using the absorption lines of indene, CO_2 vapor in air, didymium glass and the radiation lines of mercury and neon lamps, was additionally refined before this work was started. The calibration in the IR area was refined using the absorption spectra of water vapor in the atmosphere and in mica. The absorption spectrum recorded by a strip–chart recorder was

[10]Equations (23)–(25) are also used to compute ϵ_1 and ϵ_2 in the compensator method.

compared with the spectrum presented in [25]. This yielded reference points for λ=2.742, 3.419, 4.256, 5.91, 6.05, 6.57 and 6.86 μm. In the visible and near IR areas, the absorption lines of didymium glass yielded the reference points λ=0.6842, 0.7423, 0.8082, 0.8805, 1.067, 1.220, 1.517, 1.919 μm; the radiation lines of the mercury lamp was used to produce reference points at λ=1.71, 1.37, 1.129, 1.014, 0.578 μm.

The radiation sources were an incandescent lamp and globar lamp: The incandescent lamp was used in the 0.8–2.4 μm range, the globar — in the 2.8–8.5 μm range. The light exiting the monochromator struck the radiation receiver. The receiver for the 0.8–2.4 μm range was a type FSA–G2 photoconductive cell. In the middle and far IR areas, a germanium–gold photoconductive cell was used.

Before striking the mirror being studied, the radiation was modulated by a mechanical interrupter with a frequency of 215–230 Hz. The signal from the receiver was sent to the input of a V6–9 selective microvolt meter tuned to the frequency of the modulator. For visual monitoring, the output of the volt meter was connected to an S1–55 oscilloscope. One essential requirement for the radiation detector is linearity of gain over a broad dynamic range. The radiation receivers used satisfied this requirement in the range of radiation intensity measurements we used.

Film polarizers were used in the 0.8–2.4 μm spectral range. In the 2.5–8.5 μm range, we used germanium and selenium polarizers, the operation of which is based on polarization of radiation as it passes through a stack of selenium or germanium plates, with the stack placed at the Brewster angle to the incident light. The degree of polarization of these polarizers was 99.8%.

Measurements of optical constants were performed in parallel beams of monochromatic IR radiation. First the null of the polarizer was found, i.e., the polarizer azimuth at which it transmitted oscillations of the electrical vector parallel to the p direction. There should be a mark on the polarizer indicating the polarization plane. By rotating the polarizér, this mark was approximately set so that it was in the incident plane. Then, by rotating the analyzer and adjusting the polarizer, the precise position of the polarizer azimuth null was determined by complete extinction of the signal exiting the optional system. After this operation, the analyzer was rotated by 45° from the signal extinction position and fixed in place. The installation was then ready for use.

As was noted earlier, we measured four intensities of reflected radiation with four polarizer positions. The accuracy of setting of the polarizer null azimuth varied from 10′ to 40′ depending on the spectral area. The error in measurement of the optical constants depends on the metal studied. It was determined by the mean spread of data in each case individually. The error in true and imaginary parts of the complex dielectric constant, for example for NbAl, was 5–7% in the 0.75–1 μm area, 2–4% in the 1.0–2.4 μm area and 5–8% in the 2.8–7.5 μm area.

CHAPTER 3
EXPERIMENTAL RESULTS

3.1. Nb–Al

We studied a specimen of Nb–Al[9] with 80% A15 phase content, the characteristics of which are presented in Table 2. The smoothed experimental results[11] for ϵ_1 and ϵ_2 are presented in Figures 10–11. Table 4 tabulates the function $\epsilon_2|\mathrm{Im}\,\epsilon(\omega)|$ for T=295 and 670 K, which was used to determine the electronic characteristics of the Nb–Al. The experiments generated 111 points each for ϵ_1 and ϵ_2 at room temperature and 104 points each at T≈670 K.

We can see from the figure that throughout almost the entire spectral area the values of ϵ_2 (670 K)$>\epsilon_2$ (295 K). The reverse picture is observed for ϵ_1. We note that basically $\epsilon_1(\omega)$ is an increasing function, while $\epsilon_2(\omega)$ is decreasing. Against the background of this general structure, there are a number of maxima in $\epsilon_2(\omega)$. In turn, the structure of the maximum consists of the basic broad maxima, with a fine structure against this background.

At both temperatures the structure of the functions $\epsilon_1(\omega)$ and $\epsilon_2(\omega)$ is similar, but at room temperature it is more clearly seen. The positions of the singularities of $\epsilon_2(\omega)$ remain virtually unchanged with temperature. This is particularly clearly seen in the IR area of the spectrum.

It follows from Figure 11*a,b* that we can distinguish three broad interband transition zones in the function $\epsilon_2(\omega)$, each of which has a complex structure. The first band is found at energies of ~0.35 eV, the second at energies of ~1.5 eV,

[11]The smoothing algorithm is presented in Chapter 4.

the third at the edge of the spectral range at ~3 eV. At the edge of the IR area, both $\epsilon_1(\omega)$ and $\epsilon_2(\omega)$ show singularities which, apparently, are related to the interband transition zone at energies ~0.2 eV.

The major tendencies in the behavior of $\epsilon_1(\omega)$ and $\epsilon_2(\omega)$ in the IR area, as in ordinary metals, is related to the conduction electrons.

3.2. Nb–Ge

Two Nb–Ge specimens were studied[26, 27]. Their characteristics are presented in Table 2. The phase analysis of specimen No. 1 showed that it contained practically only phase A15 (cf. Figure 4). Specimen No. 2 contained phase A15 plus a σ phase and an NbO phase. The quantity A15 was over 50%. Comparison of the optical constants of these specimens shows that the presence of the other phases has a significant influence.

The smoothed values of ϵ_1 and ϵ_2 are presented in Figures 12–14. Table 5 tabulates the function $\epsilon_2(\omega)=|\mathrm{Im}\,\epsilon(\omega)|$ of Nb–Ge specimen No. 1 at T=295 and 670 K, which was used to determine the electronic characteristics of this specimen. In the figures, curves *(1)* relate to ϵ_1 and ϵ_2 of Nb–Ge specimen No. 1 for 295 K. Curves *(2)* relate to the same specimen at 670 K. Curves *(3)* were constructed for ϵ_1 and ϵ_2 of Nb–Ge specimen No. 2, T=670 K.

The dispersion of the complex dielectric constant of the compound Nb_3Ge was studied by Yao and Shnetterly. The results of processing of their data are presented in [28]. We have presented the results of their data in Figures 13*a* and 14*a*, which show the corresponding curves *(4)* for the functions $\epsilon_1(\omega)$ and $\epsilon_2(\omega)$.

Let us study Figures 12 and 13, which show the function $\epsilon_1(\omega)$ for $\hbar\omega<1$ eV. We shall study the curves relating to Nb–Ge specimen No. 1, i.e., curves *1* and *2*. We see that throughout, except for the infrared edge of the area, $|\epsilon_1(\omega, 295\ \mathrm{K})|>|\epsilon_1(\omega, 670\ \mathrm{K})|$. For the visible area of the spectrum, this conclusion is not justified. The sign of the inequality differs in different intervals of ω.

Let us now look at the figures illustrating $\epsilon_2(\omega)$ for Nb–Ge specimen No. 1 (curves *1* and *2*). In the visible area of the spectrum $\epsilon_2(\omega, 670\ \mathrm{K})>\epsilon_2(\omega, 295\ \mathrm{K})$. The function $\epsilon_2(\omega)$ is basically decreasing throughout the spectral range studied. The function $\epsilon_1(\omega)$ includes decreasing and increasing areas. The structure $\epsilon_2(\omega)$, in addition to the behavior discussed above, also includes broad maxima and a number of smaller maxima within the broader ones. As temperature rises, the positions of the maxima remain practically constant. At room temperature the maximum at ~0.33 eV is practically not seen, but at 670 K it is clearly seen. We note the positions of the basic broad maxima in $\epsilon_2(\omega)$ for Nb–Ge specimen No. 1 (curves *1* and *2*): the first main maximum is at ~0.21 eV, the second ~0.33 eV, the third at ~0.5 eV, the fourth ~1 eV and the fifth at ~2 eV.

Table 4

$\epsilon_2(\omega)$ of Nb–Al at T=295 and 670 K

ω, eV	ε₂(ω)		ω, eV	ε₂(ω)		ω, eV	ε₂(ω)	
	295 K	670 K		295 K	670 K		295 K	670 K
3,10	8.4	8.7	1.57	15.6	18.8	0.652	50.9	54.2
3,02	7.3	9.0	1.55	15.5	19.2	0.632	52.8	56.3
2.95	7.4	8.2	1.53	15.6	19.3	0.610	55.3	58.4
2,88	7.3	8.3	1,51	16.2	19.4	0.599	56.7	59.6
2,82	7.6	8.4	1.49	16.7	19.6	0.587	58.2	61.1
2.76	7.9	8.7	1.48	17.0	—	0.568	61.2	64.0
2.70	8.3	9.0	1.46	17.0	—	0.553	64.2	66.5
2.64	8.7	9.2	1.44	17.0	—	0.545	65.9	67.6
2.58	9.2	9.3	1.42	17.1	—	0.539	66.9	68.2
2.53	9.6	9.5	1.41	17.4	—	0.536	67.1	68.3
2.48	9.9	9.6	1.39	17.7	20.4	0.525	—	69.6
2.43	10.1	9.8	1.38	18.0	—	0.516	70.3	—
2.38	10.2	10.2	1.36	18.2	—	0.442	79.3	84.4
2.34	10.4	10.6	1.33	18.6	—	0.427	78.7	85.5
2,30	10.6	11.0	1.32	18.9	21.2	0.413	77.3	86.1
2.25	10.8	11.4	1.28	20.3	—	0.400	77.6	88.2
2,21	10.9	11.5	1.27	—	22.7	0.387	79.2	90.6
2,18	11.1	11.6	1.25	21.4	21.3	0.375	82.0	93.5
2.14	11.2	11.7	1.24	21.9	24.9	0.364	85.3	97.1
2.10	11.4	11.8	1.22	22.9	23.7	0.354	89.7	102.1
2,07	11.6	12.0	1.16	24.5	24.2	0.344	95.4	109.0
2.03	11.8	12.2	1.11	25.8	25.5	0.335	102	117
2.00	11.9	12.4	1.07	26.6	26.7	0.326	108	123
1,97	11.9	12.8	1.02	28.0	28.8	0.317	111	127
1,94	12.0	13.3	0.992	29.0	30.1	0.310	112	128
1,91	12.1	13.7	0.946	30.6	32.0	0.291	111	129
1.88	12.5	13.9	0.911	32.0	33.7	0.275	115	139
1,85	13.1	13.9	0.885	33.2	35.3	0,261	129	160
1.82	13.9	14.1	0.861	34.6	36.9	0.248	136	186
1.80	14.5	14.4	0.835	36.2	38.5	0.236	153	205
1.77	14.7	14.9	0.810	37.9	40.0	0,225	159	223
1.75	14.4	15.3	0.787	39.6	41.4	0,215	—	231
1.72	13.9	15.4	0.763	41.3	42.9	0,206	—	261
1.70	13,6	15.3	0.735	43.3	45.1	0.202	244	—
1.68	13.8	15.3	0.725	44.1	46.1	0,198	270	265
1.65	14.2	15.7	0.714	45.0	47.1	0.190	—	323
1.63	15.3	16.4	0.698	46.4	48.8	1.183	307	—
1.61	15.8	17.4	0.675	48.7	51.4	0.177	347	341
1,59	15.9	18.2	0.663	49.9	52.9			

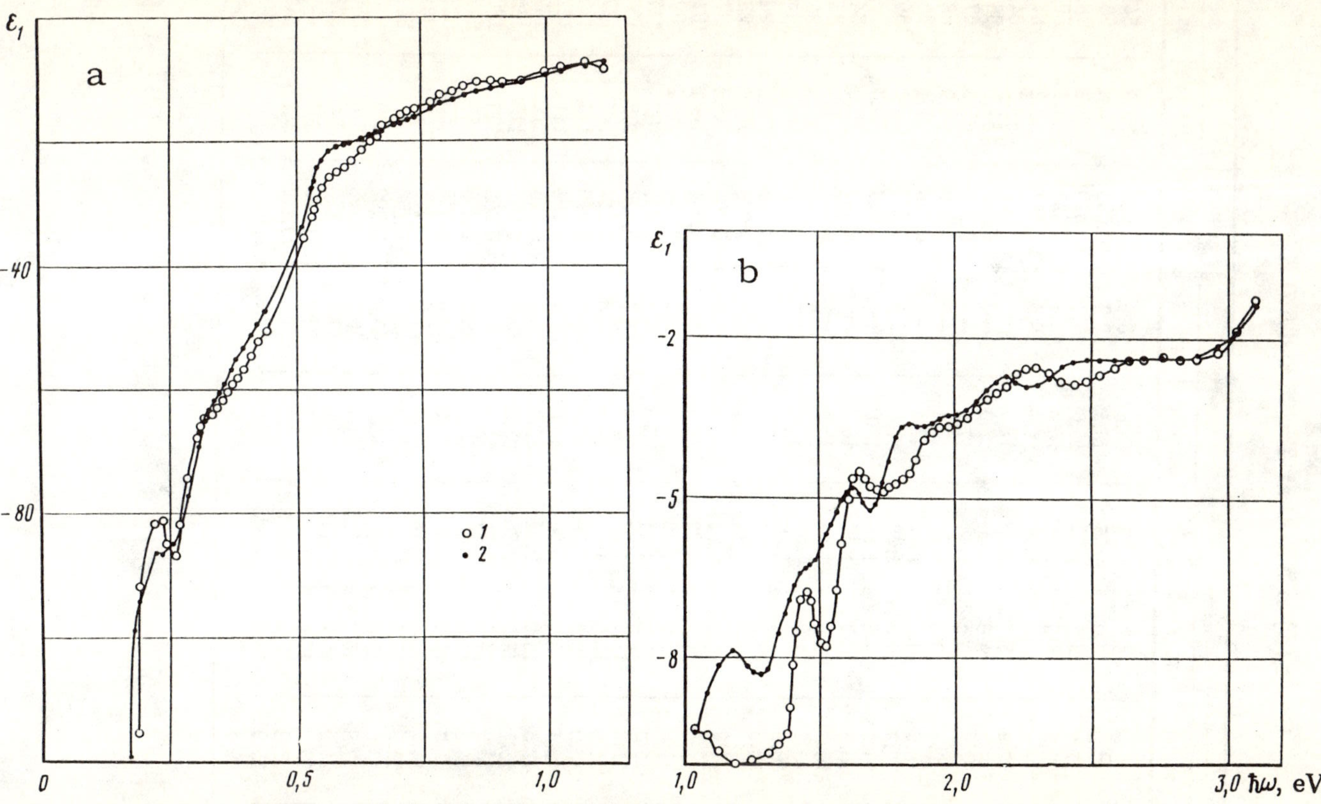

Figure 10. Real Portion $\epsilon_1(\omega)$ of Dielectric Constant of Nb–Al in IR (a) and Visible (b) Light Bands. 1 — $T \approx 670$ K; 2 — $T \approx 295$ K.

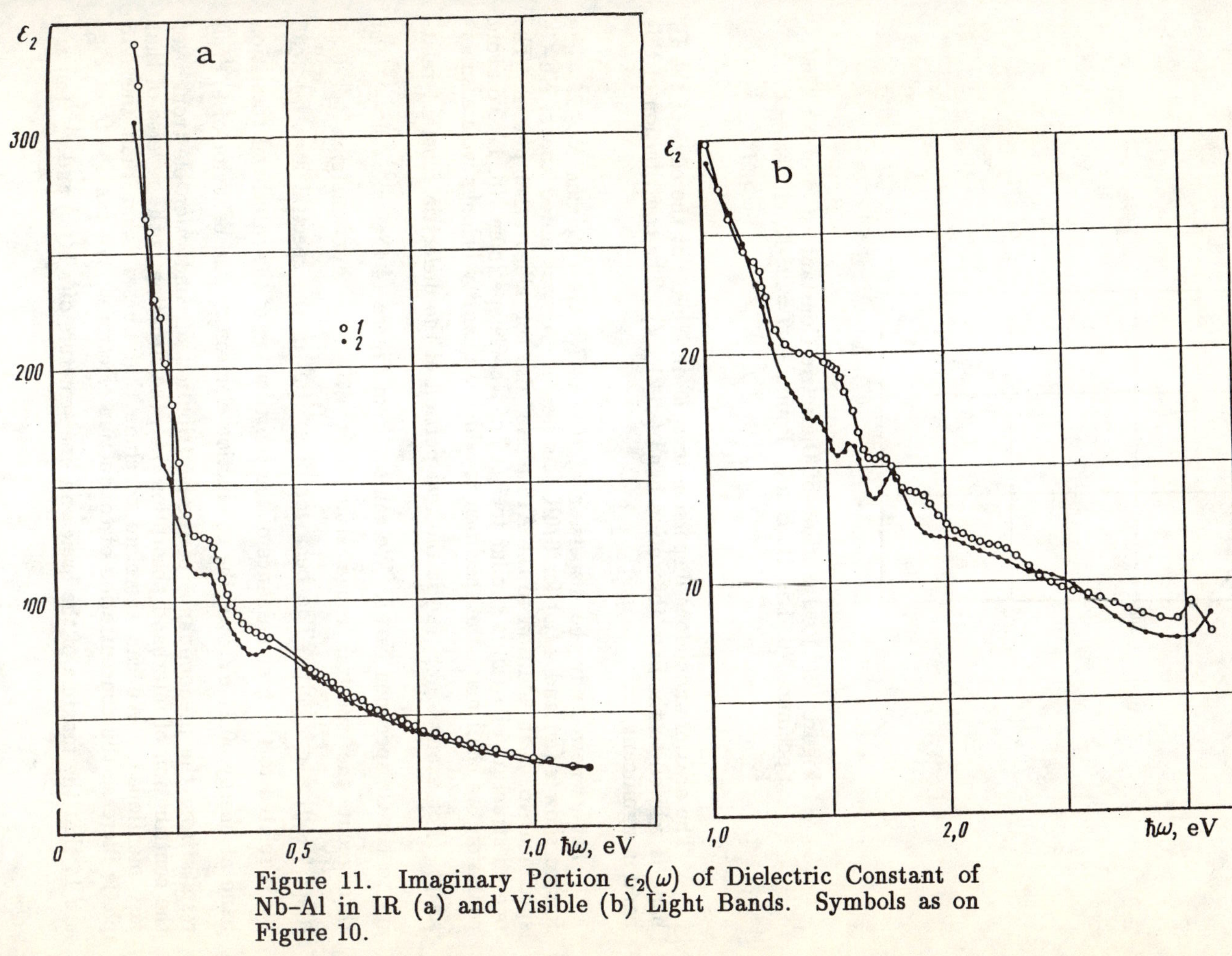

Figure 11. Imaginary Portion $\epsilon_2(\omega)$ of Dielectric Constant of Nb–Al in IR (a) and Visible (b) Light Bands. Symbols as on Figure 10.

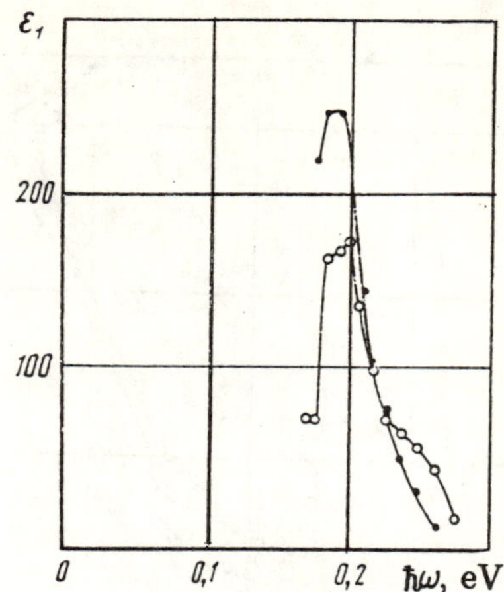

Figure 12. Real Portion of Dielectric Constant $\epsilon_1(\omega)$ of Nb–Ge Specimen No. 1 in IR Band. Points — T=295 K, circles — 670 K.

The area of anomalous dispersion in $\epsilon_1(\omega)$ in Nb–Ge at the edge of the IR band is related to the first absorption band (~0.21 eV). The contribution of this band is significant.

Let us now study the dispersion of $\epsilon(\omega)$ of Nb–Ge specimen No. 2. The figures show $\epsilon_1(\omega)$ and $\epsilon_2(\omega)$ for 670K (curves *3*). As was noted above, Nb–Ge specimen No. 2, in addition to the A15 phase, also contained other phases. The overall structure of $\epsilon_2(\omega)$ is similar to $\epsilon_2(\omega)$ of Nb–Ge specimen No. 1. To avoid confusion, we shall mark the specimen to which a quantity belongs with a superscript. For example, $\epsilon_1^{(1)}(\omega)$ is the real portion of the dielectric constant relating to Nb–Ge specimen No. 1. We can see from Figure 14 that $\epsilon_2^{(1)}(\omega)>\epsilon_2^{(2)}(\omega)$ throughout the spectral range studied at 670 K. For energies $\hbar\omega>0.35$ eV $|\epsilon_1^{(1)}(\omega)|>|\epsilon_1^{(2)}(\omega)|$. The behavior of $\epsilon_1^{(1)}(\omega)$ in the spectral area $\hbar\omega<0.35$ eV is related to the predominance of the contribution of the interband transition band at energy ~0.21 eV over the conduction electrons. The behavior of $\epsilon_1^{(2)}(\omega)$ results from the predominance of the contribution of conduction electrons over the contribution of interband absorption, i.e., the situation is the reverse of that just mentioned. We can therefore conclude that phases other than the A15 phase increase the concentration of conduction electrons. We can see from Figure 14 that the maxima of the basic and fine structure of $\epsilon_2^{(1)}(\omega)$ and $\epsilon_2^{(2)}(\omega)$ are practically the same. Further, $\epsilon_2^{(1)}(\omega) > \epsilon_2^{(2)}(\omega)$ throughout the spectral area studied. In the IR area of the spectrum, $\epsilon_2^{(2)}(\omega)$ increases sharply and at the edge of the IR area it approaches values of $\epsilon_2^{(1)}(\omega)$. In the visible area of the spectrum in $\epsilon_1^{(2)}(\omega)$ the areas of anomalous dispersion are more clearly expressed than in $\epsilon_1^{(1)}(\omega)$. In Figures 13, 14*a*, curves *(4)* relate to [28]. The data of Yao and Shnetterly[28], relating to $\epsilon_1(\omega)$ and $\epsilon_2(\omega)$, are similar to $\epsilon_1^{(2)}(\omega)$ and $\epsilon_2^{(2)}(\omega)$. This indicates that Nb–Ge specimen No. 2 is similar in its static characteristics to the specimen studied by Yao and Shnetterly[28]. Unfortunately, work[28] does not include data describing the quality of the specimen studied, but based on the optical data we can conclude that they were not studying a single-phase specimen. The measurements of Yao and Shnetterly were made with an excessive step. Determination of the contribution of interband absorption bands requires measurement with a much smaller step.

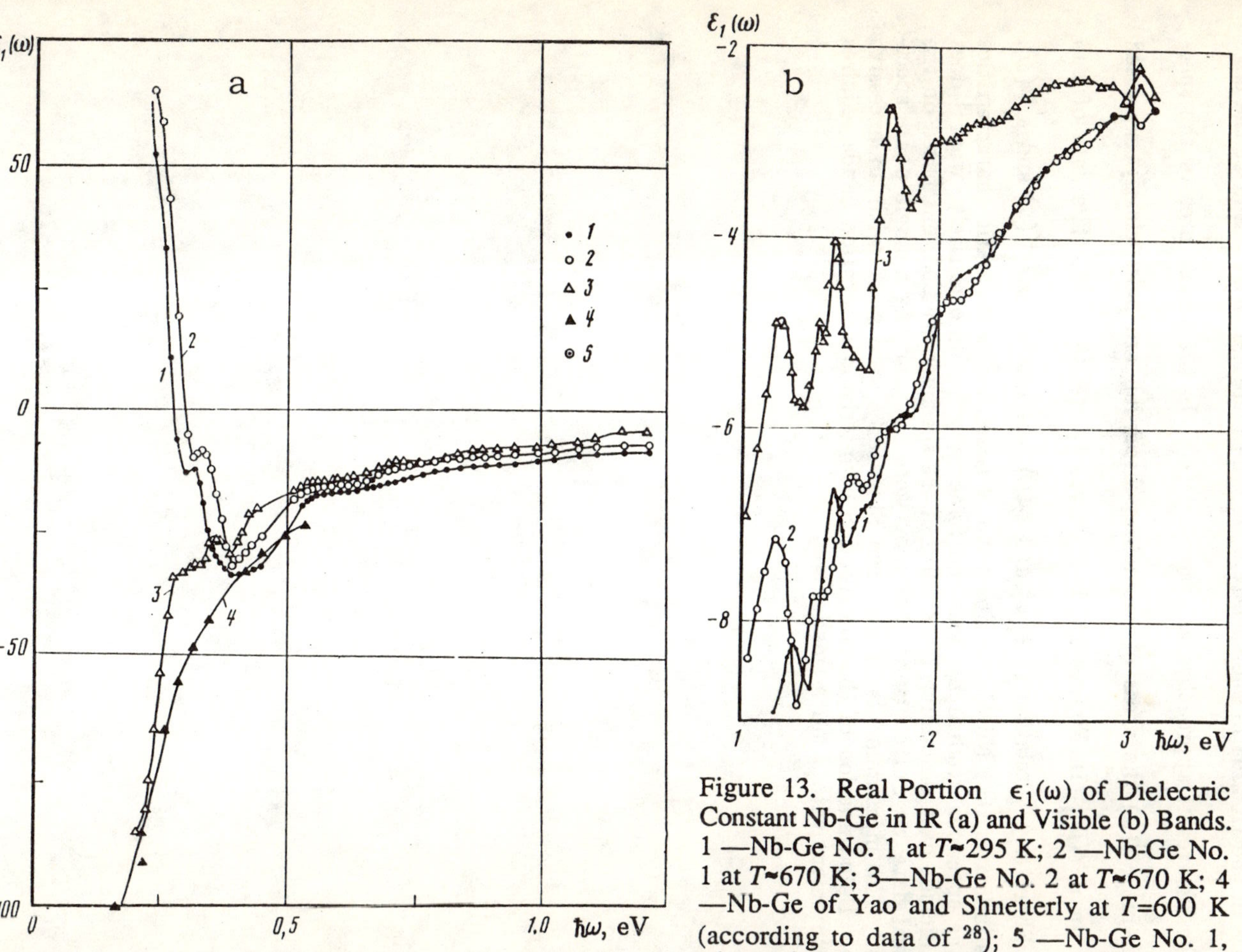

Figure 13. Real Portion $\epsilon_1(\omega)$ of Dielectric Constant Nb-Ge in IR (a) and Visible (b) Bands. 1 —Nb-Ge No. 1 at T~295 K; 2 —Nb-Ge No. 1 at T~670 K; 3—Nb-Ge No. 2 at T~670 K; 4 —Nb-Ge of Yao and Shnetterly at T=600 K (according to data of [28]); 5 —Nb-Ge No. 1, when points *1* and *2* merge.

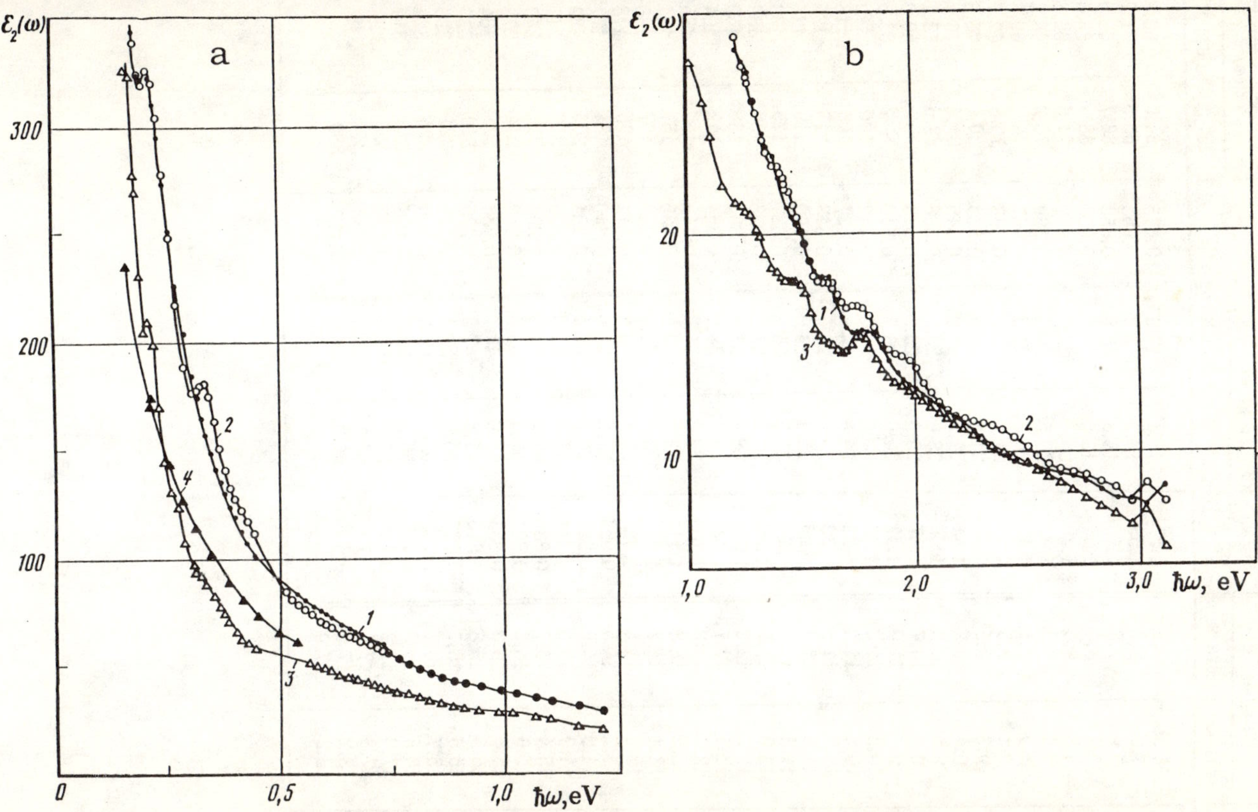

Figure 14. Imaginary Portion $\epsilon_2(\omega)$ of Dielectric Constant of Nb–Ge in IR (a) and Visible (b) Bands. Symbols same as Figure 13.

Table 5

$\epsilon_2(\omega)$ of Nb–Ge Specimen No. 1 at T=295 and 670 K

ω, eV	$\varepsilon_2(\omega)$		ω, eV	$\varepsilon_2(\omega)$		ω, eV	$\varepsilon_2(\omega)$	
	295 K	670 K		295 K	670 K		295 K	670 K
3.10	8.6	7.7	1.57	18.1	17.9	0.675	64.0	60.7
3.02	7.6	8.7	1.55	18.3	18.2	0.663	65.5	61.7
2.95	7.9	7.8	1.53	18.8	18.8	0.652	66.7	62.5
2.88	7.9	8.4	1.51	19.5	19.5	0.632	68.9	64.3
2.82	8.3	8.6	1.49	20.1	20.2	0.610	71.6	66.8
2.76	8.6	9.0	1.48	20.5	20.7	0.599	73.2	68.2
2.70	8.9	9.1	1.46	20.8	21.3	0.587	74.9	69.8
2.64	9.1	9.3	1.44	21.2	21.9	0.568	77.8	72.7
2.58	9.3	9.5	1.42	21.8	22.5	0.533	80.1	75.3
2.53	9.4	9.8	1.41	22.5	22.9	0.545	81.4	76.7
2.48	9.5	10.2	1.39	23.0	23.0	0.536	82.5	78.1
2.43	9.6	10.7	1.37	23.6	23.2	0.525	—	80.0
2.38	9.9	11.0	1.34	24.0	23.7	0.510	—	83.5
2.34	10.2	11.3	1.33	—	24.3	0.442	100	110
2.30	10.5	11.4	1.32	24.5	—	0.427	106	118
2.25	10.9	11.5	1.30	—	25.5	0.413	113	124
2.21	11.3	11.6	1.28	25.9	26.0	0.400	119	128
2.18	11.6	11.7	1.26	27.3	26.8	0.387	125	133
2.14	12.0	11.9	1.25	27.7	27.0	0.375	130	139
2.10	12.2	12.2	1.24	28.4	27.6	0.364	136	150
2.07	12.5	12.7	1.22	29.4	28.9	0.354	142	164
2.03	12.7	13.2	1.16	30.9	31.2	0.344	149	175
2.00	12.9	13.8	1.11	32.5	33.2	0.335	157	181
1.96	13.1	14.2	1.07	33.8	34.4	0.326	167	180
1.94	13.3	14.4	1.02	36.1	36.2	0.317	175	178
1.91	13.7	14.5	0.992	37.8	37.6	0.310	184	177
1.88	14.2	14.6	0.946	40.3	39.4	0.291	205	189
1.85	14.9	15.0	0.911	42.1	41.0	0.275	227	217
1.82	15.4	15.7	0.885	43.2	42.3	0.264	251	249
1.80	15.6	16.2	0.861	44.4	43.8	0.248	274	279
1.77	15.6	16.6	0.835	45.9	45.7	0.236	295	304
1.75	15.5	16.8	0.810	47.8	47.8	0.225	310	320
1.72	15.5	16.7	0.787	50.0	50.1	0.215	321	326
1.70	15.7	16.7	0.763	52.9	52.5	0.206	324	324
1.68	16.5	16.9	0.735	56.3	55.2	0.198	323	319
1.65	17.0	17.2	0.725	57.7	56.3	0.190	324	323
1.63	17.8	17.6	0.714	59.1	57.3	0.183	344	338
1.61	18.2	17.9	0.698	61.2	58.9	0.177	389	—
1.59	18.2	17.9						

CHAPTER 4
PROCESSING OF EXPERIMENTAL DATA

4.1. Smoothing of Experimental Results

Experimental values of sets $\{\epsilon_1(\omega_i)\}$ and $\{\epsilon_2(\omega_i)\}$ obtained on the frequency grid $\{\omega_i\}$ were triple smoothed by third-power polynomials, the coefficients of which were determined by the least squares method using seven unevenly spaced points. Here i=1, 2, ..., N (N is the total number of experimental points). The smoothing of the functions ϵ_1 and ϵ_2 was performed independently. Let us represent the selected function as y. The third-power approximating polynomial was determined by four coeffcients. To compute its coefficients by the method of least squares, the minimum number of experimental points considered in smoothing y_i must be 5. Furthermore, this number, which we shall represent as m, must be odd. We took m=7. For each point i, where $3<i<N-3$, the coefficients of the approximating polynomial are found, after which the smoothed value of $\tilde{y}_i$ is taken equal to the value of the polynomial for m=4. In set $\{y\}$, y_i is replaced by $\tilde{y}_i$ and the next point y_{i+1} is smoothed. Thus, we obtain a set of smoothed values $\{\tilde{y}\}$, in which three points at each end of the experimental interval remain unsmoothed. The smoothing algorithm was implemented in a BASIC program.

Coefficients of the approximating third-power polynomial were obtained for evenly spaced points for various values of m[29]. In this case, these are constant numbers, which are presented in their work. In our case, the coefficients of the polynomial must be computed where each value i

$$\frac{m-1}{2}<i<N-\frac{m-1}{2},$$

since they depend on the nonuniformity of the frequency grid $\{\omega_i\}$.

4.2. Method of Determining Electronic Characteristics [30]

It was shown in [1] that for nontransition metals at frequencies $\omega \ll \Omega_p$ (Ω_p is the plasma frequency of the valent electron), the basic contribution to $\epsilon(\omega)$ is that of the conduction electrons and electrons participating in the vertical interband transitions related to the Bragg planes. We note that $\Omega_p^2=4\pi e^2 N_{val}/m$, ($N_{val}$ is the concentration of valent electrons, e and m are the charge and mass of a free electron). In the transition metals, the picture of interband transitions is more complex, but in this case as well, as we noted in Chapter 1, the major contribution to $\epsilon(\omega)$ from interband transitions is made by vertical interband transitions, for which there is a maximum in the combined state density related to the Van–Hoff type Q_0 singularities. The latter, in turn, are related to the presence of parallel (or near-parallel) sectors of the isoenergy surfaces, of which one sector is located below the Fermi surface, the other above. This situation obtains in particular near the area of intersection of the Fermi surface by the Bragg plane. For nontransition metals, the parallel sectors of isoenergy surfaces may be on the boundary or within the Brillouine zone. In both cases it can be assumed that the form of the band of such interband transitions can be described approximately by equation (11). Equation (3) works for conduction electrons. (In both equations the one related to the vacuum is omitted, i.e., $\epsilon=4\pi\chi$, where χ is the polarizability.)

Considering that the contribution to $\epsilon(\omega)$ of various groups of electrons are additive, we write

$$\epsilon(\omega)=\sum_{j=1}^{n}\epsilon_j(\omega), \tag{26}$$

where n is the total number of intraband and interband transition zones. The subscript $j=1$ relates to the band of conductivity electrons, while other values relate to interband transition zones.

To estimate the electronic characteristics of various groups of electrons, we must distinguish their contributions to $\epsilon(\omega)$. For this purpose, we shall use $\epsilon_2(\omega)$. Selection of function $\epsilon_2(\omega)$ is determined by the following consideration. There are usually short-wave bands which fall outside the experimental interval.

These bands have little influence on $\epsilon_2(\omega)$, but make a significant contribution to $\epsilon_1(\omega)$. This is the situation with the compounds we studied, Nb–Al and Nb–Ge with A15 structure.

It follows from equations (3) and (11) that the band of conduction electrons $\epsilon_1(\omega)\equiv\epsilon_{11}(\omega)-i\epsilon_{21}(\omega)$ is described by two parameters: the square of the plasma frequency ω^2_{p1} and the relaxation frequency v_1[12]. Each main band j of interband transitions $\epsilon_j(\omega)\equiv\epsilon_{1j}(\omega)-i\epsilon_{2j}(\omega)$ is determined by three parameters: the value of the gap $\Delta j=\hbar\omega_{0j}$, the square of the plasma frequency ω^2_{pj} and the relaxation frequency v_j (j=2, 3, ..., n). (We note that ω^2_{pj}/Ω^2_p is equal to the strength of oscillators for electron group j.) The number of bands of the $(n-1)$ interband transitions is equal to the number of main broad maxima in $\epsilon_2(\omega)$.

The parameters of the bands are found by the method of least squares[30]. Their values should yield the minimum of function F

$$F=\sum_{k=1}^{N}\left[\varepsilon_2(\omega_k)-\varepsilon_2^T(\omega_k)\right]^2. \tag{27}$$

Here $\epsilon_2(\omega_k)$ is the experimental value at point ω_k, $\epsilon_2^T(\omega_k)=\sum_{j=1}^{n}\epsilon^T_{2j}(\omega_k)$ is the theoretical value, N is the number of experimental points.

The equations for $\epsilon^T_{2j}(\omega)$ are obtained from equations (3) and (11)

$$\varepsilon^T_{21}(\omega)=\frac{\omega^2_{p1}v_1}{\omega\left(\omega^2+v_1^2\right)};$$
$$\varepsilon^T_{2j}(\omega)=\frac{\omega^2_{pj}\omega_{0j}\omega v_j}{(\omega^2+v_j^2)^2}\left(\frac{\omega^2-v_j^2}{A_jB_j}-\frac{2B_j}{A_j}+\frac{2}{\omega_{0j}}\right), \tag{28}$$

[12]Relaxation frequency $v=1/\tau$, where τ is the relaxation time.

$$A_j = \sqrt{Z_j^2 + 4\omega^2\nu_j^2}; \quad B_j = \sqrt{(A_j + Z_j)/2},$$
$$Z_j = \omega_{0j}^2 - \omega^2 + \nu_j^2, \quad \hbar\omega_{0j} = \Delta_j \quad (j = 2, 3, \ldots, n). \tag{29}$$

For completeness, we present the corresponding equations for $\epsilon_{1j}^T(\omega)$:

$$\varepsilon_{11}^T(\omega) = \frac{\omega_{p1}^2}{\omega^2 + \nu_1^2}, \tag{30}$$

$$\varepsilon_{1j}^T(\omega) = \frac{\omega_{pj}^2\omega_{0j}}{(\omega^2 + \nu_j^2)^2}\left[\left(\frac{B_j}{A_j} + \frac{1}{\omega_{0j}}\right)(\omega^2 - \nu_j^2) + \frac{2\omega^2\nu_j^2}{A_jB_j}\right]. \tag{31}$$

All calculations were performed on computers. We used an iterative method. The algorithm used to find the band parameters was composed in [30], where it was implemented in FORTRAN 4+[13].

The algorithm used to seek the minimum of F is as follows [30]. The initial values $\{\omega_{0j}, v_j\}$ (ω_{01}=0) are given, the values $\{\omega_{pj}^2\}$ are found and this completes iteration zero. Determining the components of vector $\{\omega_{pj}^2\}$ is not difficult, since they are included linearly in $\epsilon_2^T(\omega)$. Then with fixed $\{\omega_{pj}^2\}$ the values of $\{\omega_{0j}, v_j\}$ are varied in the direction which decreases the sum F and the first approximation of $\{\omega_{0j}, v_j\}$ is found, etc. The direction of variation of $\{\omega_{0j}, v_j\}$ is selected by the conjugate gradient method[31]. Computations can be stopped after a sufficient number of iterations. The rate of convergence depends on the initial values of the parameters $\{\omega_{0j}, v_j\}$.

The error in determining the parameters of the bands is determined first of all by the accuracy of the model used; secondly, by the error in computing the parameters within the framework of the model.

The accuracy of the model is determined by the accuracy of equations (29). They were obtained in the first approximation using $\Delta_j/2E_F$, with higher-order terms ignored (E_F is the Fermi energy). Thus, the error is related to

---

[13]In this work, precise formulas (29), obtained in [9, 10], were used for $\epsilon_{2j}(\omega)$ rather than the approximate equations used in [30].

the accuracy of the model $\sim\Delta j/2E_F$. The error in calculating the parameters within the framework of the model used can be estimated by the equation

$$\sigma_\xi = \sqrt{\frac{F}{(N-M)\,\partial^2 F/\partial\xi^2}}. \qquad (32)$$

Here σ_ξ is the mean square deviation for the parameter ξ, M is the number of parameters defined, the other symbols are as used above. For all the compounds we studied, the error of the calculation is significantly less than the error related to the accuracy of the model. Therefore, we selected the quantity $\Delta_j/2E_F$ as the error of determination of the parameters of the bands.

4.3. Results of Processing of Dispersion of Imaginary Portion of Dielectric Constants of Nb–Al[9, 10]

We can see from Table 2 that the Nb–Al specimen studied has a lattice constant a=5.19Å. A unit cell contains 36 valent electrons. The plasma frequency of all valent electrons $\Omega_p \approx 19$ eV and the condition $\omega_{max} \ll \Omega_p$ is met. Here ω_{max}=3.1 eV is the upper limit of the spectral interval studied. Considering this, we accepted for the main interband transition zones the form defined by equations (29). The theoretical function $\epsilon_2^T(\omega)$ consists of the sum of four terms:

$$\epsilon_2^T(\omega) = \sum_{j=1}^{4} \epsilon_2^T(\omega),$$

j=1 corresponds to the zone of conductivity electrons described by equation (28), while $j \geqslant 2$ corresponds to the main interband transition zones described by equations (29). The results obtained are presented in Table 6. Agreement of the theoretical function $\epsilon_2^T(\omega)$ with the experimental function $\epsilon_2(\omega)$ at room temperature is as shown in Figure 15. We can see from this figure that the basic structure of the curve of $\epsilon_2(\omega)$ is described well by the band forms we accepted, i.e., by equations (28) and (29). Against the background of the main structure we see a weaker fine structure, the amplitude of which $\leqslant$8% of the amplitude of the main structure, and which is not described by equations (28, 29). It is apparently related to singularities in the combined state density of another type, for example lines and points of intersection of several Bragg planes.

Table 6
Electronic Characteristics of Nb–Al with A15 Structure

Band Number	Δ, eV	v, eV	ω_p^2, eV^2	Band Number	Δ, eV	v, eV	ω_p^2, eV^2
	T=295 K				T=670 K		
1	0	0.08	19	1	0	0.10	27
2	0.35	0.35	29	2	0.34	0.40	30
3	1.46±0.11	0.91±0.06	46±3	3	1.47±0.11	0.91±0.06	49±3
4	3.1±0.5	1.4±0.2	46±7	4	3.1±0.5	1.4±0.2	44±7

Note: In all cases where the error is not indicated it is less than one in the last digit.

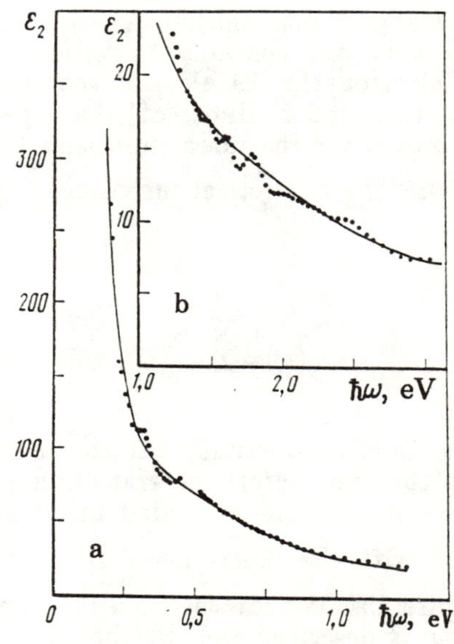

Figure 15. Comparison of Experimental Data on $\epsilon_2(\omega)$ (points) and Theoretical Curve of $\epsilon_2^T(\omega)$ (solid line) for Nb–Al at T=295 K.

The results for ϵ_1 and ϵ_2 in the IR area show the presence of interband transitions with an energy gap of ~0.2 eV. The experimental data are not sufficient to define the parameters of this band against the background of the conduction electrons. Therefore, in Table 6 the parameters of the conduction electrons are slightly distorted by the presence of this band. We shall return to this question in Chapter 5.

We note that the square of the plasma frequency of the conduction electrons $\omega_p^2=19$ eV^2 at room temperature and 27 eV^2 at $T=670$ K. Bands 2 and 3 have a fine structure which is lost as the temperature increases. The great relative width of these bands ($v/\Delta \sim 1$) may be related to the fact that each of them results from overlapping of two bands. However, it has been possible to distinguish the component bands using the band formula only for the second band at room temperature.

4.4. Results of Processing Dispersions of Imaginary Portion of Dielectric Constant of Nb–Ge

Data were processed relating to Nb–Ge specimen No. 1, since it contains practically no other phases than the A15 phase. We can see from Table 2 that the specimen had lattice constant $a=5.187\pm0.002$Å. The high accuracy of measurement of the lattice constant results from the great homogeneity of the specimen.

The Nb–Ge unit cell contains 38 valent electrons. The concentration of valent electrons $N_{val}=2.71\cdot10^{23}$cm^{-3}. The plasma frequency of the valent electrons $\Omega_p\approx19.5$ eV.

The condition $\omega_{max}\ll\Omega_p$ is also well met here. We processed the experiments for the base interband transition zones as in the case of Nb–Al compounds using a form described by equations (29). The Drude equation (28) was used for the conduction electron bands. The theoretical curve $\epsilon_2^T(\omega)$ was computed, consisting of the sum of six terms. The first term is the contribution of conduction electrons, the other terms are the contributions of the main interband transitions.

The results of this process are shown in Table 7.

Figure 16 shows the agreement of the theoretical curve determined by the parameters of Table 7 with the experimental curve $\epsilon_2^{(1)}(\omega)$[14] at $T=670$ K. We

[14]The superscript 1 with ϵ_2 means that experimental data relating to Nb–Ge specimen No. 1 were used.

can see from the figures that both in the visible and in the IR bands, the major structure of the curve of $\epsilon_2(\omega)$ is described well by the band equations which we used, i.e., equations (28) and (29). Along with the main structure in the visible area of the spectrum there is a weaker fine structure with an amplitude of $\leqslant 5\%$ of the amplitude of the main structure, not described by equations (28) and (29). Apparently, as is true for Nb–Al, this results from singularities in the combined state density of another type, for example lines and points of intersection of several Bragg planes.

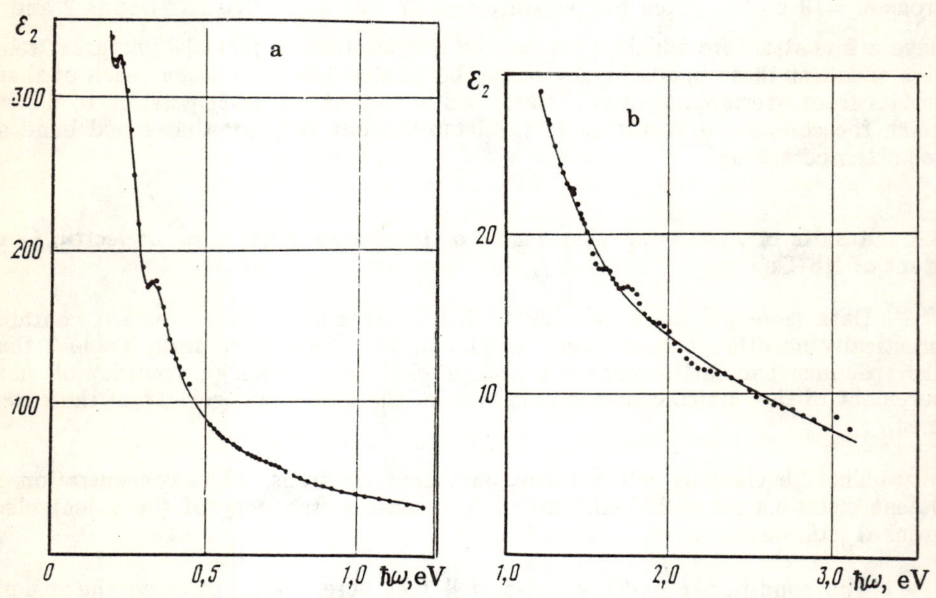

Figure 16. Comparison of Experimental Data on $\epsilon_2(\omega)$ (points) and Theoretical Curve of $\epsilon_2^T(\omega)$ (solid line) for Nb–Ge Specimen No. 1 in the IR (a) and Visible (b) Band.

Let us study the parameters of the conduction electron band (band No. 1). As temperature increases, v remains almost unchanged. The quantity ω_p^2 for this band increased from 19–22 eV^2 with increasing frequency. The parameters of the second bands changed significantly. As temperature increased, v_2 decreased by a factor of 1.65, and ω_p^2 for this band also decreased by a factor of 1.45. The quantity representing the energy gap of the interband transition Δ_2 increased slightly.

Table 7
Electronic Characteristics of the Compound $Nb_{0.8}Ge_{0.2}$ with A15 Structure

Band	Parameter			Band	Parameter		
Number	Δ, eV	v, eV	ω_p^2, eV^2	Number	Δ, eV	v, eV	ω_p^2, eV^2
	T=295 K				T=670 K		
1	0	0.098	19	1	0	0.097	22
2	0.213±0.002	0.049	8.2	2	0.217±0.002	0.030	5.7
3	0.324±0.006	0.061	1.7	3	0.332±0.006	0.034	4.1
4	0.48	0.245±0.005	22	4	0.50	0.250±0.005	19
5	0.95±0.05	0.42±0.02	23	5	0.90±0.04	0.42±0.02	24±2
6	2.0±0.2	1.0	50±5	6	2.0±0.2	1.0	57±6

Note: Where no error is given it is less than one in the last digit.

The energy gap of the third band increased. The width of the interband transition v_3 decreased by a factor of almost two, while ω_{p3}^2 increased by more than two times.

Let us now go over to the parameters of the fourth band. The energy gap Δ_4 increased, v_4 also increased. The square of the plasma frequency ω_{p4}^2 decreased with increasing temperature from 22 to 19 eV^2. The low sensitivity of the sum F to values of the parameters of bands 5 and 6 does not allow us to determine their temperature variation.

CHAPTER 5
DISCUSSION OF RESULT PRODUCED

5.1. Nb–Al

Let us turn once more to Figures 10–13 and Table 6. These figures and table indicate that the major contribution to $\epsilon(\omega)$ in the infrared band is that of the conduction electrons. At room temperature $\omega_{p1}^2(295)=19$ eV^2. The corresponding effective concentration N of conduction electrons $N_e \simeq 1.4\cdot10^{22}$, which is 3.2 times less than for pure niobium[32]. As the temperature is doubled, ω_{p1}^2 and the corresponding value of N increase by a factor of 1.4. This strong temperature variation of the square of the plasma frequency of the conduction electrons is partially related to the transfer of electrons from the base state of the second band through a small energy gap space ~0.2 eV. The presence of this gap was noted in Sections 3.1 and 4.3. The experimental data are insufficient to determine the parameters of this band. The possibility cannot be excluded that the form of this band is not described by equations (29), for example if it is related to the line (point) of intersection of Bragg planes. We suggest that the form of the band with the energy gap of ~0.2 eV is described by equations (29), and estimate the square of the plasma frequency for this band. We can conclude from Figure 10 that the asymptotic value of $\epsilon_1(0)$ for this band is ~10 (the value of the dump in the area of anomalous dispersion). We estimate v on the basis of the width of the dump in $\epsilon_1(\omega)$. We obtain $v \approx 0.1$ eV ($v/\Delta \sim 0.5$) and, substituting into equation (31), obtain $(\omega_p^2)_{0.2\ \mathrm{eV}} \sim 1$ eV^2. This band is most clearly seen at a temperature of 670 K. Thus, $(\omega_p^2)_{0.2\ \mathrm{eV}}/\omega_{p1}^2 \ll 1$, i.e., the parameters of the conduction electrons are similar to those presented in Table 6. The ratio $N_{\mathrm{eff}}/N_{\mathrm{val}}$ where T=295 K and 670 K is 0.05 and 0.07, respectively.

The relaxation frequency of conduction electrons v_1 at room temperature is $v(295)=0.08$ eV. Where $T=670$ K, $v_1(670)=0.10$ eV. If we assume that $v_1=v_{ep}+v_{ed}$, where v_{ed} is the frequency-independent relaxation temperature, related to defects, while v_{ep} is the frequency-dependent relaxation frequency, related to electron-phonon interaction, it follows from Table 6 that $v_{ep}=0.016$. We shall return to this question below.

The sum of ω_p^2 of all the bands distinguished at 295 and 670 K is 140 and 150 eV^2, respectively, which amounts to 0.39 and 0.42 times $\Omega_p^2=356$ eV^2. This fact indicates the presence of significant interband transition zones in the spectral area $\hbar\omega>3$ eV.

The adjusted state density $J(\omega)$ was presented in [33] for Nb_3Al. From a figure in this work, maxima can be seen for $J(\omega)$ at energies of ~0.5 eV, ~0.8, ~1.9 and ~3.2 eV, plus a broad maximum in the 5–6 eV area. We can see from Table 6 that only gaps of ~1.3 and 3.2 eV agree with the data in the table. In the infrared band, there is no such agreement, but theoretical computation also indicates that $J(\omega)$ has singularities in this area. In the work just mentioned, ω_p was computed for the conduction electrons, and found to be 3.83 eV. According to the data of Table 6 at room temperature $\omega_p=4.4$ eV. These values are rather similar to each other.

In [34], the parameters of the theoretical model were adjusted to fit the experimental data on $R(T)$ (resistance as a function of temperature), yielding a value of 3.7 eV for ω_p. However, the model of $R(T)$ of [34] did not consider the variation in concentration of conduction electrons N as a function of temperature, which we determined experimentally for this compound (cf. Table 6). The disagreement in the value of conduction electron parameters of [34] and the present work may also be related to the difference in specimen quality. The characteristics of the Nb_3Al specimen used in [34] are not presented.

5.2. Nb–Ge

Let us study the results of processing the experimental data relating to Nb–Ge specimen No. 1. They are presented in Table 7.

We begin with the conduction electron band. ω_{p1}^2 at $T=295$ and 670 K is 19 and 22 eV^2, respectively. The corresponding effective concentrations of conduction electrons are $N_e(295\text{ K})\approx1.4\cdot10^{22}\text{cm}^{-3}$ and $N_e(670\text{ K})=1.6\cdot10^{22}\text{cm}^{-3}$. At room temperature the effective concentrations of conduction electrons for the compounds Nb–Ge and Nb-Al coincide with an accuracy equal to the measurement error. The ratio $N(670\text{ K})/N(295\text{ K})=1.16$. It is less than that of Nb-Al, but remains significantly greater than that of ordinary metals. This change in conduction electron concentration with temperature may be partially related to the thermal shifting of electrons through the energy gap Δ_2=0.21 eV. The relaxation frequency of conduction electrons v_1 did not change as temperature T rose to 670 K considering errors, i.e., v_{ep}<0.001, where v_{ep} is the temperature-dependent portion of v_1. We shall return to this question below.

Let us now look at the interband transition zones. It follows from Table 7 that the changes in energy gaps with increasing temperature are slight. Basically all the gaps except for Δ_5 increased slightly. Gap Δ_5 at T=295 K is approximately 5% greater than at T=670 K. The accuracy of determination of parameter Δ_6 does not allow us to estimate its change with temperature.

The sum $\omega_{p1}^2+\omega_{p2}^2$ at the two temperatures studied is 27.2 and 27.7 eV^2, i.e., within the limits of error it remains constant. This agrees with assumption of transfer of electrons through gap Δ_2 into the conduction band. The width of the second band v_2 decreased with increasing temperature from 0.049 to 0.030, which is also apparently related to the emptying of the lower level of the second band.

Let us study the parameters of the third and fourth bands. We can see from Figure 15 that at 670 K the maximum of the third band seems clearer than at room temperature. Apparently, the picture is related to the imposition of an indirect interband transition over the main interband transmission (third band). For this reason, there is a change in the parameters of the third band. We note that ω_{p3}^2 at T=670 K is 2.4 times greater than at room temperature, while v_3 decreases with increasing temperature. The sum $\omega_{p3}^2+\omega_{p4}^2$ remained constant within the limits of error of determination of the parameters (it is 23.7 and 23.1 eV^2 at T=295 K and 670 K). The quantity v_4 is slightly higher at T=670 K than at T=295 K.

The low sensitivity of the sum F to the parameters of bands five and six did not allow us to determine the temperature variation of the parameters of these bands with sufficient accuracy.

The sum ω_p^2 of all interband transition zones, as well as the bands of conduction electrons at room temperature and at T=670 K is equal to 124 and 132 eV^2, 33 and 35% of Ω_p^2. Therefore, in the spectral area $\hbar\omega$>3.1 eV there are significant interband transition zones.

The positions of the maxima of $J(\omega)$ are presented in [34], as computed by Mattheis, Testardi, Yao and Shnetterly. The maxima of $J(\omega)$ have the following positions: ~0.2 eV, ~0.7, ~1.2, ~1.9 eV. Table 7 indicates that the results of calculation for three bands agree rather well with the values Δ_2=0.213 eV, Δ_5=0.95 eV and Δ_6=2.0 eV. This work presents for conduction electrons ω_p=3.7 eV ($\omega_p^2 \simeq 14$ eV^2), which is somewhat lower than the experimental values presented in Table 7.

Article [28] presents the results of processing of experimental data obtained by Yao and Shnetterly by the authors of that article. The authors of [28] obtained at 600 K a value of ω_p for conduction electrons of 2.3 eV ($\omega_p^2 \simeq 5.3$ eV^2). The value of ω_p^2 is about one-fourth our value of ω_{p1}^2 (cf. Table 7 for 670 K). This same work obtained v=0.17 eV, which is ~1.7 times greater than v_1 in Table 7.

First of all, the disagreement in conduction electron parameters of [28] with the conduction electron parameters of this work may be related to insufficient quality of the specimens studied by Yao and Shnetterly. Furthermore, long-wave interband transitions were not distinguished in [28], although the presence of these transitions in the 0.2–0.3 eV area can be established by careful analysis of the figures presented in that work. The influence of these interband transition zones on the optical properties of a compound results in deviation from linear variation of $\omega\epsilon_2$ and $-\epsilon_2$ as functions of $(\omega^2+v^2)^{-1}$. We note that the bands at energies ~0.2 and 0.33 eV agree well with results of our work (cf. Table 7). Ignoring interband transitions in the long-wave area leads to erroneous values of conduction electron parameters.

Conclusions

Optical investigation of two superconducting compounds with high critical superconducting paramers, Nb–Al and Nb–Ge, with A15 structure showed a number of common properties. Let us note that.

1. For both compounds in the 0.177–3.1 eV spectral interval reserved, in addition to the conduction electron bands, a number of basic interband transition zones corresponding to energy gaps Δ of 0.2, 0.35, 1.34 and 3.1 eV for Nb–Al (4 bands) and 0.21, 0.32, 0.48, 0.95 and 2.0 eV for Nb–Ge (5 bands).

2. The shape of the interband transition zones can be described by equations (29), obtained for Van-Hoff type Q_0 singularities of the combined state density, i.e., for parallel (or near-parallel) sheets of isoenergy surfaces in k space. The difference in energies at two points with identical quasimomenta was constant (or nearly constant) for them. This situation is apparently also valid with Bragg energy splitting.

Against the background of the structure related to the basic interband transitions, there is a weaker and finer structure, not described by equations (29). It may be determined by singularities in the combined state density of another type, for example lines and points of intersection of several Bragg planes, when the triple and quadruple energy degeneration is relieved.

3. Both compounds have long-wave interband transition zones corresponding to small gap energies, the least gap in the two compounds $\Delta \sim 0.2$ eV.

4. The squares of the conduction electron plasma frequencies in both compounds are the same at room temperature: $\omega_{p1}^2 = 19$ eV^2 (the corresponding concentration of conduction electrons $N_e = 1.4 \cdot 10^{22}$cm^{-3}), which is 3.2 times less than in pure niobium[32]. The conduction electron oscillator strengths are small[15]. The values of ω_{p1}^2/Ω_p^2 for Nb–Al and Nb–Ge are $5.3 \cdot 10^{-2}$ and $5.0 \cdot 10^{-2}$, respectively.

5. Optical measurements at the two temperatures of 295 K and 670 K showed a somewhat greater increase in ω_{p1}^2 with increasing temperature than is observed for ordinary metals. For the compounds Nb–Al and Nb–Ge, the ratios $\omega_{p1}^2(670\ \mathrm{K})/\omega_{p1}^2(295\ \mathrm{K})$ are 1.42 and 1.16. This increase in ω_{p1}^2 may be partially related to the thermal transfer of electrons from the base long-wave band state to the conduction band. This "transfer" hypothesis is supported by the decrease in the square of plasma frequency in the second band ω_{p2}^2 (in Nb–Ge) with preservation of the sum of $\omega_{p1}^2 + \omega_{p2}^2$, as well as the decrease in width of the second band by a factor of 1.6. This may be explained by emptying of the lower level of the second band. In Nb–Al, the parameters of the band with energy gaps 0.2 eV were not determined. The presence of a small gap may influence the temperature variation of static thermodynamic characteristics, and must be considered.

6. For both compounds, $\sum \omega_p^2 \ll \Omega_p^2$, where the left portion represents the sum of the squares of plasma frequencies of electrons of the interband and interband transition zones, while the right portion represents the square of the plasma frequency of valent electrons. Thus, there are significant interband transition zones in the spectral interval $\hbar\omega > 3.1$ eV.

In conclusion, we note some differences in the properties of the compounds Nb–Al and Nb–Ge with A15 structure.

---

[15]The small oscillator strength may be related either to the small concentration of conduction electrons or to the great mass of these electrons.

1. The basic difference is found in the number of interband transition zones and in the energy gap spectrum. In the 0.177–3.1 eV interval in Nb–Al there are four interband transition zones, while in Nb–Ge there are five. Only in the two longest-wave bands are the energy gaps of Nb–Al and Nb–Ge similar. In Nb–Al there is no band with Δ=0.5 eV, which is present in Nb–Ge. The two highest frequency bands are located at higher energies in Nb–Al than in Nb–Al [sic—Tr.].

2. The presence of an interband transition zone in Nb–Ge with $\Delta \approx 0.5$ eV (band 4), located near band 3 ($\Delta \approx 0.32$ eV) results in a strong temperature variation of ω_{p3}^2. As temperature rises from 295 to 670 K, ω_{p3}^2 increases by a factor of 2.4. The maximum on the curve of $\epsilon_2(\omega)$ related to the third band seems clearer at high temperature than at room temperature. Apparently, all of this is related to the superimposition on the basic third band transition of an indirect interband transition related to the fourth band. The sum $\omega_{p3}^2+\omega_{p4}^2$ remains constant. In the compounds Nb–Al, the bandwidth $\Delta \approx 0.5$ eV is absent and these effects are not seen.

3. In the spectral interval 0.155–3.1 eV for Nb–Al the sums of oscillator strengths $\sum_j \omega_{pj}^2/\Omega_p^2$ at 295 and 670 K are equal to 0.39 and 0.42 (an increase by a factor of 1.68); for Nb–Ge, 0.33 and 0.35 (an increase by a factor of 1.06). Thus, for Nb–Al these sums are approximately 20% greater than for Nb–Ge.

REFERENCES

1. Motulevich, G. P., "Optical Properties of Nontransition Metals," Trudy FIAN, Vol. 55, pp. 3–150, 1971.

2. Motulevich, G. P., "Optical Properties of Polyvalent Nontransition Metals," UFN, Vol. 97, pp. 211–256, 1969.

3. Phillips, J. C., Kleinman, L., "New Methods for Calculating Wave Functions in Crystals and Molecules," Phys. Rev., Vol. 16, pp. 127–294, 1959.

4. Ginzburg, V. L., Motulevich, G. P., "Optical Properties of Metals," UFN, Vol. 55, pp. 469–533, 1955.

5. Gurzhi, R. N., "Quantum Kinetic Equation of Electrons in Metals," ZhETF, Vol. 33, pp. 451–458, 1957.

6. Maksimov, Ye. G., Motulevich, G. P., "Determination of Electron-Phonon Bonding Constant from Optical Measurements," ZhETF, Vol. 61, pp. 414–417, 1971.

7. Golovashkin, A. I., Motulevich, G. P., "Influence of Bragg Interband Transitions on Dielectric Constant of Metals," ZhETF, Vol. 57, pp. 1054–1063, 1969.

8. Mash, I. D., "Study of Optical Properties and Electronic Characteristics of Certain Transition Metals," Trudy FIAN, Vol. 82, pp. 3–71, 1975.

9. Kuz'michev, N. D., Levchenko, I. S., Motulevich, G. P., "Determination of Electronic Characteristics of Nb-Al with A15 Lattice by Optical Method," Fizika metallov i metallovedeniye, Vol. 56, pp. 266–270, 1983. Also: Preprint FIAN No. 27, Moscow, 18 pp., 1983.

10. Kuz'michev, N. D., "Optical Properties of a Number of Niobium Compounds with A15 and B1 Lattices," Dissertation, Candidate of Physical-Mathematical Sciences, Moscow, FIAN, 142 pp., 1983.

11. Golovashkin, A. I., Shelekhov, A. L., "Determination of Electronic Characteristics of Metals and Alloys Based on Their Modulation Optical Spectra in Interband Transition Zones," Kratkye soobshcheniya po fizike, No. 5, pp. 11–17, 1978. Also: Preprint FIAN, No. 96, Moscow, 25 pp., 1981.

12. Golovashkin. A. I., Motulevich, G. P., "Linear Polarizability of Metals Related to Electron Transitions Near Bragg Planes," Preprint FIAN, Moscow, No. 76, 27 pp., 1969.

13. Golovashkin, A. I., Levchenko, I. S., Motulevich, G. P., "Characteristics of Superconducting Alloys with A15 Lattice Obtained by Evaporation in a Vacuum," Trudy FIAN, Vol. 82, pp. 72–102, 1975.

14. Golovashkin, A. I., Shubin, A. A., "Use of Simplified Electron Gun Design to Produce Layers of Refractory Metals," Preprint FIAN, No. 65, Moscow, 10 pp., 1969.

15. Mirkin, A. I., Spravochnik po rentgenostrukturnomu analizu polikristallov [Handbook on X-Ray Structural Analysis of Polycrystals], Moscow, Fizmatgiz Press, 863 pp., 1961.

16. Motulevich, G. P., "Determination of Lattice Constant of Cubic Structures," Preprint, FIAN, No. 167, Moscow, 65 pp., 1981.

17. Beatti, Y. R., "Optical Constants of Metals in the Infrared-Experimental Methods," Phil. Mag., Vol. 46, pp. 235–245, 1955.

18. Kuz'michev, N. D., "Experimental Installation for Measurement of Optical Constants of Metals and Alloys at 290–700 K in Visible and Near IR Bands," Preprint FIAN, No. 97, Moscow, 13 pp., 1982.

19. Borbam, A. M., Gorban', I. S., et al., Opticheskiye izmereniya [Optical Measurements], Kiev, Tekhnika Press, 419 pp., 1967.

20. Born, M., Bol'f, E., Osnovy optiki [Principles of Optics], Moscow, Nauka Press, 855 pp., 1970.

21. Sokolov, A. V., Opticheskiye svoystva metallov [Optical Properties of Metals], Moscow, Fizmatgiz Press, 464 pp., 1961.

22. Gorban', N. Ya., Poperenko, L. V.,"Optimized Version of Beatti–Conn Method in Study of Optical Properties of Nickel in the Visible Area of the Spectrum,:" Zhurnal prikladnoy spektroskopii, Vol. 33, pp. 706–711, 1980.

23. Bakhshiyev, N. G., "Determination of Optical Constants by Reflections," Optika i spektroskopiya, Vol. 2, pp. 818–819, 1957.

24. Kuz'michev, N. D., "Experimental Installation for Investigation of Optical Properties of Alloys in the Infrared Band at 290-700 K," Preprint, FIAN, No. 5, Moscow, 7 pp., 1983.

25. Aleksandrov, A. N., Nikitin, V. N., "Selection of Normals and Methods for Calibration of Prism IR Spectrometers," UFN, Vol. 56, pp. 1–52, 1955.

26. Kuz'michev, N. D., Levchenko, I. S., Motulevich, G. P., "Optical Properties and Electronic Characteristics of Nb–Ge Compounds with A15 Lattice," Fizika metallov i metallovedeniye, Vol. 60, pp. 488–493, 1985.

27. Kuz'michev, N. D., "Optical Properties of Nb–Ge with A15 Lattice," Kratkiye soobshcheniya po fizike, No. 11, p. 16, 1983.

28. Tutto, J., Kahn, L. M., Ruvalds, J., "Optical Properties and Electron-Phonon Interactions in Superconducting Transition Metals," Phys. Rev. B., Vol. 20, pp. 952–959, 1979.

29. Miln, V. E., Chislennyy analiz [Numerical Analysis], Moscow Foreign Literature Press, 292 pp., 1951.

30. Motulevich, G. P., "Processing of Metal–Optical Experimental Results by Nonlinear Method of Least Squares," Preprint FIAN No. 148, Moscow, 33 pp., 1980.

31. Flecher, R., Recves, C. M., "Function Minimization by Conjugate Gradients," Computer., Vol. 1, pp. 149–154, 1964.

32. Golovashkin, A. I., Leksina, I. Ye., Motulevich, G. P., Shubin, A. A., "Optical Properties of Niobium," ZhETF, Vol. 56, pp. 51–64, 1969.

33. Matterss, L. F., "APW–LCAO Band Model for A15 Compounds," Phys. Rev. B., Vol. 12, pp. 2161–2180, 1975.

34. Allen, P. B., Pickett, W. E., Ho, K. M., Cohen, M. L., "Anomalous Resistivities of A15 Metals — Insights from Band Theory," Phys. Rev. Lett., Vol. 40, pp. 1532–1534, 1978.

DETERMINATION OF ELECTRONIC CHARACTERISTICS OF NIOBIUM NITRITE BY OPTICAL METHODS

D. R. Dzhurayev, N. D. Kuz'michev and G. P. Motulevich

ABSTRACT

The dispersions of complex dielectric constant of compounds of niobium nitrite with B1 structure are measured in the 1.24–3.1 eV spectral interval, and used to determine the electronic characteristics of the compounds, as well as their variation with composition. For one of the specimens, measurements were performed at 295 and 600 K. A strong variation in plasma frequency ω_p of conduction electron as a function of specimen composition was observed. As the content of niobium increased, ω_p monotonically rose remaining, however, a small quantity. For the stoichiometric composition ω_p^2=0.7 eV^2, 100 times less than in pure niobium and 30 times less than in compounds with A15 structure. The variation of electronic characteristics with temperature manifests a number of anomalies. As the temperature rises from 295–600 K, there is an anomalously great (by 2.4 times) increase in ω_p^2 for conduction electrons and a decrease (by 1.5 times) in the corresponding value for the long-wave interband transition zone with preservation of their sum, as well as a great decrease (by 3.5 times) in the width of the long-wave bands. These anomalies may be related to thermal transfer of electrons from the lower level of the long-wave band to the conduction band. Figures 14, Tables 7, References 15.

INTRODUCTION

A niobium nitrite compound with B1 structure has a number of interesting properties, making its study important from both the scientific and practical views. Niobium nitrite defines good superconducting properties with very low static conductivity, changing but little with temperature. It is resistant to external effects, including corrosive media, and insensitive to radiation. The area of practical application of niobium nitrite is quite broad. In spite of this, studies of its optical properties were not undertaken prior to our own work[1, 2]. However, in [3, 3a], the great possibilities of metal optics were demonstrated, allowing the results of measurement of the dispersion of complex dielectric constant to be used to determine a number of electronic characteristics which define the energy band structure of the electrons. In this work, combined optical and x-ray structural studies were performed of niobium nitrite films, and their superconducting properties were also tested. Two experimental tasks were performed. The first was to study the variation of complex dielectric constants $\epsilon=\epsilon_1-i\epsilon_2$ with temperature, to determine the electronic characteristics and their variation with temperature. The second was to study the variation of ϵ and electronic characteristics with composition. Based on the results of performance of these tasks, we shall study the variation in optical and superconducting properties as functions of electronic characteristics in niobium nitrite.

CHAPTER 1
EXPERIMENTS

As was noted above, we studied compounds of niobium nitrite with B1 structure (NaCl type). Figure 1 shows a unit cell of the B1 lattice. The atoms of Nb, like the atoms of N, form cubic face-centered lattices, i.e., the B1 lattice consists of two cubic face-centered lattices located one within the other. A unit cell of the B1 lattice contains eight atoms, four each of Nb and N or four NbN molecules.

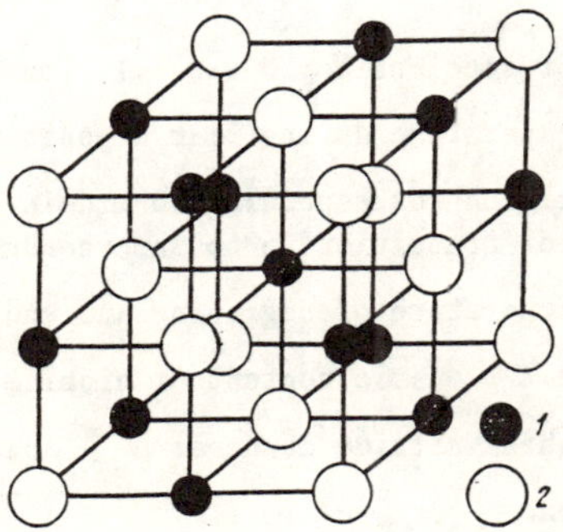

Figure 1. B1 Lattice Unit Cell (NaCl Type). 1 — Niobium atoms; 2 — nitrogen atoms.

1.1. Specimens and Their Parameters

Four specimens of niobium nitrite film with B1 structure were obtained, differing from their composition. The films were manufactured by a technology developed by Ye. A. Antonova and V. A. Sukhov[4] using their installation. The method used involved getter reactive cathodic atomization of a niobium target in a glow discharge (direct current) in an atmosphere of nitrogen and argon. Deposition of niobium nitride was performed onto heated substrates of polished quartz.

As the studies of the authors of [4] showed, niobium nitride films have a columnar structure with column growth direction perpendicular to the substrate.

Table 1
Characteristics of Specimens

Parameters	Specimens			
	1	2	3	4
T_s, °C	700	700	750	700
T_c, K	15.6	15.6	14	16
ΔT_c, K	0.5	0.5	1.8	0.3
a_0, Å	4.394±0.001	4.395±0.003	4.362±0.001	4.389±0.001
Nb, at.%	51.0±0.2	50.4±0.2	58.2±0.7	49.5±0.2
Texture	111	111	200	200
ω_p^2, eV^2	1.3	1.1	1.9	0.315

Note: Ratio R_{300}/R_{25} for specimen No. 3 was 1.1, for other specimens - 0.99. T_s - substrate temperature during film deposition; T_c - temperature of superconducting transition corresponding to middle of resistance jump; ΔT_c - temperature width of transition up to superconducting state, defined as difference between temperature of beginning and end of transition; a_0 - lattice constant; Nb, at.% - atomic content of niobium; texture - primary direction of axes of niobium nitride columns; ω_p^2 - square of plasma frecy of conduction electron.

Simultaneously with the main specimen for optical studies, two supplementary specimens were also manufactured for measurement of T_c, ΔT_c and the ratio R_{300}/R_{25} (R is the static resistance at the temperature indicated). The dimensions of the films of the main specimens were 78×20 mm^2 (specimens No. 1 and No. 2) and 60×20mm^2 (specimens No. 3 and No. 4), while the supplementary specimens measured 12×6mm^2, film thickness 4000Å The characteristics of the specimens are presented in Table 1.

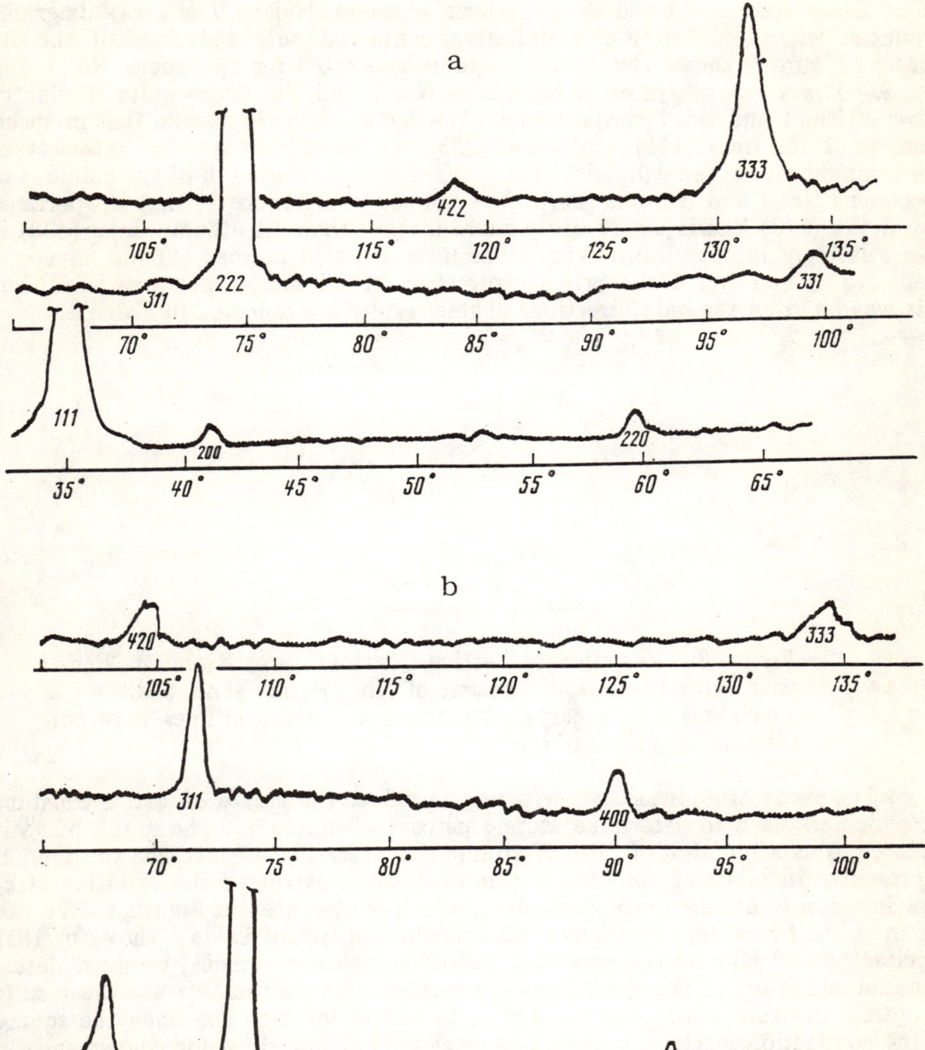

Figure 2. X-Ray Diffractogram of Niobium Nitride Specimen No. 1 (a) and No. 3 (b). Shown are indices of diffraction lines and doubled diffraction angles in degrees.

X-ray studies allowed us to perform phase analysis[1]. The x-ray diagrams produced with the DRON-2 installation contained only the lines of the B1 phase. Figure 2 shows the x-ray diagrams generated for specimens No. 1 and No. 3. The x-ray diagrams of specimens No. 2 and No. 4 are quite similar to those of No. 1 and No. 3, respectively. We can see from Figure 2*a* that in specimen No. 1 the lines (111), (222) and (333) are quite intense. The intensity of the remaining lines is negligible. Thus, the direction of growth of the columns of specimens No. 1 and No. 2 is [111]. We can see from Figure 2*b* that in specimen No. 3 the (200) line is significantly more intense than all others. Line (400) is also relatively intense. However, other lines are also present. Thus, in specimens No. 3 and No. 4 the primary direction of column growth was [100], but this was far from the only direction. These results are reflected in Table 1.

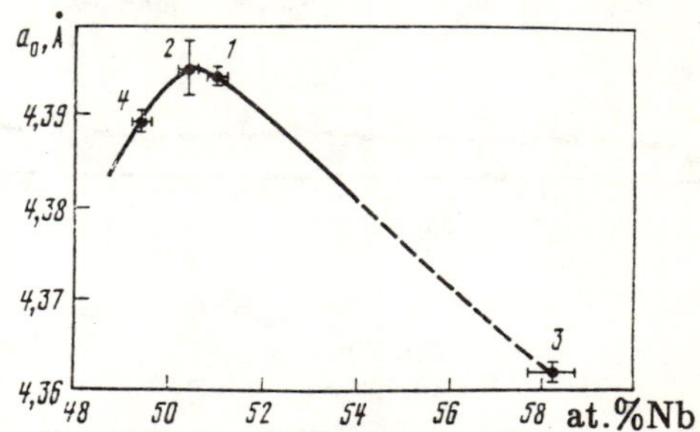

Figure 3. Variation of Lattice Constant a_0 of Niobium Nitride with Atomic Percent Content of Nb. Points show positions of specimens 1–4 on curve. Vertical and horizontal lines show corresponding values of error.

The x-ray studies also allowed us to produce the values of lattice constant a_0, which we used to determine atomic percent of niobium. The graph of variation of a_0 as a function of ratio of atomic concentrations of nitrogen to niobium is presented in [5]. Using the data of this work, we constructed the variation of a_0 as a function of atomic percent niobium, which is presented in Figure 3. We can see from the figure that this curve has a maximum, while Table 1 shows us that specimen No. 2 falls on the maximum. For the other specimens, we must determine on which leg of the curve they are located. To do this, we use the results of optical measurements, which allow us to obtain for each specimen the square of the conduction electron plasma frequency. They are given for convenience in Table 1. (Four complete electronic characteristics are presented below in Table 5.) Considering that an increase in the at.% niobium results in an increase in ω_p^2 of the conduction electrons, we see that specimens No. 1 and No. 3 must be on

[1]X-ray studies were performed by S. P. Merkulova using the same specimens which were used for the optical studies.

the right leg, specimen No. 4 — on the left leg, as is shown in Figure 3. The atomic percents of niobium for all specimens are also given in Table 1. The error in this quantity is related first of all to the error in determination of a_0, secondly to the inaccuracy of the graph of [5]. For specimen No. 3, we must use extrapolation, which increases the error. The errors in determination of a_0 and at.% Nb are presented in Table 1 and Figure 3.

1.2. Experimental Methods and Installations

At present, the most complete information concerning the complex dielectric constant $\epsilon=\epsilon_1-i\epsilon_2$ is yielded by polarimetric methods. They do not require a linear receiver-amplifier system or stable radiation source and for this reason their accuracy and reliability are greater than those of non-null methods.

We used three different versions of polarimetric methods. To measure the temperature variation of ϵ and electron characteristics we used a compensator method in the visible and near infrared areas, which is a null method; in the IR area we used the method of Beatti, which is not a null method[2].

We determined the variation of ϵ and electronic characteristics as functions of niobium content using a modulation null polarimetric method developed by G. P. Motulevich and A. A. Shubin[6]. The essence of this method is generation and detection of circular polarization of light reflected from a system of mirrors. The phase shift between the p and s components

$$\Delta=(2k+1)\frac{\pi}{2}, \text{ while } \tan\rho=R_p/R_s=\pm1,$$

k is an integer, the subscripts p and s relate to polarization of light parallel and perpendicular to the incident plane, R_p and R_s are the corresponding components of the reflected light wave electrical vector.

A schematic diagram of the installation used is presented in Figure 4. The light from source (globar, silite rod) was focused by spherical mirror L_1 on the input slit of monochromator M with an NaCl prism. After the exit slit of the monochromator, the light beam of the required wavelength was made parallel by means of two mirrors: flat mirror L_2 and cylindrical mirror L_3. The use of the additional flat mirror L_2 allowed the light beam to be directed at a slight angle

[2]These methods and the installations implementing them, allowing measurement of the optical properties of niobium nitride in the 0.155–3.1 eV spectral interval at 295 and 600 K, are described in the work of N. D. Kuz'michev, I. S. Levchenko, G. P. Motulevich, "Dispersion of Complex Dielectric Constant and Electronic Characteristics of Nb-Al and Nb–Ge Compounds with A15 Structure" (cf. this collection).

to the perpendicular to mirror L_3. This is necessary to decrease aberrations. After passing through polarizer П, the linearly polarized light struck the system of mirrors studied m_1, m_2, m_3, m_4. The incident angle of the light at all four mirrors studied was strictly the same (variation not over 2′). It was measured by a goniometer with an accuracy of 1′. The light then passed through rotating analyzer A and was focused by spherical mirror L_4 on radiation receiver B (bolometer).

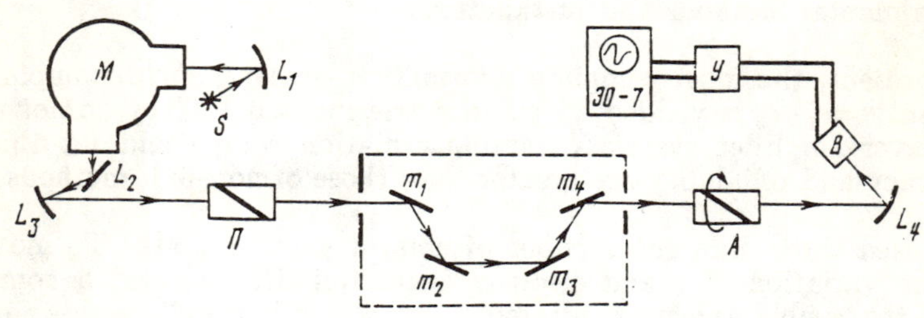

Figure 4. Schematic Diagram of Experimental Installation for Measurement of Optical Constants of Metals and Alloys by Modulation Null Polarimetric Method[3, 6].

This installation allowed us to conduct measurements in the 1–10 μm spectral interval. The measurement error of n and κ at 1.3–8.0 μm averaged 2%, increasing in the 1.0–1.3 μm and 8.0–10.0 μm intervals to 7% for n and 6% for κ.

Selenium polarizers were used to measure n and κ. The polarizers consisted of a stack of eight selenium films placed at the Bryuster angle to the incident light, achieving ≈99.8% polarization in the 0.8–10 μm area. The transmission of the polarizers was ≈40%. One of the polarizers was rotated at 4.5 Hz and used as a modulator. As a result, the bolometer received a varying light flux. The bolometer signal was amplified by a resonant amplifier. At $\Delta=(\pi/2)(2k+1)$ and $\tan\rho\cdot\tan\alpha=\pm 1$, the variable portion of the signal vanished, which was recorded visually on the oscilloscope connected to the output of the amplifier ($\tan\alpha=A_p/A_s$, A_p and A_s are the corresponding components of the incident light wave).

For good metals, all four mirrors m_1, m_2, m_3 and m_4 are identical specimens. For metals with small concentrations of conduction electrons or high relaxation frequency of these electrons the intensity of component R_p is small. In these cases, three standard mirrors of a good metal are used (in our case — gold mirrors) plus one mirror of the metal being studied.

We represent by Δ_1 and ρ_1 the values of these quantities for a single reflection of light at the fixed incident angle φ_1. Then the real and imaginary parts of the complex dielectric constant $\epsilon_1 - i\epsilon_2$ or the complex index of refraction $n - i\kappa$, where $n - i\kappa = \sqrt{\epsilon_1 - i\epsilon_2}$, are related to Δ_1 and ρ_1 by the following equations[3]:

$$\varepsilon_1 = n^2 - \varkappa^2 = n_0^2 - \varkappa_0^2 + \sin^2\varphi, \qquad \varepsilon_2 = 2n\varkappa = 2n_0\varkappa_0, \tag{1}$$

$$n_0 = \sin\varphi_1 \cdot \tan\varphi_1 \frac{\cos 2\rho_1}{1 - \sin 2\rho_1 \cdot \sin\Delta_1}, \qquad \kappa_0 = \sin\varphi_1 \cdot \tan\varphi_1 \frac{\sin 2\rho_1 \cdot \sin\Delta_1}{1 - \sin 2\rho_1 \cdot \cos\Delta_1}. \tag{2}$$

1.3. Algorithm for Computation of ϵ

The measurements were performed in two configurations: 1) with all four mirrors (m_1, m_2, m_3, m_4) standard mirrors and 2) with three standard mirrors (m_1, m_2, m_3) and a fourth (m_4) mirror consisting of a niobium nitride specimen.

The following algorithm was used to compute ϵ.

1. The frequency grid of ω_i (or wavelengths λ_i) was fixed.

2. First the data were processed relating to the four standard mirrors. For each frequency, the incident angle of the light φ_{st} and azimuth ρ_{st} were measured, corresponding to circular polarization, obtained after reflection of the light from the four standard mirrors. The phase shift Δ_{1st} and ρ_{1st} relating to the standard mirror with incident angle φ_{st} was

$$\Delta_{1st} = 90^\circ/4 = 22.5^\circ, \qquad \tan\rho_{1st} = (\tan\rho_{st})^{1/4}. \tag{3}$$

3. For the same frequency ω_i the data were processed from three standard mirrors and one niobium nitride mirror. The measured quantities φ_{exp} and ρ_{exp} were determined. They must be used to determine the contribution of the niobium nitride mirror.

---

[3]The minus sign in the denominator of equation (2) is determined by the fact that for normal incidence of light we assume $\Delta=0$.

4. For this purpose, the contribution of one standard mirror to Δ and ρ was calculated with incident angle φ_{exp} ($\varphi_{exp} \neq \varphi_{st}$). (We represent the contribution of one standard mirror to Δ and ρ with incident angle φ_{exp} as Δ'_{st} and ρ'_{st}, respectively.)

Table 2

$\epsilon_2(\omega)$ of Niobium Nitride Specimen No. 4 and T=295 and 600 K

ω, эВ	ε₂ (ω)		ω, eV	ε₂ (ω)		ω, eV	ε₂ (ω)		ω, eV	ε₂ (ω)	
	295 K	600 K		295 K	600 K		295 K	600 K		295 K	600 K
3,10	2,43	2,51	1,77	7,76	7,11	1,22	12,5	12,1	0,536	23,9	23,6
3,02	2,99	3,44	1,75	7,85	7,06	1,16	12,9	12,3	0,442	36.4	23,6
2,95	3,42	3,73	1,72	7,88	6,99	1,11	13,5	12,8	0,427	38,8	23,6
2,88	3,55	3,60	1,70	7,86	7,14	1,07	13,9	13,6	0,423	40,5	24,7
2,82	3,65	3,47	1,68	8,01	7,66	1,02	14,5	14,1	0,400	42,0	36,2
2,76	3,69	3,45	1,65	8,29	8,46	0,992	14,9	14,9	0,387	43,0	38,2
2,70	3,75	3,57	1,63	8,72	9,24	0,946	15,4	15,5	0,375	44,1	39,2
2,64	3,84	3,72	1,61	9.29	9,81	0,911	15,9	16,3	0,364	45,1	39,9
2,58	3,97	3,85	1,59	9,94	10,2	0,885	16,3	16,8	0,354	46,2	40,6
2,53	4,14	3,96	1,57	10,5	10,3	0,861	16.8	17,3	0,344	47,3	41,7
2,48	4,35	4,03	1,55	11,0	10.2	0,835	17,4	17,7	0,335	48,4	43,2
2,43	4,56	4,11	1,53	11,3	10,1	0.810	18,0	18,2	0,326	49,5	44,7
2,38	4,75	4,24	1,51	11,4	10,0	0,787	18,5	18,3	0,317	50,5	46,1
2,34	4,91	4,41	1,49	11,4	10,2	0,763	19,2	19,0	0,310	51,5	47,1
2,30	4,99	4,63	1,48	11,4	10,5	0,735	20,0	19,5	0,291	53,4	47,9
2,25	5,05	4,89	1,46	11,3	10,8	0.725	20,3	20,1	0,275	54,2	48,6
2,21	5,17	5,11	1,44	11,4	11,1	0,714	20,5	20,3	0,261	54,2	49,3
2,18	5,34	5,29	1,43	11,4	11,4	0.698	20,7	20,6	0,248	54,4	50,9
2,14	5,59	5,41	1,41	11.4	11,7	0,675	20.8	21,1	0,236	55,2	52,3
2,10	5,86	5.46	1,39	11.6	12,3	0,663	20,9	21.9	0,225	56,9	53,7
2,07	6,11	5,50	1,38	11.8	12,7	0,652	21,0	22,0	0,215	59,2	55,3
2,03	6,35	5,60	1,36	12,1	12,9	0,632	21,4	22,0	0,206	61,6	56,6
2,00	6,59	5,77	1,35	12.3	12,6	0,610	22,2	21,9	0,198	64,0	57,4
1,97	6,80	6,00	1,33	12,6	12,6	0,599	22,8	22,2	0,190	66,4	58,2
1,94	6,96	6,23	1,32	—	12,4	0,587	23,3	22,6	0,183	69,4	59,9
1,91	7,04	6,42	1,31	12.8	12,2	0,568	23,8	23,0	0,177	73,2	62,9
1,88	7,10	6.55	1,29	12.8	12,2	0,553	23,8	23,6	0.171	77,8	66,0
1,85	7,17	6,67	1,26	12,6	12.2	0,545	23,8	23,7	0,165	82,4	69,9
1,82	7,37	6,89	1,25	12,6	12,1				0,160	87,0	74,3
1,80	7,58	7,07	1,24	12,5	12,1				0,155	89,0	82,2

Note. ω is the energy of a light quantum in eV, $\epsilon_2(\omega)$ is the value of the imaginary portion of the dielectric constant at light frequency ω.

5. The contribution of the niobium nitride mirror Δ_1 and $\tan\rho_1$ is then determined by the equations

$$\Delta_1 = 90^\circ - 3\Delta'_{st}, \quad \tan\rho_1 = \frac{\tan\rho_{exp}}{\tan^3 \rho'_{st}}. \qquad (4)$$

6. The final result for ϵ_1 and ϵ_2 is obtained from equations (1) and (2).

This algorithm is discussed in more detail in [7].

Calculations were performed on a WANG-2200 computer.

CHAPTER 2
EXPERIMENTAL RESULTS

2.1. Variation of Dispersion $\epsilon(\omega)$ with Temperature

The study of the temperature variation of dispersion $\epsilon(\omega)$ of niobium nitride was performed on specimen No. 4, the characteristics of which are presented in Table 1. The specimen contained only the B1 phase. Measurements were performed on two polarimetric installations using the compensator method and the method of Beatti, at temperatures of 295 and 600 K in the spectral interval 0.155–3.1 eV (8–0.4 μm).

The smooth[4] experimental results for ϵ_1 and ϵ_2 are shown in Figures 5 and 6. Table 2 tabulates the function $\epsilon_2(\omega)=|Im\epsilon(\omega)|$ for T=295 and 600 K, which was used to determine the electronic characteristics of niobium nitride specimen No. 4.

The measurement errors were 3–6% in the 0.4–0.75 μm interval, 5–7% in the 0.75–1 μm interval, 2–3% in the 1–2.4 μm interval and 5–6% in the 2.8–8.5 μm interval. In our experiments, 117 values each were determined for ϵ_1 and ϵ_2 at 295 K, 118 each at 600 K.

We can see from Figures 5 and 6 that $|\epsilon_1(\omega, 295\text{ K})|>|\epsilon_1(\omega, 600\text{ K})|$ and $\epsilon_2(\omega, 295\text{ K})>\epsilon_2(\omega, 600\text{ K})$ with the exception of the area 1 eV and at the ends of the spectral interval. $\epsilon_2(\omega)$ is basically a decreasing function, while $\epsilon_1(\omega)$ has significant decreasing and increasing areas. The real portion of the dielectric constant vanishes twice.

---

[4]The smoothing algorithm is presented in the work of N. D. Kuz'michev, I. S. Levchenko and G. P. Motulevich mentioned above (cf. this collection).

Against the background of the general structure of the functions $\epsilon_2(\omega)$, maxima appear which can be divided into the main broad maxima and a fine structure. We shall see below that as the temperature rises, the main structure of $\epsilon_1(\omega)$ and $\epsilon_2(\omega)$ remain. The positions of the maxima of the fine structure of $\epsilon_2(\omega)$ basically shift in the direction of greater energies. The same is true for the positions of the extremes of $\epsilon_1(\omega)$.

It is significant that where $\omega<1$ eV, $\epsilon_1(\omega)$ is a large positive quantity. This indicates predominance of the contribution of interband transitions over the contribution of conduction electrons. It follows from Figure 6*a,b* that in our spectral interval there are four main interband transition zones. The first is located at ~0.26 eV, the second at ~0.74 eV, the third at ~1.3 eV and the fourth at ~2 eV. Furthermore, the behavior of the functions in the spectral area of ~0.25 eV at both temperatures indicate that presence of a fifth band at energies of ~0.15 eV. The contribution of conduction electrons to $\epsilon_1(\omega)$ and $\epsilon_2(\omega)$ is significantly less than the contribution of interband transitions. In other words, the concentration of conduction electrons in NbN is small, which agrees with data on statistical conductivity and heat capacity[5].

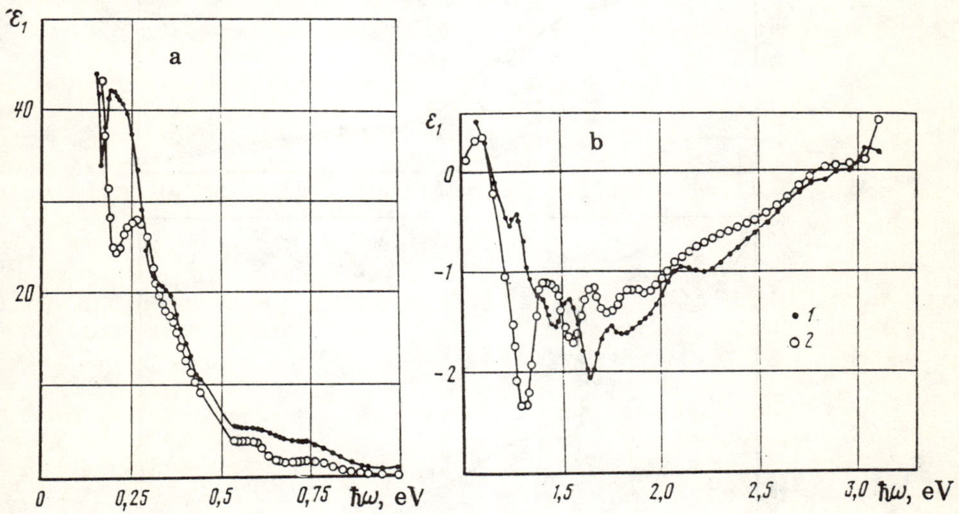

Figure 5. Real Portion $\epsilon_1(\omega)$ of Dielectric Constant of Specimen No. 4 in IR (a) and Visible (b) Bands. 1 — $T\approx295$ K; 2 — $T\approx600$ K.

2.2. Variation of Dispersion of $\epsilon(\omega)$ with Composition

The basic influence of composition on dispersion of $\epsilon(\omega)$ should be expected in the area of the band of conduction electrons and the band of interband transitions with small energy gaps Δ, i.e., the IR area of the spectrum. To solve this problem, in addition to measurement of $\epsilon(\omega)$ of specimen No. 4, presented above, we undertook measurements of $\epsilon(\omega)$ for specimens No. 1-3, the characteristics of which are presented in Table 1. Studies were performed in the 0.124–1.24 eV spectral interval (10–1 μm) at room temperature by a modulation null polarimetric method on the installation described in section 1.2. The experimental error was determined from the spread of values of φ and ρ upon repeated measurements.

The experimental results for specimens No. 1–3 are shown in Figures 7 and 8. Table 3 tabulates the function $\epsilon_2(\omega)=|\mathrm{Im}\epsilon_2(\omega)|$ of these specimens at room temperature, used to determine their electronic characteristics.

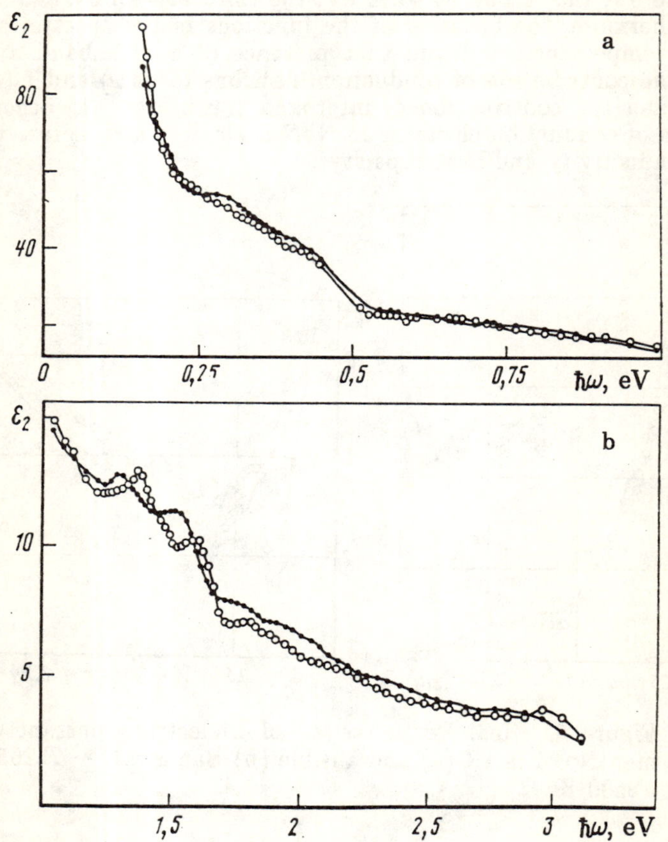

Figure 6. Imaginary Portion $\epsilon_2(\omega)$ of Dielectric Constant of Specimen No. 4 in Ir (a) and Visible (b) Bands. 1 — $T\approx 295$ K; 2 — $T\approx 600$ K.

We can see from Figures 5*a*, 6*a*, 7 and 8 that a change in composition has a significant influence on dispersion of $\epsilon_1(\omega)$ and $\epsilon_2(\omega)$ in the IR area of the spectrum. As the composition changes, the forms of the curves of $\epsilon_1(\omega)$ and $\epsilon_2(\omega)$, the positions of singularities and absolute values of ϵ_1 and ϵ_2 change, i.e., the optical properties of niobium nitride with B1 structure in the IR area are quite sensitive to the composition of the specimen. Common for all specimens is the presence of several maxima of the function $\epsilon_2(\omega)$ in the IR area, corresponding to interband transition zones. In all specimens, there are interband transitions in the area of 0.1–0.2 eV.

To estimate the number of major interband transition zones, let us study the curves of $\epsilon_2(\omega)$ and $\epsilon_1(\omega)$ for each specimen in somewhat more detail.

Table 3

Variation in Dispersion of $\epsilon_2(\omega)$ as a Function of Specimen Composition

ω, eV	ε₂ (ω)			ω, eV	ε₂ (ω)			ω, eV	ε₂ (ω)		
	Specimen №1	№ 2	№ 3		Specimen № 1	№ 2	№ 3		Specimen № 1	№ 2	№ 3
1.24	10.1	17.2	3.2	0.590	15.6	12.8	9,5	0.261	28.6	27.5	26.2
1.18	10.5	18.1	3.4	0,567	16.4	12.8	9.8	0.248	30.0	29,1	27.3
1.13	10.1	17.2	3.3	0,563	17,0	12.8	10.1	0.236	31.6	30,3	27.2
1.078	10.4	16.8	3.5	0.551	17.5	12.9	10.3	0,225	33,4	31,5	26.4
1,033	10.7	16.0	3.7	0.539	17.8	12.9	10.5	0.215	34.9	32,6	26.1
0.992	11.0	15.6	4.0	0.527	18.1	13.1	10.8	0.206	35.7	33.8	27.3
0.953	11.1	15.2	4.4	0.516	18.4	13.2	11.1	0,198	35.9	35.3	30.5
0.918	11.1	15.1	4.9	0,496	19.1	13.8	11.8	0.190	35,7	36.8	34.1
0,885	11.1	15.1	5.5	0.476	19.7	14.3	12,5	0.183	35.9	37.8	36.7
0.855	11.4	15.1	6.1	0.459	20.2	14.9	13.1	0,177	36.7	38,1	37.6
0,826	11.8	14.9	6.7	0.442	20.8	15.5	13.7	0.171	37,8	37.8	37.6
0.800	12.2	14.6	7.2	0.427	21,4	15.9	14.4	0.165	38.1	37,3	37.8
0,775	12.4	14.2	7.6	0.413	21.9	16.4	15.1	0.160	36,5	37.0	39.1
0.751	12.3	13.8	7.9	0.400	22.3	16.9	15.9	0.155	33.2	37.2	41.2
0.729	12.1	13.4	8.0	0,387	22.7	17.5	16.8	0,150	29.2	37.9	43.5
0.708	12.0	13.0	8.1	0.375	23.1	18.2	17.7	0.145	26.2	38.9	45.4
0.688	12.1	12.6	8.2	0.364	23.5	19.0	18.5	0,141	25.0	40.1	47.2
0.670	12.3	12.4	8.4	0.354	24.0	19.7	19.4	0.137	25.5	41.3	48.7
0.652	12.7	12.4	8.5	0.330	25.0	21.1	20.9	0.134	27.1	42.1	50.8
0.635	13.2	12.5	8.8	0.310	25.7	22.6	22.1	0.130	28.9	42.9	52.5
0,620	13.9	12.7	8.9	0.291	26.6	24.1	23.3	0.127	30.7	42.4	55.6
0.604	14.7	12.7	9.2	0,275	27.5	25.8	24.6	0.124	31.3	42.5	56.7

Specimen No. 1. The curve $\epsilon_2(\omega)$ in the area of 0.15–0.2 eV clearly shows two maxima, corresponding to the two interband transition zones. The curve of $\epsilon_1(\omega)$ in this area clearly shows an area of anomalous dispersion, common for both bands. Thus, we can assume the presence of two interband transition zones in the area of 0.15–0.2 eV.

In the 0.3–0.4 eV area, the shape of the curve of $\epsilon_2(\omega)$ differs significantly from the shape of the conduction electron band. On the curve of $\epsilon_1(\omega)$ we see an area of anomalous dispersion. The great width of this interval indicates the presence of two interband transition zones.

Similar arguments indicate the probability of a broad interband transition zone at 0.7–1.0 eV. Possibly it may consist of several bands, which cannot be resolved.

Specimen No. 2. At the long-wave end of the interval studied for $\epsilon_1(\omega)$ we see an area of clearly anomalous dispersion. The absolute values of $\epsilon_1>0$, meaning that there is an interband transition zone at $\omega\leqslant0.124$ eV. The curve of $\epsilon_2(\omega)$ has a clear maximum at $\omega=0.17$ eV, indicating the presence of a second band. Areas of anomalous dispersion on the curve of $\epsilon_1(\omega)$ are also seen at $\omega=0.33$ and 0.44 eV. They indicate the presence of two more interband transition zones. Finally, the curve of $\epsilon_2(\omega)$ shows a broad maximum at 1.1–1.2 eV, related to a fifth band.

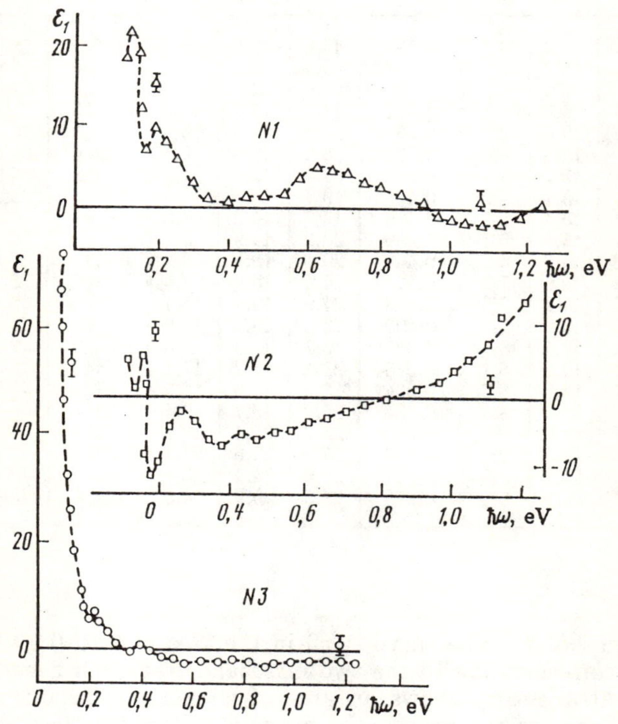

Figure 7. Experimental Curves of $\epsilon_1(\omega)$. Points show unsmoothed experimental value; dash line constructed based on smoothed values, vertical bars show experimental errors in the corresponding areas of the spectrum.

Specimen No. 3. At the long-wave end of the interval studied, ϵ_1 is a large positive value, i.e., the contribution of interband transitions is greater than that of conduction electrons. The sharp drop in ϵ_1 at this end is related to an area of anomalous dispersion. All of this indicates an interband transition zone at $\omega<0.124$ eV. A second band is detected with its maximum at $\epsilon_2(\omega)$ for $\omega=0.25$ eV. The presence of a third band can be seen from the area of anomalous dispersion at ≈0.45 eV. Finally, a fourth band is detected by the maximum on the curve of $\epsilon_2(\omega)$ in the area of 0.8 eV.

Thus, in the 0.124–1.24 eV interval, specimen No. 1 has five main interband transition zones, specimen No. 2 also has five and specimen No. 3 has four.

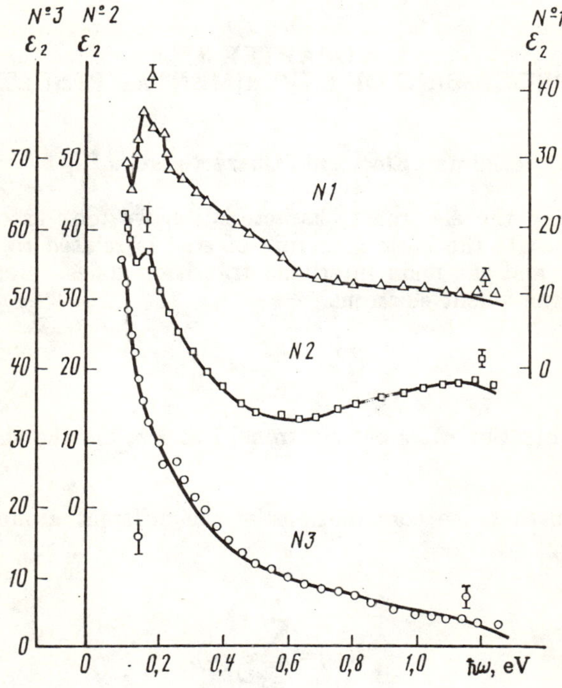

Figure 8. Experimental and Theoretical Curves of $\epsilon_2(\omega)$. Points shows smoothed experimental values; solid line shows theoretical curve. Corresponding scales on ordinate labeled with specimen numbers.

CHAPTER 3
PROCESSING OF EXPERIMENTAL RESULTS

3.1. Method of Determining Electronic Characteristics [8]

In computing the electronic characteristics, we took as our basis the fact that in the area $\omega \ll \Omega_p$ the basic structure of $\epsilon(\omega)$ is related to the band of conduction electrons and the main interband transition zones. Here Ω_p is the plasma frequency of the valent electrons.

$$\Omega_p^2 = \frac{4\pi e^2}{m} N_{\text{val}}, \tag{5}$$

N_{val} is the concentration of valent electrons, e and m are the charge and mass of a free electron.

We assume that the contributions of the different groups of electrons to $\epsilon(\omega)$ are additive, i.e.,

$$\epsilon(\omega) = \sum_{j=1}^{n} \epsilon_j(\omega), \tag{6}$$

where j is the band number, n is the total number of bands. In what follows, the subscript j=1 will be attributed to conduction electrons, other subscripts — to the main interband transition zone.

To separate the contributions of the various groups of electrons, we must know the shapes of the bands[5]. We utilize the equation of Drude for the conduction electron bands:

$$\epsilon_1(\omega)=\frac{\omega^2}{\omega(\omega-i\nu)}. \tag{7}$$

For the basic interband transition zones, we used an expression which was obtained for vertical transitions in the area of Van-Hoff singularities of combined state density of type Q_0, i.e., for "parallel" (or nearly parallel) sections of isoenergy surfaces, of which one section is located below the Fermi surface, while the other is above the Fermi surface[9-11 6].

$$\varepsilon_j(\omega)=\frac{\omega_{pj}^2}{(\omega-i\nu_j)^2}\left(\frac{\omega_{0j}}{\sqrt{\omega_{0j}^2-(\omega-i\nu_j)^2}}-1\right). \tag{8}$$

In equations (7) and (8) for electron group j, ω_{pj}^2 is the square of the plasma frequency (ω_{pj}^2/Ω_p^2 is the oscillator strength), ν_j is the relaxation frequency (or width of the corresponding band), $\omega_{0j}=\Delta_j/\hbar$, Δ_j is the energy gap ($\Delta_1=0$).

In the computations below we used the function $\epsilon_2(\omega)$, since this function contains no contribution by high-frequency interband transitions, with $\omega_0>\omega_{max}$, where ω_{max} is the upper limit of the spectral interval we used.

The calculation equations for theoretical values ϵ_2^T are [10, 11]

$$\begin{aligned}\varepsilon_{21}^T(\omega)&=\frac{\omega_{p1}^2\nu_1}{\omega(\omega^2+\nu_1^2)};\\ \varepsilon_{2j}^T(\omega)&=\frac{\omega_{pj}^2\omega_{0j}\omega\nu_j}{(\omega^2+\nu_j^2)^2}\left(\frac{\omega^2-\nu_j^2}{A_jB_j}-\frac{2B_j}{A_j}+\frac{2}{\omega_{0j}^2}\right),\end{aligned} \tag{9}$$

---

[5]In equations (7) and (8), the one related to the vacuum is omitted.

[6]The basis of this form is discussed in more detail in Chapter 1 of the work by N. D. Kuz'michev, I. S. Levchenko and G. P. Motulevich previously mentioned (cf. this collection).

$$A_j = \sqrt{Z_j^2 + 4\omega^2 \nu_j^2}, \quad B_j = \sqrt{\frac{A_j + Z_j}{2}},$$
$$Z_j = \omega_{0j}^2 - \omega^2 + \nu_j^2, \quad \hbar\omega_{0j} = \Delta_j \quad (j = 2, 3, \ldots, n). \tag{10}$$

It follows from equations (9), (10) that the conduction electron band is described by two parameters: ω_{p1} and v_1, while each band of interband transition j is described by three parameters ω_{pj}^2, v_j and ω_{0j}. The number of interband transition zones is found from the number of main broad maxima in $\epsilon_2(\omega)$.

The parameters of the bands are found by the method of least squares[8]. The function F is minimized:

$$F = \sum_{k=1}^{N} \left[\varepsilon_2(\omega_k) - \varepsilon_2^T(\omega_k)\right]^2, \tag{11}$$

$$\varepsilon_2^T(\omega_k) = \sum_{j=1}^{n} \varepsilon_{2j}^T(\omega_k). \tag{12}$$

The sum with respect to k is taken for all experimental points. N is the total number of experimental points; $\epsilon_2(\omega_k)$ are the corresponding experimental values of ϵ_2. The sum for j is taken for all bands, including the conduction electron band (j=1) and all interband transition zones (j=2, 3, 4, ...,n); n is the total number of bands.

The algorithm of the search for the minimum is described in detail in [8], where it is implemented in FORTRAN 4+. All computations were performed by computer[7].

The basic error in the values obtained for the parameters is related to the accuracy of the model, which considered only terms of the first order of magnitude with respect to $\Delta_0/2E_F$ (E_F is the Fermi energy). The computation error is significantly less. Below, in Tables 4 and 5, we estimate the relative error of the parameters of band j by the equation $\Delta_j/2E_F$.

We note that the cubic unit cell of niobium nitride contains 40 valent electrons. With lattice constant a_0=4.39Å, we obtain Ω_p=25.6 eV. Thus, $\omega_{max} \ll \Omega_p$, where ω_{max} is the upper boundary of the spectral interval used.

---

[7]In this work for $\epsilon_{2j}^T(\omega)$ we used precise equations (10) instead of the approximate equations used in [8].

3.2. Results of Processing Dispersion of Imaginary Part of Complex Dielectric Constant of Specimen No. 4

Specimen No. 4 was used to study the variation of electronic characteristics with temperature. Above, in Section 2.1, we noted that in the spectral interval 0.155–3.1 eV, the dispersion of $\epsilon_2(\omega)$ indicates the presence of four interband transition zones. Furthermore, the behavior of function $\epsilon_1(\omega)$ at $\leqslant 0.26$ eV indicates a fifth band in the frequency area below the lower boundary of the spectral interval. Thus, the total number of bands is 6, and the theoretical model of this specimen contains 6 terms:

$$\epsilon_2^T(\omega)=\sum_{j=1}^{6}\epsilon_{2j}^T(\omega).$$

The results of processing the experimental data at 295 and 600 K are presented in Table 4.

We can see from Table 4 that in specimen No. 4 there are two interband transition zones with small gaps ~0.13 and 0.27 eV. We further see that ω_p^2 for conduction electrons is a small quantity and where T=295 K $\omega_p^2/\Omega_p^2=4.8\cdot 10^{-4}$(0.048%). The results of processing show a strong temperature variation of ω_p^2 for conduction electrons. Parameter v_1 increases very little with increasing temperature. For the second band as temperature rises parameter Δ_2 increases slightly, while v_2 decreases.

Let us study the remaining bands. As the temperature rises, the quantities v_3 and v_4, as would be expected, increase. ω_{p3}^2 and ω_{p4}^2 also increase. The temperature variations of Δ_3 and Δ_4 have different signs. The insensitivity of the sum F to values of the parameters of the fifth and sixth bands prevents us from determining their temperature variation.

We shall return to the discussion of temperature variation of electronic characteristics in Chapter 4.

Table 4
Electronic Characteristics of Specimen No. 4 in Spectral Interval 0.155–3.1 eV at 295 and 600 K

Band Number	Δ, eV	ν, eV	ω_p^2, eV^2
		T=295 K	
1	0	0.32±0.02	0.31±0.02
2	0.134±0.006	0.027	1.31±0.02
3	0.271±0.014	0.150±0.06	6.9
4	0.74±0.04	0.214±0.008	7.1±0.3
5	1.28±0.06	0.27±0.02	11.6±0.7
6	2.1±0.2	0.90±0.09	18±2
		T=600 K	
1	0	0.33±0.02	0.76
2	0.142±0.007	0.008	0.86±0.02
3	0.249±0.012	0.174±0.06	7.2
4	0.74±0.04	0.217±0.008	7.7±0.3
5	1.28±0.06	0.26±0.02	10.3±0.7
6	2.1±0.1	0.90±0.09	17±2

Note: If the error is not indicated it is less than one in the last digit.

3.3. Results of Processing Dispersion of Imaginary Portion of Complex Dielectric Constant of Specimens No. 1–3

Specimens No. 1–3 were used to study the variation in electronic characteristics with composition. It was noted in Section 2.2 that in the 0.124–1.24 eV spectral interval, the dispersion of $\epsilon_2(\omega)$ indicates the presence of five, five and four interband transition zones in specimens No. 1, 2 and 3. Considering the conductivity band, we obtain for the theoretical models

$$\epsilon_2^T(\omega) = \sum_{j=1}^{n} \epsilon_{2j}^T(\omega)$$

the values of n are equal to six, six and five.

The parameters of the conduction electron bands and interband transition zones obtained by processing the dispersion of $\epsilon_2(\omega)$ in the 0.124–1.24 eV spectral interval for specimens No. 1–3 are presented in Table 5. For convenience, the same table shows the parameters of the bands of the fourth specimen for the same spectral interval.

Let us compare the results of the various specimens. We can see from Table 5 that the conduction electron bands for the different specimens have approximately the same values of v_1 and strongly differing values of ω_p^2, which is related to the different contents of niobium in the specimen.

The interband transition spectrum for all specimens is approximately identical. The interband transition zones can be divided into three groups. The first group of bands is located in the area of 0.1–0.2 eV, the second — in the area of 0.3–0.4 eV and the third — in the area of 0.7–1.0 eV.

The small gaps in all specimens have small $v \sim 0.01$–0.02 eV, thanks to which they are observed even when the corresponding values of ω_p^2 are small (~ 0.1–0.2 eV^2).

The first group of bands for specimens No. 1 and No. 2 consists of two bands, for specimens No. 3 and No. 4 — of one band. This difference, as well as the difference in the corresponding gaps, may be related to texture: for specimens No. 1 and No. 2 the texture is [111], for specimens No. 3 and No. 4 — [200].

The bands in the second group have approximately identical Δ, but differ in values of ω_p^2. Specimens No. 3 and No. 4 have greater values of ω_p^2 than do specimen No. 1 and No. 2.

The bands of the third group differ in their values of Δ and ω_p^2. For specimens No. 1 and No. 2 these quantities are significantly greater than for specimens No. 3 and No. 4. These differences may also be related to texture.

For all specimens studied $\sum \omega_p^2 \ll \Omega_p^2$ (for niobium nitride $\Omega_p^2 \approx 660$ eV^2), i.e., the most significant interband transition zones are located in the spectral area $\omega > 1.2$ eV.

Our discussion of the variation in electronic characteristics with composition will be continued in Chapter 4.

Table 5
Electronic Characteristics of Specimens No. 1–4 in Spectral Interval 0.124–1.24 eV at Room Temperature

Specimen	J	Δ, eV	v, eV	ω_p^2, eV^2	Specimen	J	Δ, eV	v, eV	ω_p^2, eV^2
	1	0	0.39	1.3		1	0	0.40	1.9
	2	0.16	0.008	0.15		2	0.11	0.02	0.18
No. 1	3	0.2	0.03	0.4	No. 3	3	0.25	0.19	1.9
	4	0.3	0.09	1.1		4	0.40	0.30	0.29
	5	0.4	0.15	0.3		5	0.74	0.40	3.8
	6	1.0	0.33	12.5					
						1	0	0.325	0.315
	1	0	0.40	1.1	No. 4	2	0.134	0.027	1.3
	2	0.123	0.01	0.14		3	0.27	0.15	6.9
No. 2	3	0.17	0.05	0.6		4	0.74	0.214	7.1
	4	0.27	0.10	0.9					
	5	0.38	0.20	0.6					
	6	1.0	0.5	34					

CHAPTER 4
DISCUSSION OF RESULTS OBTAINED

The parameters of the bands presented in Tables 4 and 5 allowed us to use equations (9), (10) and (12) to compute the theoretical values of $\epsilon_2^T(\omega)$. Comparison of the experimental curves of $\epsilon(\omega)$ with the theoretical curves of $\epsilon_2^T(\omega)$ at room temperature for all specimens is presented in Figure 8 (cf. above) and Figure 9. We can see from these figures that for specimens No. 1 and No. 2 the theoretical curves describe the dispersion of $\epsilon_2(\omega)$ quite well throughout the 0.124–1.24 eV spectral interval. For specimens No. 3 and 4 the theoretical curve describes the main structure of the dispersion of $\epsilon_2(\omega)$ well, but does not describe the fine structure. Therefore, the main interband transition zones are well described by equations (10). The fine structure cannot be described by equations (10). The method of least square averages it. Possibly, it is related to interband transitions in the area of intersection of several Bragg planes.

The data of Table 4 show the strong temperature variation of the square of plasma frequency ω_{p1}^2. As the temperature increases from 296–600 K, ω_{p1}^2 increases by a factor of 2.4. This strong increase is not observed in pure metals or in compounds with A15 structure. The reason may be thermal transfer of electrons from the base state of the second band through the corresponding energy gap into the conduction band. Thus, band 1 receives a contribution both from the conduction electrons themselves, retained at $T \cong 0$, and from a portion of the electrons of band 2, transferred through gap Δ. The relaxation frequency v_1 remains practically unchanged with increasing temperature.

Let us study the parameters of the second band. Thermal transfer of electrons from the second band into the conduction band increases ω_{p2}^2. The sum

$\omega^2_{p1}+\omega^2_{p2}$ at T=295 K and 600 K is 1.625 and 1.614, which is a constant quantity with an accuracy equivalent to the experimental error. The parameter v_2, representing the width of the interband transition, decreases significantly. As temperature doubles, v_2 decreases by a factor of 3.4.

The sum of the squares of plasma frequencies $\sum_{j=1}^{6} \omega^2_{pj}$ for T=295 K and 600 K is 45.7 and 43.7 eV^2, i.e., remains practically unchanged.

Let us return to the variation of ω^2_{p1} as a function of specimen composition. Based on the data of Tables 1 and 5, we have constructed the variation of ω^2_p as a function of atomic percent Nb (Figure 10). This figure indicates that ω^2_p varies strongly with atomic percent niobium. As the content of niobium increases from 49.5 at.% to 58.2 at.%, ω^2_p rises monotonically from 0.32 to 1.9 eV^2. The growth rate is greatest at small values of ω^2_p, decreasing as atomic percent noibium increases. In spite of the strong variation with composition, ω^2_{p1} remains a small quantity.

Let us compare the concentration of conduction electrons in the B1 structure niobium nitride compound we studied with the corresponding values for compounds with A15 structure and pure niobium. For a film of stoichiometric composition, the conduction electron band has ω^2_{p1}=0.71 ev^2 (cf. Figure 12). The corresponding effective concentration of conduction electrons N_e=$4.4\cdot10^{20}$ cm^{-3}, which is 100 times less than in pure niobium[12] and 30 times less than in compounds with the A15 structure[10].

We know that the minimum value obtained by us ω^2_p=0.32 eV^2 corresponds to an effective conduction electron concentration of $2.3\cdot10^{20}cm^{-3}$.

It also follows from Tables 1 and 5 that the change in ω^2_p of conduction electrons is related primarily to the composition of the specimen, not its texture. Actually, specimens No. 3 and No. 4, which have the greatest differences in value of ω^2_p, have the greatest difference in atomic percent niobium, but identical texture (preferential direction of column axes [200]).

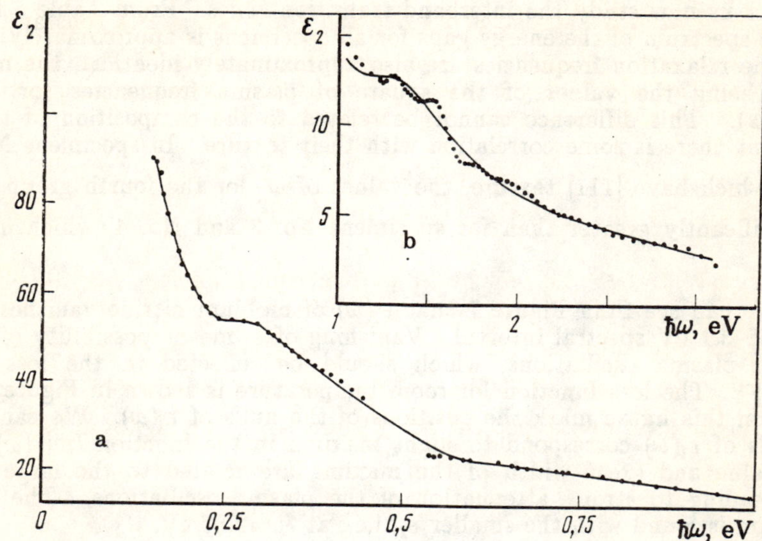

Figure 9. Comparison of Experimental Curve of $\epsilon_2(\omega)$ (points) with Theoretical $\epsilon_2^T(\omega)$ (solid line) for Specimen No. 4 in IR (a) and Visible (b) Areas of the Spectrum. T=295 K.

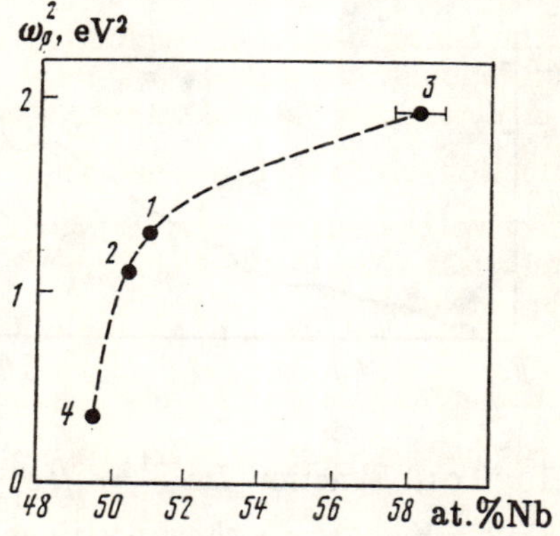

Figure 10. Variation in Square of Plasma Frequency of Conduction Electrons with Atomic Percent Niobium, in Niobium Nitride Compound. Error in atomic percent niobium for specimen No. 3 indicated by horizontal line, for all other specimens error less than size of point.

Let us now study the interband transition zones. From Table 5 it follows that the spectrum of the energy gaps for all specimens is approximately the same, the relaxation frequencies are also approximately identical, the major difference being the values of the square of plasma frequencies (or oscillator strengths). This difference cannot be related to the composition of the specimens, but there is some correlation with their texture. In specimens No. 1 and No. 2, which have [111] texture, the values of ω_p^2 for the fourth group of bands are significantly greater than for specimens No. 3 and No. 4, which have [200] texture.

We can see from Figure 7 that $\epsilon_1(\omega)$ of niobium nitride vanishes twice in the 0.155–3.1 eV spectral interval. Vanishing of ϵ_1 means possibility of development of plasma oscillations, which should be reflected in the loss function $\mathrm{Im}[\epsilon(\omega)]^{-1}$. The loss function for room temperature is shown in Figure 11. The arrows on this figure mark the positions of the nulls of $\epsilon_1(\omega)$. We can see that the nulls of $\epsilon_1(\omega)$ correspond to slight maxima in the function $\mathrm{Im}[\epsilon(\omega)]^{-1}$. The small value and great width of the maxima are related to the large value of $\epsilon_2(\omega)$, leading to strong attenuation of the plasma oscillations. The strongest maximum is found with the smaller ϵ_2, i.e., at $\hbar\omega$=2.87 eV.

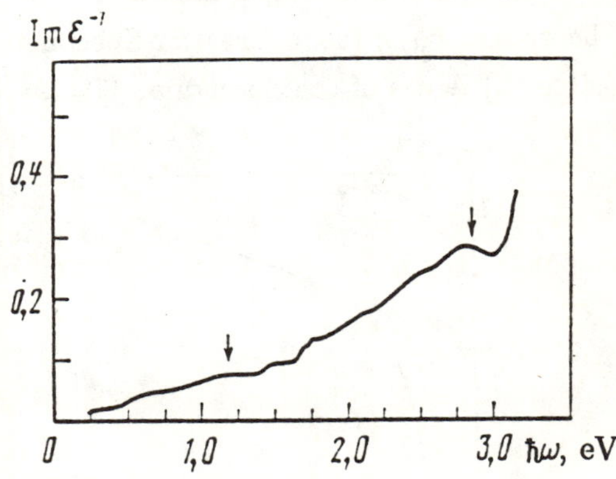

Figure 11. Loss Function $\mathrm{Im}^{\epsilon-1}=\epsilon_2/(\epsilon_1^2\_\epsilon_2^2)$ on Niobium Nitride at T=295 K. Arrows show positions of nulls in $\epsilon_1(\omega)$.

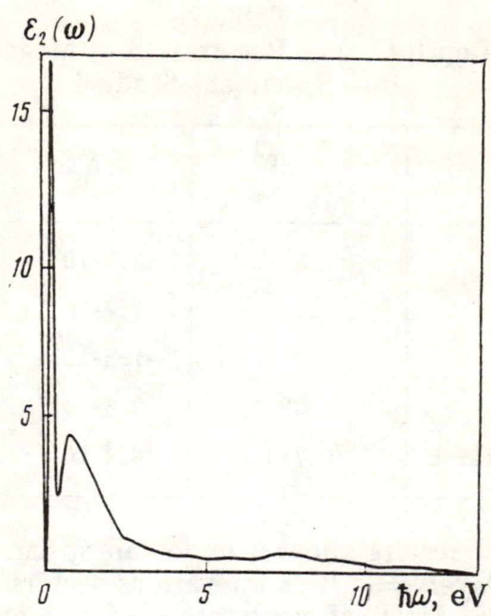

Figure 12. Variation of $\epsilon_2(\omega)$ Calculated by Empirical Pseudopotential Method of Fong and Cohen[13].

Theoretical computations of $\epsilon_2(\omega)$ for direct interband transitions of niobium nitride using the empirical pseudopotential method were undertaken in [13]. The result is shown in Figure 12. The figure shows two main maxima. In the text their positions are given as ~0.1 and ~0.6 eV (on the figure the positions of the maxima are ~0.2 and 0.8 eV). In addition to these maxima, the figures also show maxima at ~1.5, ~4 and ~7 eV. The data of Table 4 show that the positions of some bands computed in [13] do not differ greatly from the experimental positions.

Table 6
Depth of Skin Layer of Specimens Studied in IR Band

λ, μκ	δ, Å			
	Specimen № 1	№ 2	№ 3	№ 4
1	700	800	750	600
3	1500	1700	1200	1300
5	2300	2300	2100	1800
7	3200	3200	2200	2300
9	4200	4300	3400	—

Table 7
London Depth (λ_L) of Penetration of Magnetic Field Into Specimens Studied

Specimen	ω_p^2, eV^2	N_e,cm^{-2}	λ_L, Å
No. 1	1.3	$8.9 \cdot 10^{20}$	1800
No. 2	1.1	$7.7 \cdot 10^{20}$	1900
No. 3	1.9	$1.3 \cdot 10^{21}$	1500
No. 4	0.32	$2.3 \cdot 10^{20}$	3800
Stoichiometric Composition	0.71	$4.4 \cdot 10^{20}$	2500

The experimental results allowed us for all specimens studied to estimate the depth of the skin layer $\delta = \lambda_0 / 2\pi\kappa$ for various sectors of the spectral interval, as well as the London depth of penetration of the magnetic field $\lambda_L = (mc^2/4\pi e^2 N_e)^{1/2}$. Here λ_0 is the length of a light wave in a vacuum, κ is the imaginary portion of the complex index of refraction, N_e is the concentration of conduction electrons, e and m are the charge and mass of a free electron, c is the speed of light in a vacuum. These quantities are presented in Tables 6 and 7.

It follows from Table 6 that in the IR area of the spectrum $a_0 \ll \delta \lesssim d$ (a_0 is the lattice constant, d is the thickness of the specimens; d=4000Å). However, for niobium nitride in this case the influence of light reflected from the film–quartz boundary can be ignored. As the wavelength of light increases, δ increases. No clear variation with composition is observed.

It follows from Table 7 that the London depth of penetration of a magnetic field λ_L for niobium nitride films is anomalously great, which is related to the small concentration of conduction electrons. For niobium nitride specimens with atomic percent niobium 49.5–58.2, the value of λ_L is in the 3800–1500Å interval. This high value of λ_L must be considered when experiments are performed in magnetic fields.

CONCLUSIONS

Optical studies of four niobium nitride specimens with A15 structure and different composition allows us to draw the following conclusions.

1. The structure of dispersion variations of dielectric constant $\epsilon(\omega)$ indicates the presence of contributions to $\epsilon(\omega)$ both by intraband transitions (conduction electron band) and by interband transitions (interband transition zones).

The presence of singularities in the graph of $\epsilon_2(\omega)=\text{Im}\epsilon(\omega)$ indicates that the interband transition spectrum in the 0.11–1.2 eV spectral interval for all four specimens is approximately the same. The interband transition zones can be divided into three groups located at 0.1–0.2, 0.3–0.4 and 0.7–1.0 eV (cf. Table 5). In addition to these bands, measurement of the dispersion $\epsilon_2(\omega)$ of specimen No. 4, performed over a broader interval up to 3.1 eV, shows the presence of two more bands in the 1.3 and 2.1 eV areas (cf. Table 4).

We should particularly note the long-wave interband transition zones which correspond to small energy Δ~0.1–0.2 eV. The minimum energy Δ=0.11 eV is found in specimen No. 3 with maximum niobium content and 200 texture.

2. The shape of the basic interband transition zones can be described by equations (10), obtained for type Q_0 Van–Hoff singularities of combined state density, i.e., for vertical transitions between parallel (or near parallel) sectors (sheets) of isoenergy surfaces in k space (k is the wave number), of which one sector is located below the Fermi surface, while the other is above it.

Against the background of the structure related to the main interband transitions there is a weaker and finer structure, the amplitude of which represents <10% of the amplitude of the main structure. This additional structure is not described by equations (10). It is apparently related to other types of singularities, for examples lines or points of intersection of several Bragg planes. (See the comparison of experimental and theoretical curves of $\epsilon_2(\omega)$ in Figures 11 and 12.)

3. The plasma frequency of conduction electrons ω_{p1} of all the specimens studied is small. For the stoichiometric composition ω_{p1}^2=0.71 eV^2 (cf. Figure 10), corresponding to the effective concentration N_e=4.4·10^{20}cm^{-3}, 100 times less than in pure niobium[12], 30 times less than in compounds with A15 structure[10]. The oscillator strength ω_p^2/Ω_p^2=1.1·10^{-3}, i.e., the response of the conduction electron to the electromagnetic field of a light wave is small. However, we cannot answer the question whether this is related to the low concentration of conduction electrons or to their greater mass without additional data.

4. There is a strong variation of plasma frequency of conduction electrons with specimen composition. As the content of niobium increases, ω_{p1}^2 increases monotonically. The growth rate is greatest at small values of ω_p^2 and decreases with increasing atomic percent niobium (cf. Figure 10). The texture does not influence ω_{p1}^2.

5. The measurements of dispersion of $\epsilon_2(\omega)$ of specimen No. 4 at temperatures 295 and 600 K reveal the number of effects in the temperature variation of electronic characteristics of the conduction electron band and the long-wave interband transition zone (band 2). As the temperature rises from 295 K to 600 K, we observed: a) anomalously large changes in ω_{p1}^2 (increase by 2.4 times) and ω_{p2}^2 (decrease by 1.5 times) while conserving the sum $\omega_{p1}^2+\omega_{p2}^2$; b) a great reduction in relaxation frequency v_2 (3.5 times); c) a slight increase in energy gap Δ_2 (cf. Table 4). All of these effects are well explained by the hypothesis of thermal transfer of electrons from the base state of the second band through a small energy gap into the conduction band. The thermal transfer of conduction electrons through a small energy gap can lead to anomalies in the temperature variation of other static characteristics.

6. The sum of the squares of plasma frequencies for specimen No. 4, which was studied in the interval of 0.155–3.1 eV, $\sum_j \omega_{pj}^2$ is practically independent of temperature. The value of $\sum_j \omega_{pj}^2 \ll \Omega_p^2$, indicating that there are major interband transition zones where $\omega>3.1$ eV.

7. The depth of the skin layer δ in the IR area of the spectrum is great. Where $\lambda=7$ μm for our specimens it is 2200–3200Å. Explicit variation with composition was not observed (cf. Table 6).

The London depth λ_L of penetration of a magnetic field into the niobium nitride film is also great and depends on the content of niobium in the specimen. For the stoichiometric composition $\lambda_L\approx 2500$Å (cf. Table 7). The greater values of δ and λ_L are related to smaller concentration of conduction electrons. Estimates of δ and λ_L must be considered in organizing the corresponding experiments.

8. In niobium nitride compounds, high values of superconducting transition temperature T_c are combined with small values of the square of conductivity electron plasma frequency ω_{p1}^2.

Comparison of ω_{p1}^2 with the superconducting characteristics T_c and ΔT_c shows that as ω_{p1}^2 decreases, T_c increases, while ΔT_c drops (cf. Table 1). A similar correlation has been observed for other superconductors[14-15]. It follows from this that in order to increase T_c of niobium nitride, it is desirable to decrease ω_{p1}^2, by approaching the electron localization boundary.

REFERENCES

1. Kuz'michev, N. D., Motulevich, G. P., "Determination of Electronic Characteristics of Niobium Nitride by Optical Methods," ZhETF, Vol. 84, pp. 2316–2322, 1983.

2. Dzhurayev, D. R., Motulevich, G. P., "Optical Properties and Electronic Characteristics of Niobium Nitride Films with B1 Structure," FTT, Vol. 27, pp. 2640–2650, 1985.

3. Motulevich, G. P., "Optical Properties of Polyvalent Nontransition Metals," UFN, Vol. 97, pp. 211–256, 1969.

3a. Motulevich, G. P., "Optical Properties of Nontransition Metals," Trudy FIAN, Moscow, Nauka Press, Vol. 55, pp. 3–150, 1971.

4. Antonova, Ye. A., Sukhov, V. A., "Relationship Between Structure and Electrophysical Properties of Niobium Nitride Films," Fizika nizkikh temperature, Vol. 7, pp. 1002–1009, 1981.

5. Tot, L., Karbidy i nitridy perekhodnykh metallov [Transition Metal Carbides and Nitrides], Moscow, Mir Press, 294 pp., 1974.

6. Motulevich, G. P., Shubin, A. A., "Determination of Optical Constants of Metals in the Infrared Band," Optika i spektroskopiya, Vol. 2, pp. 633–636, 1957.

7. Dzhurayev, D. R., "Optical Properties of Niobium Nitride Films in the 1–10 μm Spectral Range," Preprint FIAN, No. 26, Moscow, 23 pp., 1985.

8. Motulevich, G. P., "Processing of Metal–Optic Experimental Results by Nonlinear Method of Least Squares," Preprint FIAN SSSR, No. 148, Moscow, 33 pp., 1980.

9. Golovashkin, A. I., Motulevich, G. P., "Influence of Bragg Interband Transitions on Dielectric Constant of Metals," ZhETF, Vol. 57, pp. 1054–1063, 1969.

10. Kuz'michev, N. D., Levchenko, I. S., Motulevich, G. P., "Determination of Electronic Characteristics of Nb–Al with A15 Lattice by Optical Method," Fizika metallov i metallovedeniye, Vol. 56, pp. 266–270, 1983.

11. Kuz'michev, N. D., "Optical Properties of a Number of Niobium Compounds with A15 and B1 Lattices," Dissertation, Candidate of Physical–Mathematical Sciences, Moscow, FIAN, 142 pp., 1983.

12. Golovashkin, A. I., Leksina, I. Ye., Motulevich, G. P., Shubin, A. A., "Optical Properties of Niobium," ZhETF, Vol. 56, pp. 51–64, 1968.

13. Fong, C. Y., Cohen, M., "Pseudopotential Calculations of the Electronic Niobium Nitride," Phys. Rev. B., Vol. 6, pp. 3633–3637, 1972.

14. Beasley, M. R., Geballe, T. H., "Superconducting Materials," Phys. Today, Vol. 37, pp. 60–68, 1984. Same, UFN,Vol. 148, pp. 347–362, 1986.

15. Golovashkin, A. I., Levchenko, I. S., Motulevich, G. P., "Correlation of Critical Temperature of Superconductors with Their Electronic and Phonon Characteristics," FTT, Vol. 23, pp. 3061–3064, 1978.

TEMPERATURE VARIATION OF CRITICAL MAGNETIC FIELD AND ELECTRONIC CHARACTERISTICS OF SUPERCONDUCTING ALLOYS WITH A15 STRUCTURE

A. I. Golovashkin, Ye. V. Pechen', N. P. Shabanova

ABSTRACT

Measurements of temperature variation of the upper critical magnetic field H_{c2} are performed for Nb_3Sn films with varying degrees of disorder. An anomalous variation of d_{Hc2}/dT (where $T \approx T_c$) as a function of film resistivity ρ is discussed. Electronic characteristics and parameters of electron-phonon interaction of specimens of Nb_3Sn and Nb_3Ge with varying degrees of disorder are determined. A decrease in electron state density at the Fermi level by a factor of four is found for Nb_3Sn, by a factor of two for Nb_3Ge, as ρ increases to 150μOhms·cm. The results are explained by means of a model of electron structure with a peak of state density. Figures 9, Tables 3, References 53.

1. Introduction

A number of unusual phenomena are observed in high-temperature superconducting compounds with the A15 lattice: strong degradation of the transition temperature to the superconducting state T_c upon bombardment with energetic particles[1, 2], anomalous temperature variation of resistivity ρ [3], strong temperature variation of magnetic susceptibility[3, 4]. The corresponding characteristics depend strongly on electron state density at the Fermi level $N(0)$ and electron-phonon interaction. The anomalous properties of type A15 superconductors are related to the presence of a peak of electron state density $N(E)$ near the Fermi energy E_F and indicate its "spreading" with temperature or when defects are introduced[5].

In order to explain the specific causes of the phenomena observed and their relationship to high critical parameters in such superconductors, we must investigate the properties of specimens with various degrees of structural disorder. Studies of the electronic characteristics of Nb_3Ge [6-8] have shown that they actually depend on the degree of disorder, which can be characterized, for example, by the mean free path of electrons l. When l changes in Nb_3Ge, such characteristics as $N(0)$, the concentration of conduction electrons N, and speed of electrons at the Fermi surface v_F also change, though these characteristics in ordinary metals are practically independent of both temperature and electron free path length, as long as it is significantly greater than the interatomic distance. The fact of variation of these electronic characteristics with the degree of disorder should lead to an anomalous variation of these characteristics with temperature.

In this article, we study the properties of films of the superconducting compound Nb_3Sn with various degrees of order. We investigated the critical magnetic fields of specimens as a function of their resistivity. Based on these data, we obtained information on the electronic structure of Nb_3Sn. Some additional data were also produced for Nb_3Ge.

2. Production of Specimens and Their Parameters

Varying degrees of crystalline lattice perfection can be obtained in specimens by varying their conditions of preparation. In our case, in order to produce Nb_3Sn films with various degrees of ordering, we changed the substrate temperature during deposition and the film annealing temperature.

The Nb_3Sn films were synthesized by simultaneous evaporation of their components in a vacuum of 10^{-6}–10^{-7} Torr and precipitation on sapphire on substrates. To form the ordered A15 phase, precipitation and annealing were performed at 900°C (series no. 3 and 10). To produce a disordered A15 structure, we used substrate temperatures below 600°C (series No. 12). Subsequently, after measurements were performed, some films in series No. 12 were annealed at temperatures from 600–800°C and measurements were repeated.

Films of Nb_3Ge were prepared by dc cathode sputtering. The method of producing such films is described in detail in [8]. Specimens with various degrees of ordering were obtained by varying the discharge parameters, substrate temperature and argon pressure in the chamber. The basic parameters of the Nb_3Ge films were published in [6-8].

The parameters of the Nb_3Sn specimens studied in series 3, 10 and 12 are presented in Table 1. Here we see the values of film thickness D; resistivity of ρ_n at temperature T slightly above the superconducting transition temperature T_c; the ratio of resistance at room temperature $\rho(300)$ to ρ_n; T_c; and the widths of the temperature interval of transition to the superconducting state ΔT_c. The table also shows the slope of the linear portion of the variation of upper critical magnetic field H_{c2} as a function of temperature $-dH_{c2}/dT$.

Table 1
Parameters of Nb_3Sn Films

Parameters	Specimens								
	10D	10	3E	12G	12D	12Z	12Zh	12E	12GG*
D, μm[1]	0.32	0.32	0.45	0.52	0.47	0.39	0.32	0.29	0.52
$\rho(300)/\rho_n$	4.2	3.7	3.7	1.0	1.0	1.0	1.0	1.0	1.2
ρ_n, μOhm·cm	20±2	23±2	23±2	215±22	206±22	203±20	197±20	198±20	147±15
T_c, K[2]	17.8	17.8	17.2	3.8	4.2	3.1	2.4	2.4	5.2
ΔT_c, K	0.07	0.12	0.19	1.00	0.60	0.55	---	---	0.75
H'_{c2}, kOe/K[3]	22	23	21	24	27	22	23	24	19

* Specimen 12 G1 was produced by annealing specimen 12G in a vacuum at 600-650°C for five minutes. [1] Error ±0.03; [2] error ±0.1; [3] error for specimens 12Z and 12Zh ±2, for others ±1.

3. Methods and Results of Measurement

Measurements of the temperature variation of upper critical magnetic field H_{c2} of the superconducting films were performed in a magnetic field parallel $H_{\parallel}$ and perpendicular $H_{\perp}$ to the surface of the films. The temperature was measured with an Allen-Bradley thermometer and by measuring the pressure of helium vapor.

Figures 1 and 2 show characteristic results of measurement of H_{c2} as a function of temperature T for Nb_3Sn specimens in series 10 and 12. Curves are shown for the beginning, middle and end of the superconducting transition, determined by the $0.05R_n$, $0.5R_n$ and $0.95R_n$ levels, respectively (R_n is the resistance in the normal state near the transition). Data for Nb_3Ge are presented in [6, 7].

The resistivity of films ρ was determined by a four-contact measurement method. Thicknes D was measured on a type MII-4 interference microscope.

Critical temperature T_c was determined from the middle of the resistive transition. The width of the transition ΔT_c corresponds to the temperature interval determined from the beginning and end of the transition. In Table 1, $H'_{c2}=-dH_{c2}/dT$ where $T=T_c$ is shown for $H_\perp$.

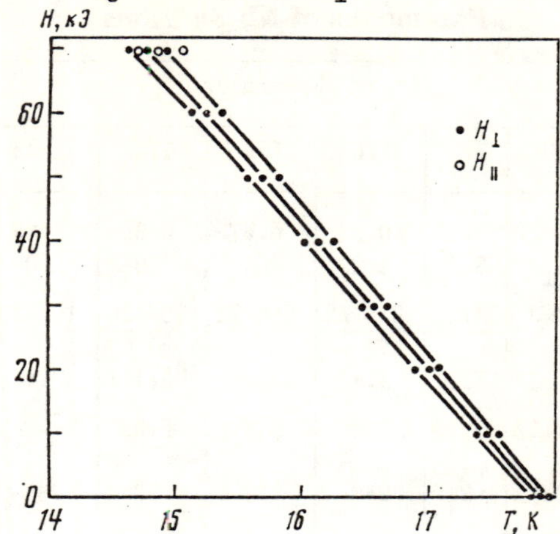

Figure 1. Variation of H_{c2} with Temperature for Nb_3Sn Specimen 10E. Resistive transitions measured with fixed field.

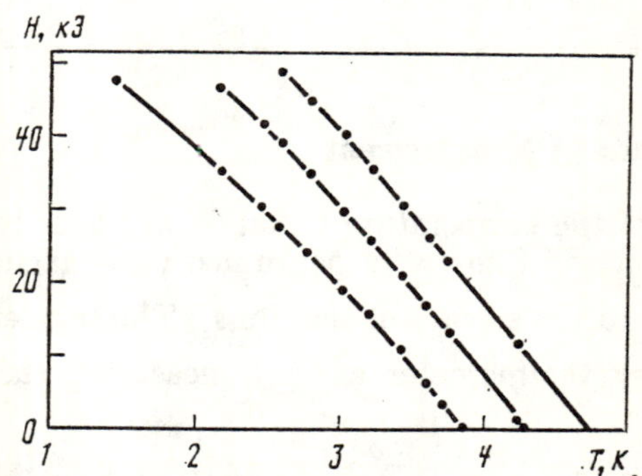

Figure 2. Temperature Variation of H_{c2} for Nb_3Sn Specimen 12D. resistiive transitions measured at fixed temperatures in $H_{||}$

Figures 3 and 4 show H'_{c2} and T_c for our Nb_3Sn and Nb_3Ge specimens, together with data from [9-11], as functions of ρ_n. In the area of measurement of ρ_n from 50–150 μOhms·cm, T_c drops sharply with increasing ρ_n for both compounds.

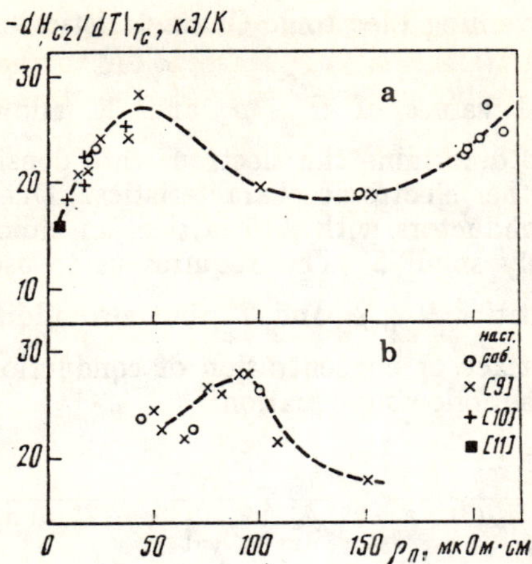

Figure 3. Variation of $H'_{c2}=-dH_{c2}/dT$ for $T=T_c$ as a Function of ρ_n for Nb_3Sn (a) and Nb_3Ge (b).

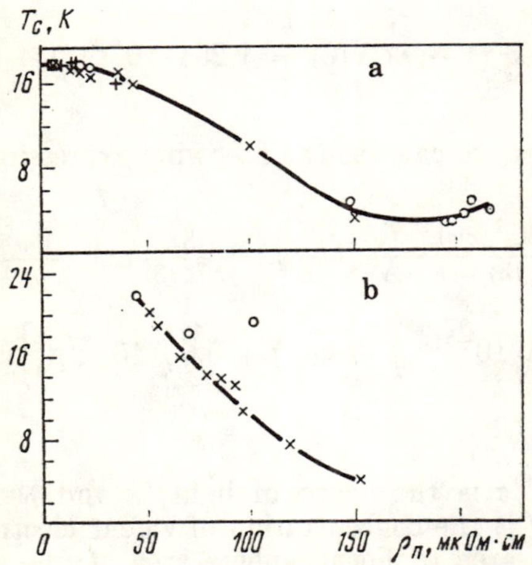

Figure 4. Variation of T_c as a Function of ρ_n for Nb_3Sn (a) and Nb_3Ge (b). Hollow points show data of this work.

4. Algorithm for Determining Electronic Characteristics

The experimental values of H'_{c2}, ρ_n and T_c allow us, using the Ginsburg–Landau theory, to determine the electron state density near the Fermi energy $N(0)$, as well as other electronic characteristics. Due to the short length of coherence ξ^* in superconductors with A15 lattice, an almost "clean" limit is observed even at relatively small l. This requires us to use in the calculation of $N(0)$ not only the quantities H'_{c2}, ρ_n and T_c, but also additional data on the area of the actual Fermi surface or concentration of conduction electrons N. In the first case, we can use the following equation[10]

$$H'_{c2} = \frac{\eta_{H_{c2}}(T_c)}{R(\lambda_{tr})}\left[\frac{24}{7\zeta(3)}\left(\frac{3}{\pi}\right)^{2/3}\frac{\hbar c}{k^2 e}T_c\frac{\gamma^{*2}}{N_{\text{val}}^{4/3}S^2}+\frac{12}{7\zeta(3)\pi}\frac{ec}{k}\frac{\rho_n}{9\cdot 10^{11}}\gamma^*\right] =$$
$$= \frac{\eta_{H_{c2}}(T_c)}{R(\lambda_{tr})}\left[9{,}55\cdot 10^{24}\,T_c\frac{\gamma^{*2}}{N_{\text{val}}^{4/3}S^2}+5{,}26\cdot 10^4\rho_n\gamma^*\right], \tag{1}$$

where the electron heat capacity coefficient

$$\gamma^* = \frac{2\pi^2}{3}k^2(1+\lambda)N(0) = 1{,}254\cdot 10^{-31}(1+\lambda)N(0); \tag{2}$$

while in the second case we can use the following expression[6]

$$H'_{c2} = \frac{\eta_{H_{c2}}(T_c)}{R(\lambda_{tr})}\left[\frac{16\pi^2}{7\zeta(3)}\frac{cmk^2}{e\hbar}\frac{T_c}{N}(1+\lambda)+\frac{8\pi}{7\zeta(3)}eck\frac{\rho_n}{9\cdot 10^{11}}\right](1+\lambda)N(0) =$$
$$= \frac{\eta_{H_{c2}}(T_c)}{R(\lambda_{tr})}\left[1{,}93\cdot 10^{-11}\frac{T_c}{N}(1+\lambda)+6{,}60\cdot 10^{-27}\rho_n\right](1+\lambda)N(0). \tag{3}$$

In equations (1)–)(3), c is the speed of light, k and $\hbar$ are the Boltzman and Planck constants, N_{val} is the concentration of valent electrons (in cm^{-3}), S is the ratio of Fermi surface area to Fermi sphere area of free electrons at concentration N_{val}, λ is the electron–phonon interaction constant, $\eta_{H_{c2}}$ is an additional correction for the strong coupling of the quantity H_{c2}[12],

$$\lambda_{tr} = \pi e^{-v}\xi_0^*/(2l) \approx 0.882\xi_0^*/l, \tag{4}$$

$e^{v} \approx 1.781$, $\zeta(3) \simeq 1.202$, $R(\lambda_{tr})$ is a coefficient on the order of one, while

$$R(\lambda_{tr})=\begin{cases} R(0)=1 \text{ where } l\to\infty \text{ ("clean" limit)} \\ R(\infty)=\frac{\pi^2}{7\zeta(3)}\approx 1.173 \text{ where } l\to 0 \text{ ("dirty" limit).} \end{cases}$$

The dimensions are: ρ_n — Ohms·cm, γ^* — erg/cm^{3}K^2, $N(0)$ — states/(cm^3·erg·spin).

We used data for N obtained by optical measurements on Nb_3Sn[13, 14] and Nb_3Ge[15]. These measurements were undertaken at room temperature. However, as the temperature drops to $T=T_c$, N may change significantly, since, due to a decrease in spreading of the peak of $N(E)$, the value of $N(0)$ is altered. Considering the contribution from this effect to the temperature variation to be the major contribution, we use for calculation of the Fermi surface area of Nb_3Sn specimens with known experimental values of H'_{c2} and T_c, the value of N from [13, 14], obtained for films with the resistivities. (In our calculations we used specimens with ρ_n=90–100 μOhms·cm.) In other words, we assumed agreement of electron concentration N at various temperatures for specimens with identical resistivity ρ. This is correct on the assumption that the measure of the influence on $N(0)$ and N of both temperature and structural defects is the relaxation time, which determines ρ.

In computing the electronic characteristics of Nb_3Ge, no correction was made to N for temperature variation, since even in the best specimens of this compound it is relatively small[8, 16].

We should note that the difference in the real Fermi surface from the free electron sphere was considered as the change in N in comparison to the concentration of valent electrons N_{val}, not the "optical" mass m_{opt} in comparison to the free electron mass m. Both approaches are equivalent where $N_{val}/m_{opt}=N/m$. The ratio $N/N_{val}=m/m_{opt}$ for Nb_3Sn~0.03–0.1, for Nb_3Ge ~0.02.

The electronic characteristics, normalized considering the strong electron–phonon interaction, were determined by the following equation[12]

$$\kappa^*(A)=\eta_\kappa \kappa(A^*), \tag{5}$$

where $\kappa(A^*)$ is the BCS equation for κ, which depends on the parameter A, A^* is the normalized value of parameter A, η_κ is the strong–coupling correction for the value of κ.

The normalized electron state density $N^*(0)$ can be determined from (3) by the equation

$$N^*(0) \equiv N(0)(1+\lambda) = \eta_{H_{c2}}^{-1} R(\lambda_{tr}) \left[1{,}93\cdot 10^{-11} T_c N^{-1}(1+\lambda) + + 6{,}60\cdot 10^{-27}\rho_n\right]^{-1} H'_{c2} \qquad (6)$$

or from (1)-(2) by the equation

$$N^*(0) = -2{,}20\cdot 10^{10}\left(N_{\text{val}}^{2/3} S\right)^2 \rho_n T_c^{-1} + \left\{\left[2{,}20\cdot 10^{10}\left(N_{\text{val}}^{2/3} S\right)^2 \rho_n T_c^{-1}\right]^2 + + 6{,}66\eta_{H_{c2}}^{-1}(T_c)\, 10^{36} H'_{c2}\left(N_{\text{val}}^{2/3} S\right)^2 R(\lambda_{tr})\, T_c^{-1}\right\}^{1/2}, \qquad (7)$$

where

$$S = (N/N_{\text{val}})^{1/2}[N(0)/N_0(0)]^{1/2}, \qquad (8)$$

where the free electron state density at spin

$$N_0(0) = \frac{1}{2\pi}\left(\frac{3}{\pi}\right)^{1/3} \frac{m}{\hbar^2} N_{\text{val}}^{1/3} = 1{,}28\cdot 10^{26} N_{\text{val}}^{1/3}. \qquad (9)$$

The mean value of electron speed at the Fermi surface

$$v_F^* = [3N/(2mN(0))]^{1/2}(1+\lambda)^{-1} = \pi(\pi/3)^{1/3}(k^2/\hbar)\, N_{\text{val}}^{2/3} S/\gamma^* = = 5{,}77\cdot 10^{-5} N_{\text{val}}^{2/3} S/\gamma^*. \qquad (10)$$

The electron free path length

$$l = mv_F^*(1+\lambda)/(e^2 N\rho_n) = \frac{\pi^2 k^2}{e^2}\frac{1}{v_F^*\gamma^*}\frac{1}{\rho_n} = (3\pi^2)^{1/3}\, 9\cdot 10^{11}\frac{\hbar}{e^2}\left(N_{\text{val}}^{2/3} S\right)^{-1}\rho_n^{-1} = = 1{,}27\cdot 10^4 \left(N_{\text{val}}^{2/3} S\right)^{-1}\rho_n^{-1}. \qquad (11)$$

The diffusion factor

$$D^* = v_F^* l/3 = 9\cdot 10^{11}\left[2e^2 N^*(0)\rho_n\right]^{-1}. \qquad (12)$$

The coherence length

$$\xi_0^* = \hbar v_F^*/(\pi^2 e^{-\nu} k T_c) = 1{,}38\cdot 10^{-12} v_F^*/T_c. \tag{13}$$

The Ginsburg–Landau coherence length

$$\xi_{GL}(T) = [\hbar c/2eH_{c2}(T)]^{1/2} = \left[\hbar c/\left(2eH'_{c2}\cdot(T_c - T)\right)\right]^{1/2} = \pi e^{-\nu}\eta_\xi(T_c)\,\xi_0^* \times$$
$$\times\,[7\zeta(3)\,R(\lambda_{tr})/(48\,(1+\lambda_{tr})\,(1 - T/T_c))]^{1/2} = \xi^0_{GL}(1 - T/T_c)^{-1/2}, \tag{14}$$

where ξ^0_{GL}=1.81·10$^{-4}(T_c H'_{c2})^{-1/2}$. The depth of penetration of the magnetic field into a superconductor

$$\lambda_{GL}(T) = \eta_\lambda(T_c)\,\lambda_{BCS}(0)/[2^{1/2}(1 - T/T_c)^{1/2}] \equiv \lambda^0_{GL}/(1 - T/T_c)^{1/2}, \tag{16}$$

where $\lambda_{BCS}(0)$ is the depth of penetration according to the weak-coupling theory,

$$\lambda^0_{GL} = \eta_\lambda(T_c)\,\lambda_L\,[(1+\lambda_{tr})/(2R(\lambda_{tr}))]^{1/2}. \tag{17}$$

The London penetration depth

$$\lambda_L = (9\pi)^{1/6}\,\hbar c\gamma^*\Big/\left(2\pi k e N_{val}^{2/3} S\right) = 1{,}32\cdot 10^8 \gamma^{*1/2}\Big/\left(N_{val}^{2/3} S\right). \tag{18}$$

The Ginsburg–Landau parameter

$$\varkappa = \lambda_{tr}(T)/\xi_{GL}(T) = (24/7)^{1/2} e^\nu \eta_\lambda(T_c)\,(1+\lambda_{tr})\,\lambda_L\Big/\left[\pi\,\zeta(3)^{1/2}\,\eta_\xi(T_c)\,R(\lambda_{tr})\,\xi_0^*\right] =$$
$$= 0{,}957\eta_\varkappa(T_c)\,\lambda_L(1+\lambda_{tr})/\left(R(\lambda_{tr})\,\xi_0^*\right). \tag{19}$$

The jump in heat capacity where $T = T_c$ [17]

$$c_s - c_n = [12/(7\zeta(3))]\,\eta_c T_c\gamma^* = 1{,}43\eta_c T_c\gamma^*. \tag{20}$$

The thermodynamic critical field near T_c

$$H_c(T) = [8/(7\zeta(3))]^{1/2} e^{\nu}\left[\eta_{H_c}(T_c)/\eta_{H_c}(0)\right] H_c(0)(1 - T/T_c) = \\ = 1{,}74\left[\eta_{H_c}(T_c)/\eta_{H_c}(0)\right]\cdot H_c(0)(1 - T/T_c), \tag{21}$$

where T=0

$$H_c(0) = (6\pi)^{1/2} e^{-\nu}\eta_{H_c}(0)\,T_c\gamma^{*1/2} = 2{,}44\eta_{H_c}(0)\,T_c\gamma^{*1/2}. \tag{22}$$

Near T_c

$$H_c' \equiv -(dH_c/dT)_{T=T_c} = 4\,[3\pi/(7\zeta(3))]^{1/2}\,\eta_{H_c}(T_c)\,\gamma^{*1/2} = 4{,}23\eta_{H_c}(T_c)\gamma^{*1/2}. \tag{23}$$

At the "dirty" limit (where $k\xi_0^*$), ignoring the paramagnetic limitation, the variation of upper critical field with temperature is given by the equation[18]:

$$\ln(T_c/T) = \psi\,[1/2 + (e^{-\nu}/4)(T_c/T)\,H_{c2}(T)/H_{c2}(0)] - \psi(1/2). \tag{24}$$

Near T_c, considering strong-coupling corrections

$$H_{c2}(T) = (8e^{\nu}/\pi^2)\left[\eta_{H_{c2}}(T_c)/\eta_{H_{c2}}(0)\right] H_{c2}(0)(1 - T/T_c). \tag{25}$$

From this equation we can find

$$H_{c2}(0) = (\pi^2/(8e^{\nu}))\left[\eta_{H_{c2}}(0)/\eta_{H_{c2}}(T)\right] T_c H_{c2}' = 0{,}693\left[\eta_{H_{c2}}(0)/\eta_{H_{c2}}(T_c)\right] T_c H_{c2}', \tag{26}$$

or

$$H_{c2}(0) = (\pi^2/2)\,[7\zeta(3)]^{-1/2}\left[\eta_{H_{c2}}(0)\Big/\big(\eta_{\varkappa}(T_c)\,\eta_{H_c}(0)\big)\right]\varkappa(T_c)\,H_c(0). \quad (27)$$

At the "clean" limit $(l \gg \overset{*}{\xi}_0)$[18]

$$H_{c2}(0) = 0{,}74\left[\eta_{H_{c2}}(0)\big/\eta_{H_{c2}}(T_c)\right]T_c H'_{c2}. \quad (28)$$

With an arbitrary relationship between l and $\overset{*}{\xi}_0$, a good approximation for determination of $H_{c2}(0)$ is yielded by the equation[19]

$$H_{c2}(0) = \eta_{H_{c2}}(0)\left[7{,}1\cdot 10^{24}\gamma^{*2}T_c\left(N_{\text{val}}^{2/3}S\right)^{-2} + 3{,}1\cdot 10^{4}\gamma^{*}\rho_n\right]T_c. \quad (29)$$

The use of this equation to determine the characteristics in the square brackets based on data on $H_{c2}(0)$ can lead to significant error (10–20%), since the individual contribution of each of the components of $H_{c2}(0)$ for an arbitrary ratio between l and $\overset{*}{\xi}_0$ is not precisely reflected.

Ignoring spin-orbital scattering, the upper critical field at T=0, limited by the paramagnetic limit, can be found by the equation[20]

$$\overset{*}{H}_{c2}(0)=H_{c2}(0)(1+\alpha^2)^{-1/2}, \quad (30)$$

where the Maki parameter

$$\alpha=2^{1/2}H_{c2}(0)/H_{\text{P}}(0) \quad (31)$$

and Pauli field limit[21]

$$H_{\text{P}}(0) = 2^{1/2}\pi e^{-\nu}\eta_{H_c}(0)\,(kmc/(e\hbar))\,(1+\lambda)\,T_c = 1{,}86\cdot 10^{4}\eta_{H_c}(0)\,(1+\lambda)\,T_c. \quad (32)$$

Considering the spin-orbital interaction, represented by the scattering parameter

$$\lambda_{so} = 2\hbar/3\pi k T_c\tau_{so} = 2\pi e^{-\nu}\overset{*}{\xi}_0/(3l_{so}) = 1{,}18\overset{*}{\xi}_0/l_{so}, \quad (33)$$

(τ_{so} is the relaxation time, $l_{so}=\tau_{so}v_F$ is the effective electron free path length in the spin-orbital interaction), decreases the significance of the paramagnetic limitation[22]. For $\lambda_{so} \gg 1$, the effective limiting field

$$H_{so} \simeq 1.33 \lambda_{so}^{1/2} H_p(0). \tag{34}$$

The values of corrections η_K were calculated in this work by the equations[17]

$$\eta_{H_{c2}}(T_c) = 1 + (\pi T_c/\omega_0)^2 \, [0{,}6 \ln(\omega_0/T_c) - 0{,}26], \tag{35}$$

$$\eta_\xi(T_c) = \eta_{H_{c2}}(T_c)^{-1/2}, \tag{36}$$

$$\eta_{H_c}(T_c) = 1 + (\pi T_c/\omega_0)^2 \, [1{,}1 \ln(\omega_0/T_c) + 0{,}14], \tag{37}$$

$$\eta_{H_c}(0) = 1 + 2{,}3 \, (T_c/\omega_0)^2 \ln(\omega_0/T_c), \tag{38}$$

$$\eta_\lambda(T_c) = \eta_{H_{c2}}(T_c)^{1/2} / \eta_{H_c}(T_c), \tag{39}$$

$$\eta_\varkappa(T_c) = \eta_{H_{c2}}(T_c) / \eta_{H_c}(T_c), \tag{40}$$

$$\eta_c(T_c) = 1 + (\pi T_c/\omega_0)^2 \, [2{,}2 \ln(\omega_0/T_c) + 0{,}28]. \tag{41}$$

The characteristic frequency of the phonon spectrum ω_0 can be estimated from the experimental data on the energy gap Δ and T_c using the equation for η_Δ of[23]

$$\eta_\Delta = \Delta/(\pi e^{-\gamma} k T_c) = 1 + 5{,}3 \, (T_c/\omega_0)^2 \ln(\omega_0/T_c). \tag{42}$$

We determined the electron-phonon interaction constant using the equations of McMillan[24]

$$T_c = (\langle\omega\rangle/1{,}20 \exp\{-1{,}04 \, (1+\lambda)/[\lambda - \mu^* (1 + 0{,}62\lambda)]\} \tag{42}$$

or Allen and Dynes[25]

$$T_c = (f_1 f_2 \omega_{\log}/1{,}20) \exp\{-1{,}04 \, (1+\lambda)/[\lambda - \mu^* (1 + 0{,}62\lambda)]\}, \tag{43}$$

yield similar results where $\lambda<2$. In expressions (42)–(43)

$$\langle\omega\rangle = \frac{2}{\lambda} \int_0^\infty \alpha^2 F(\omega) \, d\omega, \tag{44}$$

$$\omega_{\log} = \exp\left[\frac{2}{\lambda} \int_0^\infty \frac{d\omega}{\omega} \, d^2 F(\omega) \ln\omega\right], \tag{45}$$

$$f_1 = \left[1 + (\lambda/\Lambda_1)^{3/2}\right]^{1/3}, \quad f_2 = 1 - \frac{\lambda^2\left(1 - \langle\omega^2\rangle^{1/2}/\omega_{\log}\right)}{\lambda^2 + \Lambda_2^2}, \tag{46}$$

$$\Lambda_1 = 2{,}46\,(1 + 3{,}8\mu^*), \quad \Lambda_2 = 1{,}82\,(1 + 6{,}3\mu^*)\left(\langle\omega^2\rangle^{1/2}/\,\omega_{\log}\right), \tag{47}$$

while μ^* is the Coulomb pseudopotential.

Our calculations of λ for Nb_3Ge and Nb_3Sn show that where μ^*=0.1–0.16, $\langle\omega\rangle=\omega_0$ is a sufficient approximation. For the best specimens of Nb_3Ge, $2\Delta/kT_c$=4.2±0.1[26], yielding values of λ=1.8±0.2, μ^*=0.1, ω_0=(14.2±0.2) meV. For Nb_3Sn, $2\Delta/kT_c$=4.2–4.4[27], λ=1.8±0.2, μ^*=0.13, ω_0=(10.8±0.9) meV.

We note that the algorithm presented above does not consider effects of anisotropy, which may appear at the "clean" limit. A theory of H_{c2} for anisotropic superconductors of the second kind, considering scattering anisotropy at the "clean" limit, was developed in[28].

5. Characteristics of Nb_3Sn and Nb_3Ge Films

Characteristics of all Nb_3Sn films were calculated on the basis of the algorithm outlined above. Table 2 presents the characteristics of two typical specimens with widely differing values of ρ_n. The characteristics presented in Table 2 were calculated for S=0.3±0.1. This value of S was obtained using experimental data on N. The characteristics of Nb_3Ge are presented in [6-8]. Figures 5 and 6 show the variation of $N^*(0)$ with ρ_n and l for Nb_3Sn and Nb_3Ge, including the results of this study and values calculated on the basis of the data of [9-11]. Figure 7 shows the variation of $N^*(0)$ for both compounds as a function of temperature, obtained on the assumption of identical relaxation time resulting from electron–phonon scattering and scattering of electrons on structural defects.

The error in the computed characteristics is determined by the experimental error of ρ, H'_{c2} and N. For specimens with small ρ_n ("clean" limit), the error for S, $N^*(0)$, γ^*, C_s-C_n, l, λ_{tr} is ~30%, for λ_L, $H_c(0)$~15%, ξ_0^*, ξ_{GL}^0, v_F^*, κ, λ_{GL}, $H_{c2}(0)$ less than 10%. In the area of great ρ_n ("dirty" limit), the error in the values of $N^*(0)$, λ_{GL}, C_s-C_n, γ^* is 10–15%, in S, v_F^*, l, ξ_0^*, λ_L — 30%, λ_{tr} — 60%, in κ, $H_c(0)$, $H_{c2}(0)$, ξ_{GL}^0 less than 10%.

Table 2
Electronic Characteristics of Nb_3Sn Specimens

Characteristic	Specimen 10D	12Z	Characteristic	Specimen 10D	12Z
$N^*(0)$, 10^{34} $\frac{\text{states}}{\text{erg}\cdot\text{cm}^3\cdot\text{spin}}$	7.2±2	2.0±0.2	l, Å	52±17	5
$N(0)$, 10^{34} $\frac{\text{states}}{\text{erg}\cdot\text{cm}^3\cdot\text{spin}}$	2.5±0.8	1.2±0.2	ξ_0^*, Å	60±6	1300±400
λ	1.8±0.2	0.6±0.2	λ_{tr}	1.0±0.3	220±130
v_F^*, 10^7 cm/s	0.8±0.1	2.8±0.9	ξ_{GL}^0, Å	28±1	69±4
v_F, 10^7 cm/s	2.2±0.3	4.6±1.4	λ_{GL}^0, Å	770±80	5200±500
γ^*, mlJ/cm^{3}K^2	0.9±0.3	0.28±0.04	λ_L, Å	1050±160	540±170
$C_s - C_n$, mlJ/cm^{3}K	42±14	1.0±0.2	κ	27±2	75±4
N, 10^{22} cm^{-3}	0.7±0.4	1.6±0.8	$H_{c2}(0)$, kOe	290±20	48±3
			$H_c(0)$, kOe	4.6±0.7	0.4±0.1

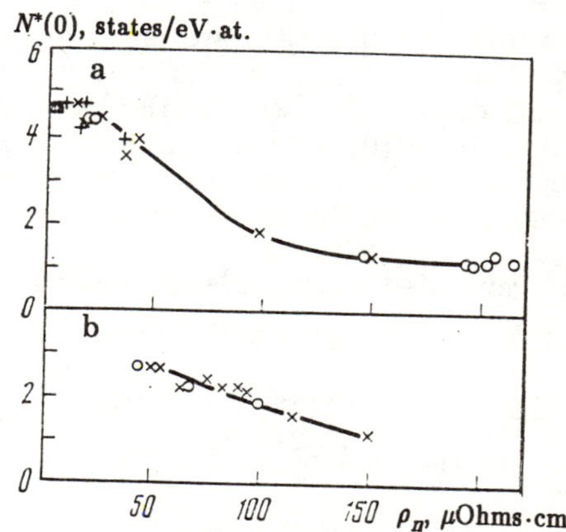

Figure 5. Variation of $N^*(0)$ with ρ_n for Nb_3Sn (a) and Nb_3Ge (b). Hollow points show data of this study.

Works[29, 30] presented measurements of $N(0)$ for Nb_3Sn based on heat capacity, while work[4] presented similar data based on magnetic susceptibility. The results were obtained with specimens with small values of ρ_n. In Table 3, the data of [4, 29, 30] are compared with the result of the present study, relating to specimens with $\rho_n \leqslant 20$ μOhms·cm. The agreement of all data within the limits of measurement is obvious.

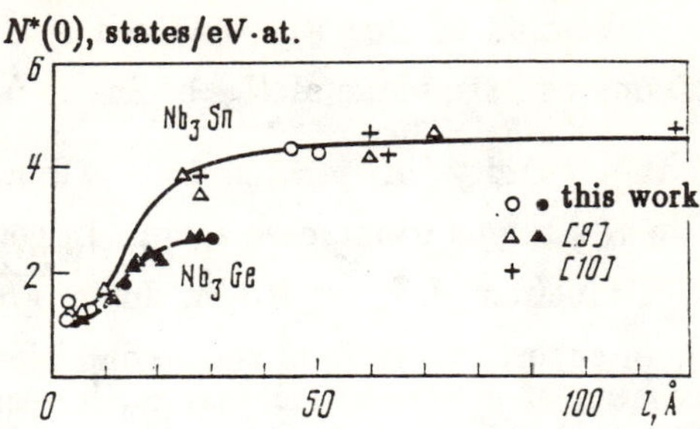

Figure 6. $N^*(0)$ as a Function of l. Results for Nb_3Sn obtained for S=0.3.

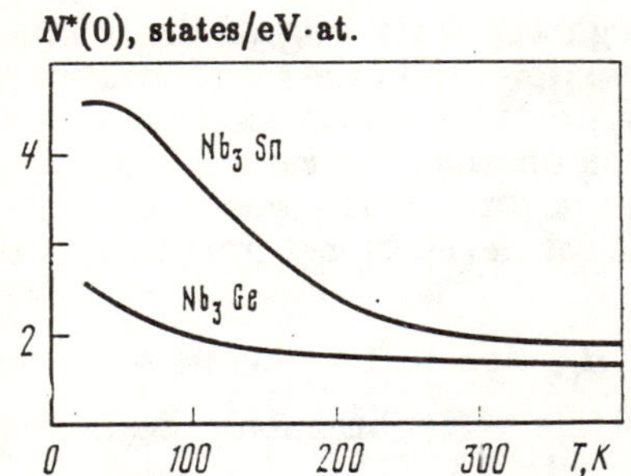

Figure 7. Variation of $N^*(0)$ with Temperature.

Table 3
Comparison of Results for N(0) for Nb_3Sn

Method	$N(0)$, states/(eV·at)	Reference
Heat Capacity	2.01	29,30
Heat Capacity	1.35±0.1	30
Susceptibility	1.5	4
$H_{c2}(T)$	1.6±0.5	This work

6. Discussion

1. As follows from Figure 3, the variation of H'_{c2} with ρ_n for Nb_3Sn is anomalous. The first indication of this was obtained in [31]. After near-linear growth up to $\rho_n \approx 50$ μOhms·cm, the value of H'_{c2} begins to fall. The minimum of H'_{c2} is observed in the vicinity of ρ_n=100–150 μOhms·cm. At ρ_n>150 μOhms·cm, the value of H_{c2} begins to increase again. In general, the experimental variation of H'_{c2} as a function of ρ_n for Nb_3Sn differs greatly from the linear variation characteristic of superconductors of the second kind. A similar anomalous form, with a maximum of ρ_n=35–40 μOhms·cm, is seen in the variation of H_{c2} as a function of ρ_n for each temperature.

The variation of H'_{c2} as a function of ρ_n for Nb_3Ge also has a maximum at $\rho_n \approx 90$ μOhms·cm, although the scattering of the points on the curve is greater than for Nb_3Sn. Unfortunately, the interval of measurements of ρ_n is smaller for Nb_3Ge. The value of H_{c2} decreases with increasing ρ_n throughout this interval. In [32] for Nb_3Ge specimens obtained by cathode sputtering and chemical precipitation from the gas phase, a drop in H_{c2} was also observed from 333 kOe to 234 kOe with a decrease in T_c of the specimen from 21.3 K to 18.6 K.

The linear rise in H'_{c2} at small ρ_n agrees with the Ginsburg–Landau theory, according to which this variation should be observed with constant $N(0)$. The presence of a maximum on the curves of H'_{c2} and H_{c2} as a function of ρ_n for Nb_3Sn was previously observed in [33]. In [34], the GLAG [Ginsburg–Landau–Abrikosov–Gor'kov] theory was extended to the case of superconductors with fine structure of state density near E_F. It was predicted that a maximum would be found on the variation of $H_{c2}(0)$ as a function of concentration, with a deviation from a linear rule for H'_{c2}.

Data on critical fields in Nb_3Sn in the area of $\rho \gtrsim 100$ μOhms·cm are sparse. According to [9], H'_{c2} changes little with increasing ρ_n under the influence of radiation from 100–150 μOhms·cm. The data obtained in the present work for films with $\rho_n \simeq 200$ μOhms.cm indicate a new increase in H'_{c2} in this area.

As was noted in [33], regularity of the change in H'_{c2} and T_c with ρ_n in Nb_3Sn is observed to a great extent, regardless of the nature of the defects. This is also confirmed by our results, which fall on the common curves. It was shown in [35] that the variation of $H_{c2}(0)$ with ρ, with a maximum at $\rho_n \approx 35$–40 μOhms.cm, is also preserved where the value of ρ_n is altered by introducing a small quantity (a few atomic percent of the impurities Ti, Ta or Ga to Nb_3Sn).

However, we should note that the scattering of points on the curves presented in the present work is still somewhat greater than the limits indicated by the authors of [9-11] for experimental error. We might assume that this deviation of the points from the mean curve still results from the influence of the nature of A15 structural defects.

To explain the drop in T_c with increasing disorder in superconductors with A15 lattice, in addition to the spreading of the state density peak near E_F, other mechanisms have also been suggested, such as Andersen localization or the influence of the quasiunidimensional electron spectrum. In [36], an explanation was suggested related to the development of a correlation gap and an expression was derived for T_c which directly contained the resistance of the specimen ρ. In addition to ρ, this equation for T_c also includes the Debye frequency, electron–phonon interaction constant λ, Coulomb repulsion potential and other parameters as constants. Unfortunately, these parameters are themselves dependent on the disorder, i.e., on ρ, in A15 materials with state density peak near E_F.

2. The electronic characteristics of Nb_3Sn and Nb_3Ge which have been obtained depend essentially on resistivity ρ_n. As ρ_n increases in Nb_3Sn from 5 to 200 μOhms·cm, $N^*(0)$ decreases by several times. In Nb_3Ge, $N^*(0)$ decreases by more than a factor of two as ρ_n changes from 50 to 150 μOhms·cm. A significant contribution to this variation is made by the decrease in electron–phonon interaction constant λ with increasing ρ_n. It is responsible for about one half of the total change in $N^*(0)$. However, the variation of $N^*(0)$ and other electronic characteristics represents more than a change in the electron–phonon interaction. The zonal electronic characteristics $N(0)$, v_F, and N are also dependent on ρ_n.

The variation in $N(0)$ as a function of perfection of Nb_3Sn specimens is confirmed by direct experimental studies of the electronic energy losses [37, 38]. The variation in electronic characteristics with ρ_n can be transformed by the equation suggested in [35] to a variation as a function of long-range order.

3. The variation of electronic characteristic of Nb_3Sn and Nb_3Ge as a function of ρ_n is equivalent to their variation as a function of electron collision frequency or the inverse of electron relaxation time $\tau^{-1}=e^2\rho_n N/m$. Assuming that the change in state density $N(0)$ is related to the spreading of the peak of $N(E)$, located near E_F, we can estimate its width δE based on $\hbar/\tau$, corresponding to a significant change in $N(0)$. Estimates based on $\hbar/\tau$ corresponding to the maximum of the derivative of $N(0)$ with respect to ρ_n yield for Nb_3Sn $\delta E \sim 0.01$ eV, for Nb_3Ge $\delta E \sim 0.01$–0.02 eV.

4. In the algorithm used to determine electronic characteristics, the value of S for Nb_3Sn was computed with respect to N and $N(0)$ considering their variation with temperature. In [31], a method is suggested for determining S and γ^* for specimens with small ρ_n using the linear variation of the initial section of the curve of $H'_{c2}(\rho_n)$ without involving additional experimental data. The values thus determined according to the data of Figure 3 are $S=0.3\pm0.1$, $N^*(0)=4.5\pm1.5$ states/(eV·at), $N(0)=1.6\pm0.5$ states/(eV·at). They relate to an Nb_3Sn specimen with $l\to\infty$ and agree with quantities obtained in the present work using optical data for N. In the case of Nb_3Ge, due to the great scattering of points in the area of small ρ_n, this method can be used only to produce rough estimates, which agree qualitatively with the results obtained earlier.

Assuming that with large ρ_n (after spreading of the peak of $N(E)$ in the area of near growth of H_{c2}) the electronic characteristics have reached saturation, a similar estimate of the mean value γ^* can be obtained by the same method.

The values obtained in the present work for the area of the Fermi surface S agree well with the data of [10] and also indicate correctness of the correlation of T_c and S [6, 7] (Figure 8).

We can see from Figure 8 that the value of T_c increases with decreasing relative area of the Fermi surface S. However, we know[39] that in the area of semiconductor concentration of electrons $N<10^{21}cm^{-3}$ the value of T_c decreases (Figure 9). Thus, the maximum values of T_c are reached at the boundary of the transition from metallic phases to semiconductors around $S<0.2$ or $N\lesssim5\cdot10^{21}cm^{-3}$. (The optimal concentration N is $(2\text{–}4)\cdot10^{21}cm^{-3}$.) It is possible that the phenomena recently observed in Nb–Ge–Al–O [40] and Ba–La–Cu–O [41] systems at $T\approx30\text{–}40$ K are related to the formation of high-temperature superconducting phases with optimal values of the parameters S and N. For such compounds and alloys the interval of compositions in which the optimal values of parameters are observed is usually quite narrow and difficult to achieve in practice[42]. The complexity of these systems, the presence of many phases, interphase boundaries and composition gradients in specimens facilitates the formation and stabilization of such high-temperature phases in certain portions of the specimens. Oxygen can in this case act as a catalyst[43]. It is possible that in [44] superconductivity was also observed in a certain complex sulfur-based compound formed due to the presence of large quantities of impurities in the specimens at high pressures. Doubtless, fluctuation expansion of the transition was of some significance in the increase in observed temperature at which the transition of the films of these heterogeneous systems started. The answer to the question, just which phase or interface was responsible for the phenomenon observed in [40, 41] at such high temperatures is difficult to obtain, due to the complexity of the systems, resulting from the method used in this work to produce the phases required.

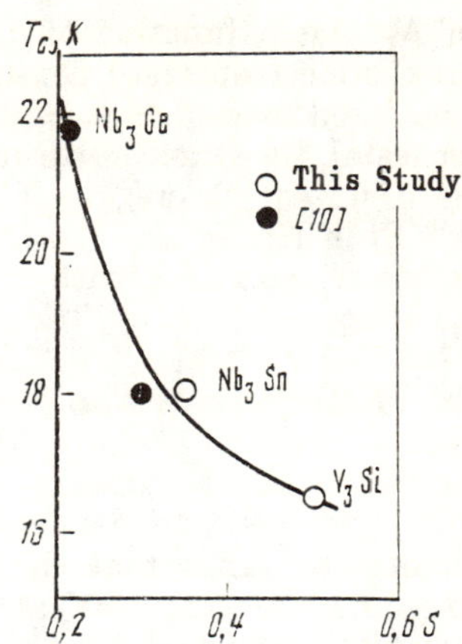

Figure 8. Correlation of T_c with Relative Area of Fermi Surface S in Superconductors with A15 Lattice.

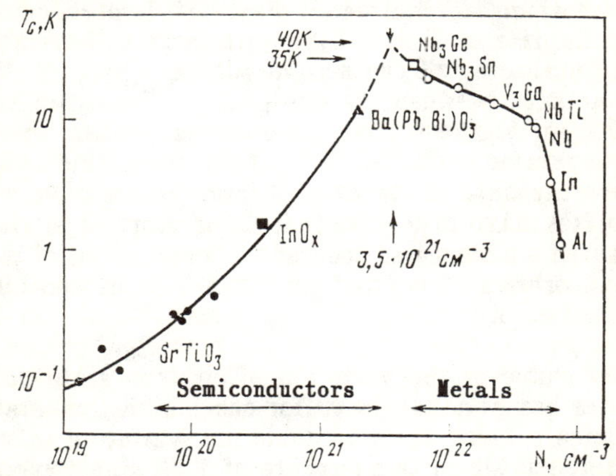

Figure 9. Variation in Maximum Values of T_c with Electron Concentration N. Data for semiconductors from [39].

5. The variation in electronic characteristics as a function of τ should cause them to vary with temperature T with a change in $\tau(T)$. This is apparently confirmed by optical measurements at room temperature[45]. However, verification requires direct temperature measurements of the characteristics. The temperature variation of the fundamental electronic characteristics and electron–phonon interaction may result not only from spreading of the electron state density peak. However, apparently in "narrow-band" superconductors this is the main factor.

6. The variation of $N(0)$ as a function of ρ_n (or τ) obtained allows us to estimate the parameters of the electron state density model. Studying the variation of $N(E)$ near E_F, considered as a narrow peak (the shape of which is unimportant) on a constant "pedestal," we can find from Figure 5: the height of the peak $N^{max}(E)$ in Nb_3Sn is 5 states/(eV·at), in Nb_3Ge — 3 states/(eV·at), the height of the pedestal $N^{min}(E)$ in Nb_3Sn and Nb_3Ge is 1 state/(eV·at). In these estimates it was assumed that E_F coincides with the position of the center of the peak[1]. Given more detailed data on the function $H'_{c2}(\rho_n)$, information can be obtained both on the shape of the peak of $N(E)$, and on the position of E_F relative to its center.

The values of $N^{max}(E)$, $N^{min}(E)$ and δE agree satisfactorily with the data of other works[37, 38]. We note the result that the values of $N(0)$ for both compounds approach the same value of 1 state/(eV·at) with decreasing l. Theoretical calculations[46, 47] predict for Nb_3Sn and Nb_3Ge practically identical values of the pedestals.

Our results contain an indirect indication of more complex structure of $N(E)$ near E_F. If the rise in T_c in Nb_3Sn with large ρ_n is related to the increase in $N(0)$ caused by spreading of the second (deeper) peak of $N(E)$, we can estimate the energy distance between the two peaks. This estimate yields for Nb_3Sn a value of $\Delta E \approx 0.2$ eV. Theoretical calculations predict the existence of two peaks in Nb_3Sn, separated by 0.19 eV[46]. Of course, this agreement may be a chance result, since the data in the area of large values of ρ_n are insufficient for quantitative estimates. We note also that T_c is more sensitive to various influences and is measured with greater accuracy. Direct proof of the presence of two peaks in Nb_3Sn and other A15 compounds[46] can be provided by optical measurements in the infrared area.

The complex shape of the variation of electron state density with energy near the Fermi level has been suggested for other high-temperature superconductors with A15 lattice. Thus, the nonmonotonic variation in magnetic susceptibility and structural transition temperature of V_3Si single crystals as a function of T and the ratio of resistances at 300 K and 18 K $\rho(300)/\rho(18)$ observed in the works of V. M. Pan, et al., (reviewed in [48]) is related to the presence of a fine structure in $N(E)$.

7. The question of the type of superconductivity occurring in such "narrow-band" superconductors as materials with A15 structure is also interesting. A number of works[49-52] have suggested models of polar (phonon or electron) constriction of the zone in the presence of a strong electron-phonon or electron-electron (from different zones) interaction. This narrowing leads to an automatic

---

[1]Unfortunately, the accuracy of available data does not allow us to answer the question as to whether $N(0)$ increases with increasing ρ_n at small ρ_n.

sharp increase in electron state density at the Fermi surface $N(0)$ (if the Fermi level is in this band) and an increase in T_c where the possible negative contribution from changes in the phonon system and electron–phonon interaction does not exceed the positive effect. The polar superconductivity[49, 52] arising in this case should lead to slightly different variations in the characteristics as functions of ρ, l and T. However, apparently, the difference between these variations and those observed with the ordinary (phonon) type of superconductivity should not be very great. Therefore, to determine the type of superconductivity (polar or ordinary electron type) from, for example, $H_{c2}(\rho)$ requires detailed theoretical calculations. This apparently requires a knowledge of the precise electron structure of the material. The theoretical problem here is to predict the values of ρ, l and T at which a significant deviation from the equations of the ordinary theory can be expected (at great ρ and small l, when the peak of $N(E)$ is practically spread, the ordinary equations must doubtless be valid).

The situation is otherwise if as ρ decreases or l increases in the material and peak $N(E)$ narrows, a new type of superconductivity arises — "localized bipolaron superconductivity"[53]. This should lead to sharp and qualitative differences in the variations of the characteristics. For example, the theory predicts a decrease in T_c with increasing electron–phonon interaction constant λ. Variations should then be observed on the curves of electron characteristics as functions of ρ, l (upon transition from one type of superconductivity to another). The problem of the theory is to predict the values of ρ, l and T at which these sudden changes should occur in various classes of materials, in which the appearance of this superconductivity can be expected. (Incidentally, these discussions are also valid for systems with heavy Fermions.) In particular, such an area in our case of Nb_3Sn and Nb_3Ge might be the area of $\rho \sim 50$ μOhms·cm, where a sudden increase in the parameters $(H_{c2}, dH_{c2}/dT)$ is observed with a decrease in ρ. Unfortunately, little accurate experimental information is available in this area of ρ.

In the present study, we have explained all variations in a consistent manner based on the GLAG theory, starting with ordinary phonon superconductivity. This consistency, of course, rests on the fact that the changes in electron state density $N(0)$ which we obtained can be achieved at the given values of ρ (i.e., τ). Precise confirmation of this must be based either on theoretical calculations or on measurement of the variation of $N(0)$ as a function of τ in independent experiments.

Thus, to determine the type of superconductivity in these materials we must use various independent experiments yielding the variation of superconductor characteristics with such parameters as τ, l and ρ.

7. Conclusions

The study of the properties of Nb_3Sn and Nb_3Ge over a broad range of disorders has yielded a number of interesting results. This method apparently allows investigation of the detailed form of electron state density near E_F. It is important that the method of introducing defects has relatively little influence on the change in H_{c2}, $N(0)$ and other characteristics.

Data have been obtained indicating an increase in H'_{c2} with ρ for Nb_3Sn not only in the areas of small $\rho \leqslant 50$ μOhms·cm, but also where $\rho \geqslant 150$ μOhms·cm. In both cases, the increase in H_{c2} is related to an increase in ρ with relatively small changes in $N(0)$, while the decrease in H'_{c2} in the area of $\rho \simeq 50$–100 μOhms·cm is related to a sharp drop in $N^*(0)$.

With disorder, $N(0)$ approaches the same value in both Nb_3Sn and Nb_3Ge, which agrees with the theoretical calculations of $N(E)$.

Thus, the results of our work experimentally confirm the existence of a peak in electron state density near E_F in Nb_3Sn and Nb_3Ge, with width of 0.01–0.02 eV.

The variations in fundamental electron characteristics of high-temperature superconductors with A15 lattice as a function of their disorder (i.e., λ, ρ, τ) can be used to optimize the parameters of a number of superconducting microelectronic elements — resonators, superlattices, weak couplings.

REFERENCES

1. Svidler, A. R., Snid, K. L., Koks, D. Ye., "Influence of Radiation on Properties of Superconducting Materials," Metallurgiya sverkhprovodyashchikh materialov [Metallurgy of Superconducting Materials], Moscow, Metallurgiya Press, 256 pp., 1984.

2. Dynes, R. S., Poate, J. M., Testardi, L. R., Strom, A. R., Hammon, R. H., "Superconductivity Defects and Stoichiometry in A15 Materials," IEEE Trans. Magn., Vol MAG-13, No. 1, pp. 640—643, 1977.

3. Testardi, L., Weger, M., Gol'dberg, I., Sverkhprovodyashchiye soyedineniya so strukturoy β-vol'frama [Superconducting Compounds with β-Tungsten Structure], Moscow, Mir Press, 436 pp., 1977.

4. Shamray, V. F., Vol'f, G., Bomkhammel', K., Padurets, L. N., "Magnetic Susceptibility and Superconducting Transition Temperature of Nb_3SnH_x Variable–Composition Phase," Fizika metallov i metallovedeniya, Vol. 54, pp. 459–464, 1982.

5. Yelesin, V. F., Aleksandrov, A. S., "Theory of Influence of Radiation Defects on Normal-Phase Properties and Critical Parameters of Superconductors with A15 Structure," Voprosy atomnoy nauki i tekhniki: Fizika radiatsionnykh povrezhdeniy i radiatsionnoye materialovedeniye [Problems of Atomic Science and Technology: Physics of Radiation Damage and Radiation Materials Science], No. 3(31), pp. 1-118, 1984.

6. Golovashkin, A. I., Pechen', Ye. V., Shabanova, N. P., "Temperature Variation of Critical Magnetic Fields and Electronic Characteristics of Nb_3Ge Films," ZhETF, Vol. 82, pp. 850–861, 1982.

7. Golovashkin, A. I., Pechen', Ye. V., Shabanova, N. P., "Critical Magnetic Fields and Electronic Characteristics of Nb_3Ge Films," Preprint FIAN, Moscow, No. 238, 23 pp., 1981.

8. Golovashkin, A. I., Pechen', Ye. V., "Superconducting Compounds Nb_3Ge and Nb_3Si," Trudy FIAN, Vol. 151, pp. 63–136, 1984.

9. Wiesmann, H., Gurvitch, M., Ghoah, A. K., Lutz, H., Kammerer, O. F., Strongin, M., "Estimation of Density of State Changes with Disorder in A15 Superconductors," Phys. Rev. B., Vol. 17, No. 1, pp. 122–125, 1978.

10. Orlando, T. P., McNiff, E. J., Js., Foner, S., Beasley, M. R., "Critical Fields, Pauli Paramagnetic Limiting, and Material Parameters of Nb_3Sn and V_3Si," Phys. Rev. B., Vol. 19, No. 9, pp. 4545–4561, 1979.

11. Lim, K. S., Thompson, I. D., Webb, G. W., "Electronic Density of States and T_c in Nb_3Sn Under Pressure," Phys. Rev. B., Vol. 27, No. 5, pp. 2781–2787, 1983.

12. Rainer, D., Bergmann, G., "Temperature Dependence of H_{c2} and x_1 in Strong Coupling Superconductors," J. Low Temp. Phys., Vol. 14, No. 5/6, pp. 501–519, 1974.

13. Golovashkin, A. I., Donner, Ye. D., Levchenko, I. S., Motulevich, G. P., "Optical Properties and Electronic Characteristics of Niobium–Tin Alloys," ZhETF, Vol. 59, pp. 1967–1974, 1970.

14. Golovashkin, A. I., Donner, Ye. D., Levchenko, I. S., Motulevich, G. P., "Optical Properties in the Infrared Band and Electronic Characteristics of Niobium–Tin Alloys," Preprint, FIAN, Moscow, No. 6, 68 pp., 1971.

15. Mattheiss, L. F., Testardi, L. R., Yao, W. W., "Plasma Energies for A15 Compounds," Phys. Rev. B., Vol. 17, No. 12, pp. 4640–4643, 1978.

16. Golovashkin, A. I., Pechen', Ye. V., Shavanova, N. P., "Characteristics of Nb_3Ge and Nb_3Sn as Functions of Free Path Length of Electrons and Temperature," Materialy 2-go nauchnogo seminara 'Metallofizika sverkhprovodnikov' [Materials of Second Scientific Seminar on "Metal Physics of Superconductors"] (October, 1983), Kiev, pp. 92–93, 1983.

17. Masharov, N. F., "Properties of Superconductors with Strong Coupling in an Arbitrary Magnetic Field Near the Critical Temperature," FTT, Vol. 16, No. 8, pp. 2342–2349, 1974.

18. Helfand, E., Werthamer, N. R., "Temperature and Purity Dependence of the Superconducting Critical Field, H_{c2}," Phys. Rev., Vol. 147, No. 1, pp. 288–294, 1966.

19. Suenaga, M., Welch, D. O., Sabatini, R. L., Kammerer, O. F., Okuda, S., "Superconducting Critical Temperatures, Critical Magnetic Fields, Lattice Parameters, and Chemical Compositions of 'Bulk' Pure and Alloyed Nb_3Sn Produced by the Bronze Process," J. Appl. Phys., Vol. 59, No. 3, pp. 840–853, 1986.

20. Maki, K., "The Magnetic Properties of Superconducting Alloys," Physics, Vol. 1, No. 2, pp. 127–143, 1964.

21. Foner, S., McNiff, E. J., Jr., "Upper Critical Fields of Cubic and Tetragonal Single Crystal and Polycrystalline Nb_3Sn in dc Fields to 30 Tesla," Sol. St. Comm., Vol. 39, No. 9, pp. 959–964, 1981.

22. Wertamer, N. R., Helfand, E., Hohenberg, P. C., "Temperature and Purity Dependence of the Superconducting Critical Field, H_{c2}. Electron Spin and Spin-Orbit Effects," Phys. Rev., Vol. 147, No. 1, pp. 295–302, 1966.

23. Geylikman, B. G., Kresin, V. Z., "Critical Temperature for Ordinary and Anomalous Superconductors," FTT, Vol. 7, No. 11, pp. 3294–3301, 1965.

24. McMillan, W. L., "Transition Temperature of Strong-Coupled Superconductors," Phys. Rev., Vol. 167, No. 2, pp. 331–344, 1968.

25. Allen, P. B., Dynes, R. C., "Transition Temperature of Strong-Coupled Superconductors Reanalyzed," Phys. Rev. B., Vol. 12, No. 3, pp. 105–922, 1975.

26. Rowell, J. M., Schmidt, P. H., "Energy Gaps of Ordered and Disordered A15 'Phases' in Nb_3Ge Measured by Tunneling," Appl. Phys. Lett., Vol. 29, No. 9, pp. 622–625, 1976.

27. Moore, D. F., Zubeck, R. B., Rowell, J. M., Beasley, M. R., "Energy Gaps of A15 Superconductors Nb_3Sn, V_3Si and Nb_3Ne Measured by Tunneling," Phys. Rev. B., Vol. 20, No. 7, pp. 2721–2738, 1979.

28. Youngner, D. W., Klemm, R. S., "Theory of Upper Critical Field in Anisotropic Superconductors," Phys. Rev. B., Vol. 29, No. 9, pp. 3890–3898, 1980.

29. Junod, A., Miller, J., Rietschel, H., Schneider, E., "Chaleur Specifique et Transformation Martensitique dans le Systeme $Nb_{1-x}Sn_x$," J. Phys. Chem. Solids, Vol. 39, pp. 317–328, 1978.

30. Stewart, G. R., Cort, B. S., Webb, G. W., "Specific Heat of A15 Nb_3Sn in Fields to 18 Tesla," Phys. Rev. B., Vol. 24, No. 7, pp. 3841–4846, 1981.

31. Golovashkin, A. I., Rodin, V. V., Shabanova, N. P., "Anomalous Variation of Critical Magnetic Field with Resistivity in Nb_3Sn," 24–ye Vsesoyuz. soveshch. po fizike nizkikh temperatur (Tbilisi, 8–10 sent. 1986): Tez. dokl. [24th National Conference on Low-Temperature Physics (Tbilisi, 8–10 Sep 86): Abstracts of Reports], Tbilisi, Metsniyereba Press, Part 1, pp. 92–93, 1986.

32. Suzuki, M., Anayama, T., Kido, G., Nakagawa, Y., "Upper Critical Fields of Superconducting Nb_3Ge Films," J. Appl. Phys., Vol. 59, No. 3, pp. 975–977, 1986.

33. Orlando, T. P., Alexander, J. A., Bending, S. J., et al., "The Role of Disorder in Maximizing the Upper Critical Field in the NbSn System," IEEE Trans. Magn., Vol. MAG–17, pp. 368–369, 1981.

34. Aleksandrov, A. S., Yelesin, V. F., Kotov, M. A., et al., "Equations of the Ginsburg–Landau–Abrikosov–Gor'kov Theory for Superconductors with Fine Structure of State Density and Radiation Effect on Upper Critical Field of Nb_3Sn," ZhETF, Vol. 87, No. 4(10), pp. 1432–1441, 1984.

35. Fahnle, M., "Atomic Disorder and Superconductivity in A15 Materials," J. Low Temp. Phys., Vol. 46, No. 1/2, pp. 3–26, 1982.

36. Belitz, D., "Correlation Gap Mechanism for T_c Degradation in High–Temperature Superconductors," J. Phys. F: Met. Phys., Vol. 15, pp. 2315–2331, 1985.

37. Muller–Heinzerling, Th., Fink, J., Weber, W., "Electron Energy Loss Spectra of High — T_c A15 Materials," Physica., Vol. 135B, No. 1–3, pp. 347–349, 1985.

38. Muller-Heinzerling, Th., Fink, J., Weber, W., "Direct Observation of the Peaked Density of States in High — T_c A15 Superconductors by Electron Energy-Loss Spectroscopy," Phys. Rev. B., Vol. 32, No. 3, pp. 1850–1852, 1985.

39. Bisli, M. R., Geboll, T. Kh., "Superconducting Materials," UFN, Vol. 148, No. 2, pp. 347–362, 1986.

40. Ogishi, T., Osono, Y., "Superconductivity in Nb–Ge–Al–O Films Above 44K," Appl. Phys. Lett., Vol. 48, pp. 1167–1168, 1986.

41. Bednorz, J. G., Muller, K. A., "Possible High T_c Superconductivity in the Ba–La–Cu–O System," Z. Phys. B. — Condensed Matter, Vol. 64, pp. 189–193, 1986.

42. Golovashkin, A. I., Kuznetsova, T. I., "Relationship of Critical Temperature to Physical Characteristics of Superconductors," Trudy FIAN, Vol. 151, pp. 3–245, 1984.

43. Golovashkin, A. I., Pechen', Ye. V., "Stabilization of High–Temperature Superconducting Phase of Nb_3Ge," Fizika nizkikh temperatur (FNT), Vol. 6, No. 4, pp. 454–464, 1980.

44. Stepanov, G. N., Yakovlev, Ye. N., "Superconductivity of Sulfur at High Pressures," Pis'ma v ZhETF, Vol. 31, No. 11, pp. 657–661, 1980.

45. Golovashkin, A. I., Levchenko, I. S., Shelekhov, A. L., "Study of Electron Structure of Nb_3Al and Nb_3Sn by Heat Reflection," ZhETF, Vol. 86, pp. 2229–2238, 1984.

46. Klein, B. M., Boyer, L. L., Papaconstantopoulos, D. A., Mattheiss, L. F., "Self Consistent Augmented–Planewave Electronic Structure Calculations for A15 Compounds V_3X and Nb_3X, X=Al, Ga, Si, Ge and Sn," Phys. Rev. B., Vol. 18, No. 12, pp. 6411–6438, 1978.

47. Mattheiss, L. F., Weber, W., "Electronic Structure of Cubic V_3Si and Nb_3Sn," Phys. Rev. B., Vol. 25, No. 4, pp. 2248–2269, 1982.

48. Shevchenko, A. D., Sverkhprovodniki s nestabil'noy reshetkoy [Superconductors with Unstable Lattice], Moscow, Metallurgiya Press, 105 pp., 1985.

49. Aleksandrov, A. S., Yelesin, V. F., "Polaron Model of Electron Spectrum and Superconductivity of Compounds with A15 Structure," FTT, Vol. 25, pp. 456–464, 1983.

50. Aleksandrov, A. S., Yelesin, V. F., Elektron-fononnoye vzaimodeystviye i anomal'nyye svoystva uzkozonnykh vysokotemperaturnykh sverkhprovodnikov [Electron-Phonon Interaction and Anomalous Properties of Narrow-Band High-Temperature Superconductors], Moscow, MIFI Press, 40 pp., 1985.

51. Kagan, Yu., Prokof'ev, N. V., "Electron Polaron Effect and Quantum Diffusion of a Heavy Particle in a Metal," ZhETF, Vol. 90, No. 6, pp. 2176–2195, 1986.

52. Soda, T., "On the Mechanism of Heavy-Fermion Superconductivity," J. Phys. Soc., Japan, Vol. 55, No. 5, pp. 1728–1740, 1986.

53. Alexandrov, A., Ranninger, J., "Bipolaronic Superconductivity," Phys. Rev. B., Vol. 24, No. 3, pp. 1164–1169, 1981.

GAP SINGULARITIES OF VOLT-AMPERE CHARACTERISTICS OF SUPERCONDUCTING JUNCTIONS WITH DIRECT CONDUCTIVITY

A. I. Golovashkin, A. N. Lykov

ABSTRACT

This article reviews works dedicated to singularities of gaps in the current-voltage characteristics of junctions with direct conductivity. Significant attention is given to analysis of the specifics observed in bridges and tunnel contacts with shorts consisting of superconductors with A15 lattice. They are explained by introducing phenomenological theoretical models. Experimental data obtained to date are partially contradictory and do not always agree with the models. Figures 6, Table 1, References 41.

1. Introduction

A number of physical and technological singularities have been distinguished among various types of weak superconducting junctions with direct conductivity. This type of junction includes superconducting structures such as superconductor–constriction–superconductor, with the constriction consisting of a short between two superconducting films, various types of bridges between superconducting "banks," point and and pressure junctions also in many cases acting as weak couplings with direct conductivity. They all have different and quite interesting properties. For structures with intentional shorts, an important parameter is the size of the short, which should be as small as possible.

Many works have been published on properties of weak connections with direct conductivity. The steady-state properties of such structures are now rather well known (cf. review[1]). However, studies of their unsteady properties remain timely even today. Quite interesting and attractive effects arise in the area of energies comparable to the gap width Δ in the spectrum of elementary excitations of a superconductor. They are manifested as singularities on the current-voltage characteristics of such structures.

The gap singularities appear most clearly in superconductive junctions. They are observed as a sharp increase in current I with potential differences $V=2\Delta/e$ across a tunnel junction made of two identical superconductors (where e is the charge of an electron). In the structures of interest to us, when the banks on either side of the shorting connection are two identical superconductors, gap singularities act as local maxima in the variation of incremental resistance $R_d=dV/dI$ as a function of voltage V. Usually there is not only a maximum at $V=2\Delta/e$, but also a number of singularities, the positions of which are determined by the equation $V_n=2\Delta/ne$, where n is an integer. These singularities are called the subharmonic gap structure. The nature of the gap singularities in weak connections with direct conductivity is still not fully understood. At present, there are a number of phenomenological models, which will be described below.

In addition to the subharmonic gap, a number of works have observed harmonic gaps as well, i.e., singularities on the current-voltage characteristic at voltages of $V_m=m\Delta/e$, where m is an integer.

Gap singularities can be observed on the current-voltage characteristics of junctions by more complex methods. We are speaking here of the so-called Riedel peak. A variation of critical Josephson tunnel junction current with frequency f was predicted theoretically in [2]. This function has a narrow peak at $hf=2\Delta$ (h is the Planck constant). The existence of this peak was experimentally confirmed in [3].

The importance of the question of gap singularities in the characteristics of weak connections extends beyond what we have discussed. The quantity Δ appears in the power spectrum of Josephson junction current fluctuations[4]. This is particularly important for the high-frequency properties of weak connections where $hf\approx 2\Delta$. The presence of current or voltage fluctuations across a junction means that the spectral width of the Josephson junction radiation line also has a singularity at $V=2\Delta/e$[4].

An energy gap and Reidel singularity have frequently been observed in the characteristics of point contacts. This is usually related to the possibility of achieving tunneling in such junctions. However, we know that point contacts are very complex structures with metallic shorts, and their physical configuration is indefinite. It is therefore possible that there may be other mechanisms causing the gap singularities in the current–voltage characteristics of such junctions.

We have not mentioned certain other possibilities of determining Δ in the characteristics of weak connections. For example, let us note the well known variation of critical Josephson junction current $I_c \sim \Delta / R_N e$ (R_N is the resistance of the junction in the normal state), or the voltage jump of $2\Delta / e$ in the resistive state of superconducting films when phase slip lines or elementary thermal domains appear. The quantity Δ may also appear in junction–receiver characteristics in the detection of recombination phonons radiated by a tunnel junction generator[5]. We note, furthermore, that we shall limit ourselves here to analysis of current–voltage characteristics, and shall not touch upon other characteristics of weak connections, which may also depend on Δ.

Of particular interest is a study of gap singularities in shorts made with superconductors having the A15 or B1 lattice. In addition to the record high critical temperatures achieved in such superconductors, they also have a number of anomalies. Tunneling measurements performed in such superconductors show that there are several gaps in the elementary excitation spectrum. There are two explanations for this fact: first, their occurrence is related to the multiple phases of the specimen, or existence of an upper defect layer with a smaller gap; second, the singularities of the electron spectrum in such semiconductors. A study of shorted junctions on this basis may yield additional information for determination of the electron spectrum of superconductors with A15 lattice.

In this article, we shall study experimentally observed gap singularities in current–voltage characteristics, and shall present models explaining their nature. Significant attention shall be given to singularities observed in bridges and shorts of superconductors with A15 lattice.

2. Experimental Method

1. In many cases, the gap singularities of weak connections with direct conductivity are expressed in the current–voltage characteristics quite weakly. In these cases, subharmonic gap structure and harmonic gap structure, the Rowell resonant system is used to measure $d^2I/dV^2(V)$ [6]. Considering its importance, let us discuss one version of this system which we developed[7] in more detail. The installation was designed to record the second derivative of the current–voltage characteristic by measuring the second harmonic voltage arising across a junction when purely sine–wave signals were applied across it. It can operate with a base frequency voltage across the junction of less than 100 μV.

A block diagram of the installation is shown in Figure 1. The signal at the base frequency of 190 Hz is applied by generator *1* (GZ–33) through separating transformer Tp1 to the input of the reactive circuit shown within the dot–dash line in the figure. The second harmonic signal is fed from the junction studied to the input of preamplifier *2,* from its output to resonance amplifier *3* (U2–6). The amplified signal is converted by synchronous detector *4* (V9–2) and sent through an H–pad attenuator to the "y" input of dual-coordinate recorder *5* (PDS–021M). The reference voltage fed to the synchronous detector is a 380 Hz signal, formed by a dual-junction diode doubler. Current scanning is performed by means of a generator which produces a linearly changing dc current *6,* based on three transistors[7].

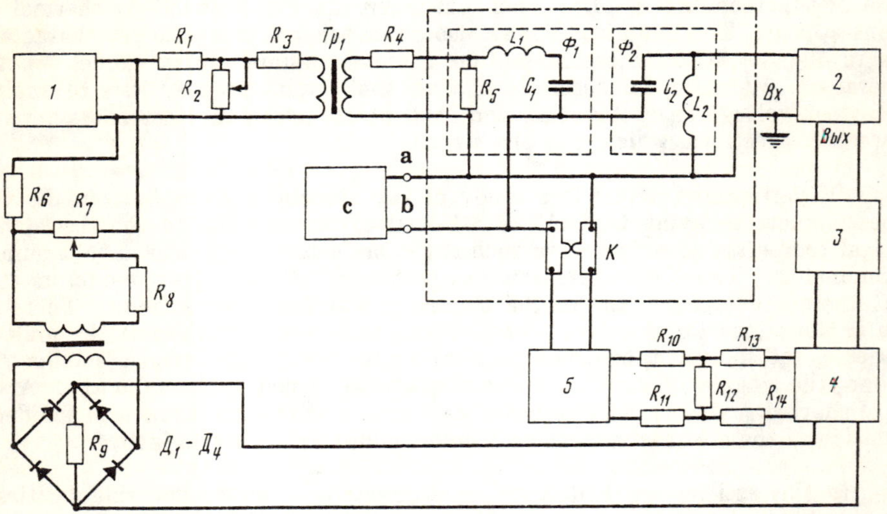

Figure 1. Diagram for Measurement of Second Derivative of d^2I/dV^2 as a Function of V.

The basic installation for measurement of the second harmonic voltage is a reactive circuit which performs signal filtration and impedance conversion. The circuit consists of two LC filters — Φ_1 and Φ_2, connected by means of the junction being studied K. The input filter Φ_1, which is tuned to 190 Hz, passes the signal at the base frequency while suppressing the second harmonic. In our case, this suppression was by a factor of 230. The need for this very careful base frequency signal filtration results from the fact that the level of noise and parasitic harmonics determines the sensitivity of the installation. Output filter Φ_2, tuned to 380 Hz, attenuates the base frequency signal by a factor of 3 while amplifying the second harmonic voltage by a factor of 170. This prevents overloading of the first stages of the resonance amplifier and reduces the influence of internal preamplifier noise.

One significant parameter of the circuit is the resistance of the junction being studied. The optimal incremental resistance R_g is determined by the equation[6]

$$R_g^3 - (R_{L1} + R_5)(2R_{L2} + R_{L1} + R_5) R_g - 2(R_{L1} + R_5)^2 R_{L2} = 0,$$

where R_{L1} and R_{L2} are the active resistances of the windings in the first and second inductance coils. The circuit developed in [6] was designed for the study of tunnel junctions with incremental resistance about 30 Ohms. Considering the low resistance of shorted junctions, we selected the parameters of the circuit so that it could measure d^2I/dV^2 with sufficient accuracy in the range of $R_g \approx 2.5$–3 Ohms. When the incremental resistance varies more broadly, the second harmonic voltage is no longer proportional to the second derivative of the current–voltage characteristic. However, the basic singularities of d^2I/dV^2 coincide with the singularities in the behavior of the second harmonic amplitude.

A schematic diagram of the preamplifier used in the circuit to measure d^2I/dV^2 is presented in [7]. The design of the preamplifier was selected so as to assure high input impedance, ~10 MOhms. This is necessary in order not to decrease the quality of circuit Φ_2 excessively.

2. In our work, we turned particular attention to junctions with direct conductivity made of superconducting compounds with A15 structure. The record value of transition temperature to the superconducting state T_c=24 K was achieved in films of Nb_3Ge[8]. Other compounds of this type have also achieved high T_c, over the triple point of hydrogen. One distinguishing feature of high-temperature superconducting compounds is the instability of their structure[9, 10]. The clearest manifestation of structural instability is the presence of structural transitions in a number of compounds. When the temperature in these compounds drops, a transition occurs from the cubic phase to a tetragonal phase. In V_3Si, this transition occurs at 21 K, in Nb_3Sn at 43 K. This structural instability makes the manufacture of certain superconducting compounds with A15 lattice a nontrivial task.

There are several methods for producing such compounds (most common are compounds based on niobium and vanadium): metallurgical (arc melting); diffusion, in which a second element (such as Sn or Ga) is precipitated onto a surface of niobium or vanadium, then annealing is performed in a vacuum or inert gas; various methods of joint evaporation of the component elements; the method of precipitation from the gas phase, consisting of mixing chlorides of the elements with their simultaneous reduction in hydrogen. Unfortunately, there is no one method suitable for manufacturing films of all compounds with A15 and B1 lattice. Usually, each compound has its own specific method which yields the best results.

We studied junctions based on films of Nb, Nb_3Sn, Nb_3Ga, Nb_3Ge and V_3Si. The technology of manufacturing these films was described previously (for example, cf. [7, 8, 11]). Here we note simply that niobium films and V_3Si films were prepared by electron-beam evaporation in a vacuum. Films of Nb_3Sn and some of the Nb_3Ga films were produced by separate evaporation of the components in a vacuum with condensation on a substrate in the required proportions. Another portion of the Nb_3Ga films was produced by a diffusion method. Films of Nb_3Ge were made by cathode sputtering in an atmosphere of argon.

3. Junctions with direct conductivity can be divided into two main types. The first is the bridge junction, i.e., superconductor — micron-sized constriction — superconductor, the second is the tunnel junction with a short. There are many methods for the creation of microscopic bridges, and we used in particular the method of scratching a film with a diamond tool[7]. Tunnel junctions were created by the usual method[11, 12]. The oxide layer was created by exposure in air for 0.5–2 hours of either freshly precipitated or electropolished films of the high-temperature superconductor. The electropolishing etchant used was a mixture of 10 H_2O+5H_2SO_4+4HNO_3+1HF. The top electrode (usually of lead) was precipitated in a vacuum of $1 \cdot 10^{-5}$ mmHg through a tin foil mask. The presence of shorts in a junction was determined by the shape of the current-voltage characteristic.

3. Experimental Results

Gap singularities at $V_n=2\Delta/ne$ were first detected in [13] for tunnel junctions. The only singularity clearly observed was at n=2. Rowell (cf. commentary[14]) reported the subharmonic structure $2\Delta/n$ right up to n=12. Apparently, the first work in which the subharmonic gap structure was observed in the current-voltage characteristic of tunnel junctions with shorts was the work of Yanson, Svistunov and Dmitrenko[15]. In this work, the $2\Delta/n$ gap structure was found in the current-voltage characteristics of Sn-SnO_2-Sn junctions with great supplementary current. The subharmonic gap structure was later observed in [16-19]. Detailed studies of the subharmonic gap structure in the current-voltage characteristic of tunnel junctions with shorts were undertaken by Rowell and Feldman[20]. Typical current-voltage characteristics observed in such junctions are shown in Figure 2. The authors of [20] noted the following features.

1. The current at $V<2\Delta/e$ may be comparable to the current at $V>2\Delta/e$. In an ordinary tunnel junction, the current at low voltage is proportional to $\exp(-\Delta/kT)$ (k is Boltzman's constant) and approaches zero as the temperature approaches zero. 2. The current at $V<2\Delta/e$ is independent of temperature. 3. The current is insensitive to magnetic fields. 4. The structure of the current-voltage characteristic in the area of $V\sim\Delta/e$ is manifested not as an increase in current ($\max dI/dV$) as in ordinary tunnel junctions, but rather as a decrease in conductivity ($\min dI/dV$). 5. The current decreases smoothly as V changes from $2\Delta/e$ to 0, and a structure of the same form as at Δ/e is observed at $2\Delta/3e$, $2\Delta/4e$, ...

The singularities noted in points 1, 2, 3 and 5 are characteristic of all superconducting tunnel junctions with shorts, this being how they differ from classic tunnel junctions.

An example of a current–voltage characteristic and its derivative for an asymmetrical Sn–Pb tunnel junction with a short is shown in Figure 3, taken from [20]. The following are characteristic for this type of junction.

1. The singularity in the vicinity of $(\Delta_{Pb}+\Delta_{Sn})/e$ appears as a peak, characteristic for a single-frequency tunnel current, against the background of a general decrease in conductivity. 2. There is a singularity at $V=\Delta_{Pb}/e$, but there is no similar singularity at $V=\Delta_{Pb}/ne$, where n is an integer greater than one. 3. There are no singularities at $V=(\Delta_{Pb}+\Delta_{Sn})/ne$. 4. Singularities are clearly seen at $V=2\Delta_{Sn}/me$, where m is an even number; for odd m (except $m=3$) this series is not observed.

Similar characteristics were also observed in microbridge junctions based on such simple superconductors as In and Sn (Figure 3*a*)[21-23]. In addition to the subharmonic gap structure and $V=2\Delta(T)/ne$, the authors of [22] also observed singularities at $V=4\Delta/e$; $6\Delta/e$, i.e., harmonics of the gap, with temperature variation of the position of the singularities similar to the variation of the energy gap as a function of temperature.

A detailed analysis of the experimental data was undertaken in [24], where a number of generalizations were made. 1. The subharmonic gap structure is not observed if one of the banks of the short or microbridge is in the normal state, although excess current is observed. 2. The subharmonic gap structure and excess current are present only at metal junctions. 3. The position of singularities (peaks) and the excess current magnitude are determined by the size of the gap in the spectrum of elementary excitation. 4. The presence of defects scattering electrons weakens the structure. 5. The amplitude of the singularities decreases monotonically with increasing n. Gap subharmonics of very high order ($n>10$) may be observed.

There are also works which do not satisfy these regularities. Interesting results in this respect were obtained in [25], which studied Pb–Nb–Pb bridge junctions (i.e., banks of lead, the bridge itself made of niobium). The dimensions of the bridges (length 0.8–4.5 μm, width 1.2–2 μm) were much greater than the coherence length. Furthermore, in order to suppress the order parameter, ions were implanted in the bridges. As a result of this operation, the critical temperature of the niobium bridge dropped to a value less than the critical temperature of lead, while the free path length was greatly reduced. In the current–voltage characteristics of these bridges, the authors observed singularities at $V=2\Delta(T)/e$, where $\Delta(T)$ is the gap in niobium. In contrast to all previous works, which reported observation of gap singularities in the current–voltage characteristics, the junctions studied in this work involved the movement of Abrikosov eddies. Furthermore, their bridges had a large number of defects, which did not prevent observation of gap singularities. In this case, in contrast to the previous works, the singularities were related to the superconductive gap formed by the microbridge, not to the banks.

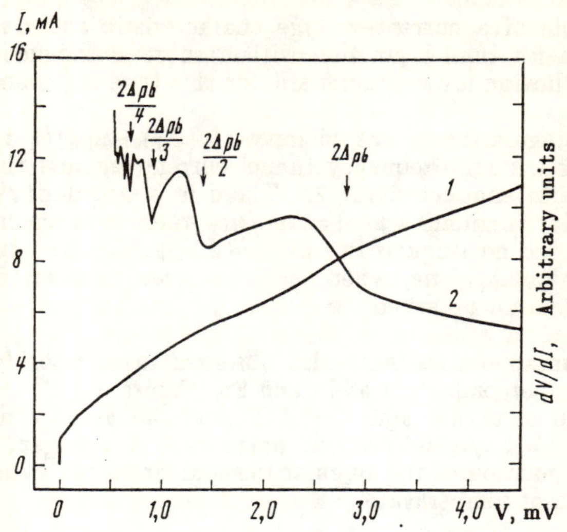

Figure 2. Current–Voltage Characteristic (1) and $dV/dI(V)$ (2) of Shorted Pb–Pb Tunnel Junction at T=1 K[20].

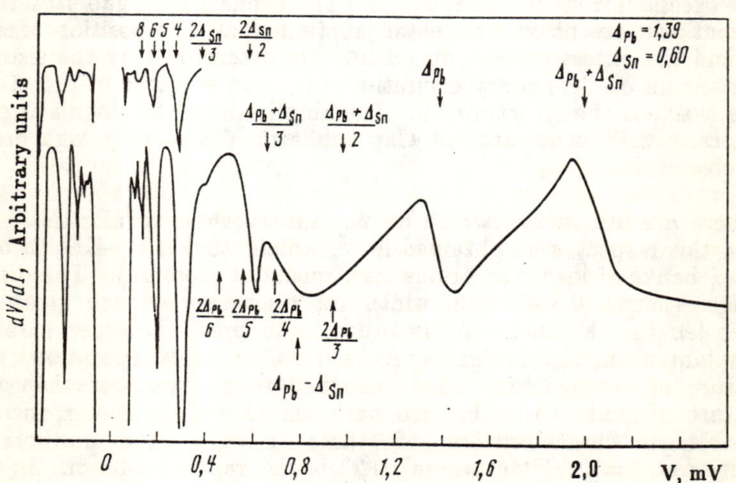

Figure 3. Incremental Resistance as a Function of Voltage in Shorted Pb–Sn Tunnel Junction at T=1 K[20].

The results presented above indicate that the subharmonic gap structure in the current–voltage characteristics of junctions with direct conductivity can be used to produce important information on junctions, particularly to determine their energy gaps. In many cases the subharmonic gap structure has also been used in more precise measurements. For example, in [26] it was used in an experimental determination of the superheating of tin whisker junctions as a function of the dissipated power.

Interesting experiments were performed by the authors of [22, 27]. In these works, the influence of external microwave radiation on the current–voltage characteristic either of microbridges of In and Sn [22] or of pressure point Nb–Nb junctions in the area of gap singularities and their subharmonics were studied. The authors found that rf energy had a strong influence on the form of the current–voltage characteristic (Figure 3*b*). As the microwave radiation power increased, the main peaks at voltages $2\Delta/me$ decreased and additional singularities appeared at voltages $(2\Delta \pm nhf)/me$, where f is the frequency of the external electromagnetic radiation. The variation in amplitude of these singularities with microwave power follows the Bessel rule $J_n^2(meV_f/hf)$, where V_f is the rf voltage applied to the junction. A similar rule of change in amplitude as a function of radiation power is observed in ordinary Josephson junctions at voltages of $nhf/2e$. This indicates similarity of the phenomena involved.

As we see, the experimental facts listed above, in part contradicting each other, indicate the complex nature of the phenomenon. Still more contradictory are the experimental results obtained in junctions of high–temperature superconducting compounds. High–temperature compounds are those superconductors with critical temperature $T_c > 15$ K. They almost all have A15 or B1 crystalline lattice.

Superconducting compounds with A15 and B1 lattice have long attracted great interest. This is a result of the high temperature of transition to the superconducting state of some of these compounds. However, there is still a great deal which is unclear in the nature of the superconductivity of these compounds. Suffice it to say that in some works, for example [28], superconductivity in these compounds is explained by a polaronic mechanism. Therefore, any new methods and results of studying such superconductors are quite useful.

Let us now list the basic experimental facts obtained in studies of junctions with direct conductivity based on high–temperature superconductors. First of all we note that the harmonics and subharmonics of the energy gap have been observed both in tunnel junctions with shorts[11, 29–31], and in bridge junctions[37]. In [29], shorted tunnel junctions were studied in which the lower electrode was Nb_3Sn, the upper electrode was Pb, Sn or Al. In the low voltage area, maxima were observed in the incremental conductance (subharmonic gap structure) at voltages $V_n=(\Delta_{eff}+\Delta_2)/ne$, where Δ_2 is the energy gap of the second metal, $\Delta_{eff}=0.95$ meV is the effective gap in Nb_3Sn, determined from the position of the main tunnel state density maximum. In this work, four values of gap in Nb_3Sn were observed simultaneously at 2.35, 1.12, 0.75 and 0.18 meV. The amplitudes of even subharmonics were greater than the amplitudes of odd subharmonics. In addition to the subharmonics of the effective gap, making up the main structure, several primary subharmonics of other gaps were also observed. Furthermore, in [29, 30] in Nb_3Sn–Sn, Nb_3Sn–Pb and Nb_3Sn–Nb_3Sn tunnel junctions with shorts, in many cases gap harmonics were observed. The variation of d^2I/dV^2 as a function of V manifested periodic minima. The major period and fine-structure period of these junctions are presented in the table below, taken from [30].

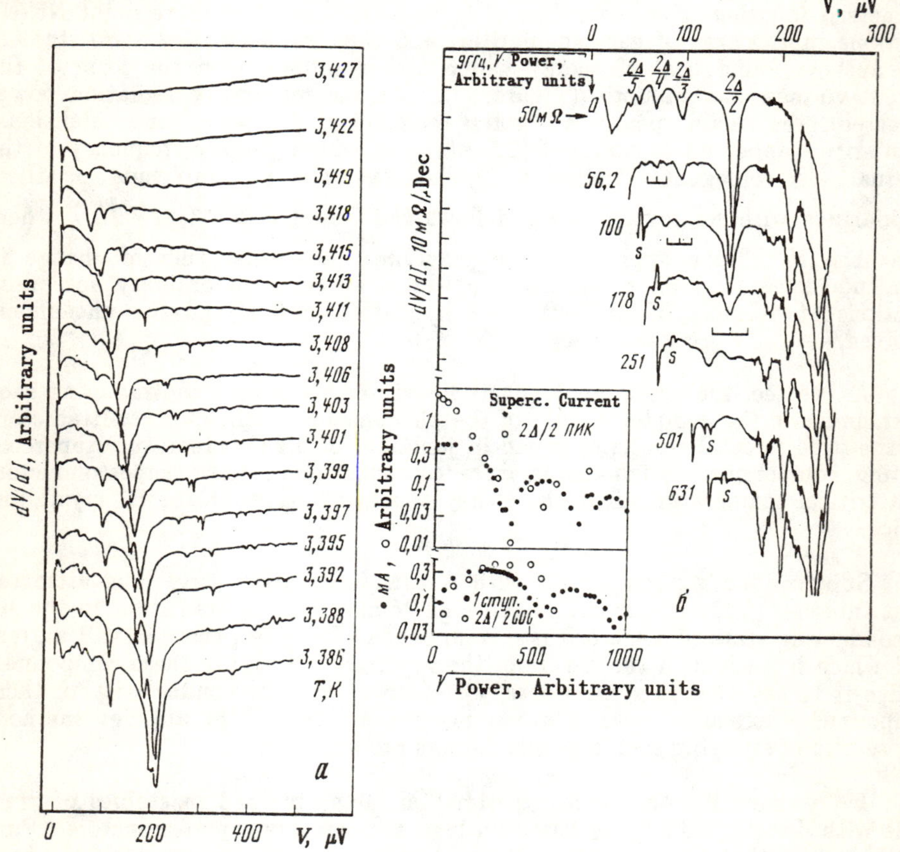

Figure 4. Incremental Resistance as a Function of Voltage in Microbridge Based on Indium Film at Various Temperatures[22] (a) and Various Levels of Microwave Power (b). Inset shows variation in critical current and singularity amplitudes as a function of microwave radiation power[22].

Subharmonic gap structures and harmonic gap structures were observed in [29, 30] in tunnel junctions with shorts having noticeable superconducting current. However, in [11] the same structure was observed in Nb_3Ga–Pb shorted tunnel junctions, which did not superconduct. Figure 5 taken from this work shows the current–voltage characteristic and its derivatives dV/dI and d^2/dV^2 for an Nb_3Ga–Pb junction with a microshort, in which two gap singularities are observed: a sharper one corresponding to $\Delta_l+\Delta_{Pb}$, where V=3.5 mV and a broader one corresponding to $\Delta_s+\Delta_{Pb}$ in the area of 2 mV (more clearly seen in the first derivative), where Δ_l and Δ_s are the large and small gaps in the excitation spectrum of Nb_3Ga, Δ_{Pb} is the gap in the excitation spectrum of lead.

Periods of Oscillations of Conductivity in Tunnel Junctions[30]

Specimen	Tunnel Junction Type	*T*, K	Main Period	Fine Structure Period
No. 1	Nb_3Sn-Sn	3	0.53	---
No. 1	Nb_3Sn-Sn	1.9	0.55	0.14
No. 2	Nb_3Sn-Sn	2	0.58	---
No. 3	Nb_3Sn-Sn	2	0.43	---
No. 4	Nb_3Sn-Sn	2	0.44	---
No. 5	Nb_3Sn-Sn	4.2	0.97	---
No. 6	Nb_3Sn-Nb_3Sn	4.2	1.0	0.25
No. 7	Nb_3Sn-Pb	4.2	0.95	---
No. 8	Nb_3Sn-Pb	4.2	0.98	---
No. 9	Nb_3Sn-Pb	4.2	1.0	---
No. 10	Nb_3Sn-Pb	4.2	1.32	0.3

The presence of two or more singularities in the current–voltage characteristics of tunnel junctions without shorts, which can be interpreted as gaps in the spectrum of elementary excitations, have been repeatedly observed for intermetallic superconducting compounds with A15 lattice[12, 31, 32]. There are several hypotheses to explain the complex nature of the current–voltage characteristics of tunnel junctions based on these materials. Apparently the most acceptable is the explanation of the complexity of the current–voltage characteristic as due to complexity of the phases of the specimens[32]. The large gap is related to the presence of a high–temperature A15 phase. We know that under certain conditions (textured specimens), gap anisotropy may appear[12]. There are also theoretical models for the presence of two gaps in materials with complex electron structure[33] or in heterogeneous systems[34]. All these explanations are based on the electron–phonon mechanism of superconductivity. It would be interesting to explain the presence of several gaps in the excitation spectrum of intermetallic superconducting compounds with A15 lattices on the basis of the polaronic mechanism of superconductivity.

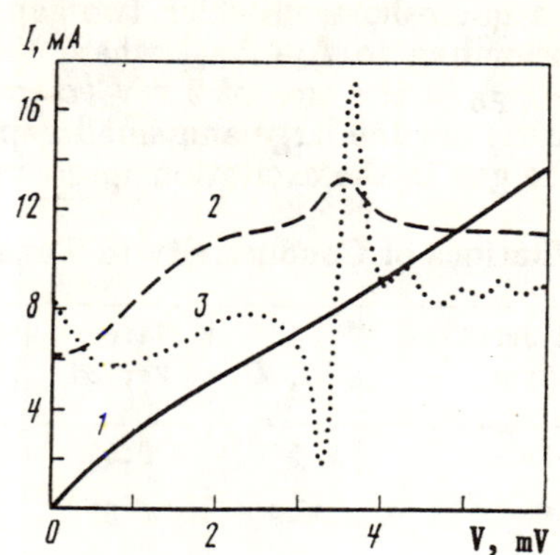

Figure 5. Current–Voltage Characteristic (1) and Its Derivatives dV/dI (2) and d^2I/dV^2 (3) of Nb_3Ga–Pb Junction (T=4.2 K). Scale of ordinate for derivatives arbitrary. Nb_3Ga film produced by joint precipitation on niobium foil, T_c^H=17 K.

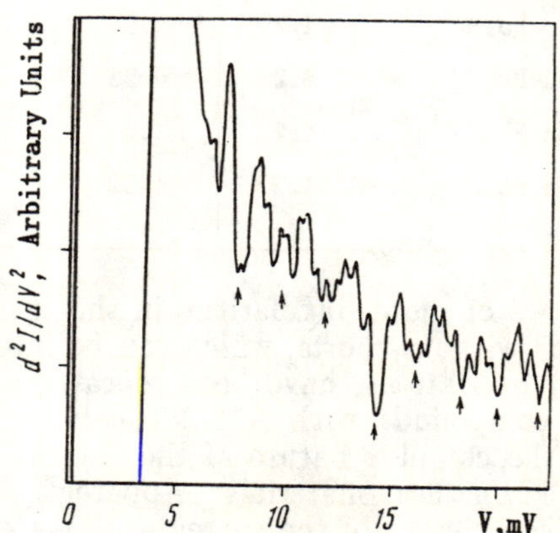

Figure 6. Example of $d^2I/dV^2(V)$ in the Area of High Voltages for Nb_3Ga–Pb Specimen. T_c^H=16.7 K, ΔT=3.5 K.

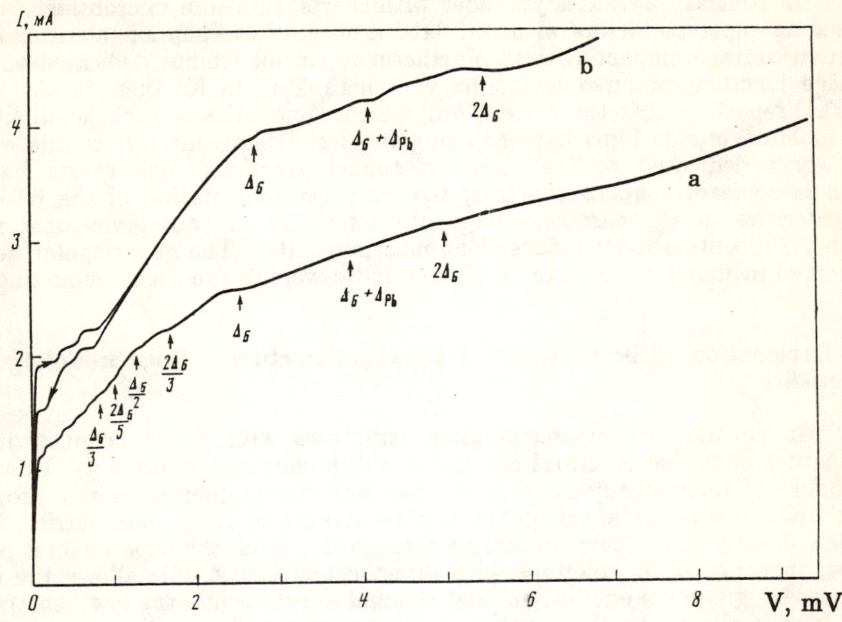

Figure 7. Current–Voltage Characteristics of Nb_3Ge–Pb Tunnel Junctions with Microshorts (T=4.2 K). Current scale for curve *b*: 1 division =2 mA.

As in [30], in [11] oscillations with a period of 0.6 mV (Figure 5) were observed in the curve of d^2I/dV^2 as a function of V in the high voltage area (where $eV>\Delta_1+\Delta_{Pb}$). We note that this period is close to the value of the small gap. Figure 6 shows another example of the variation of d^2I/dV^2 as a function of V in a shorted Nb_3Ga-Pb junction[11]. This function also has a complex periodic shape. The position of the singularity, noted in the figure by the arrows, on the voltage scale consists of multiples of 2 mV, i.e., multiples of one of the gaps of Nb_3Ga, the fine structure period is 0.6 mV. Thus, as in [29, 30], small and large gap harmonics are observed in the high-temperature superconductor.

For understandable reasons, works dedicated to the study of Nb_3Ge are of particular interest today. This includes tunnel studies of this material. Figure 7, taken from [35], shows current–voltage characteristics of Nb_3Ge–Pb tunnel junctions with microshorts. Even without measuring the derivatives we can see the singularities related to the harmonics and subharmonics of the large gap in Nb_3Ge. Thus, tunnel junctions with microshorts based on various intermetallic compounds with A15 lattice have common properties. Usually the current–voltage characteristics of these junctions contain subharmonic gap structures and harmonic gap structures related to various gaps in the elementary excitation spectra of these semiconductors.

In contrast to tunnel junctions with shorts, thin-film microbridge junctions based on superconductors with A15 lattice do not have gap singularities in their current–voltage characteristics. Furthermore, in all studies dedicated to microbridge junctions of superconductors with high T_c(>15 K) there is but a single work[36] reporting that the current–voltage characteristics of such junctions based on niobium nitride films have gap singularities. Unfortunately, in this work, as in works dedicated to the study of tunnel junctions with shorts based on high-temperature superconductors, the temperature variation of the position of singularities in the current–voltage characteristics are not given or compared with $\Delta(T)$ obtained by independent measurements. The gap singularities were observed in junctions in which both electrodes were in the superconducting state.

4. Explanation of the Existence of the Gap Structure in Current–Voltage Characteristics

In contrast to superconducting structures with direct conductivity, the properties of tunnel junctions are rather well understood at present. This is true not only of their steady-state properties, but also of their unsteady properties. The difference is explained by the low penatrance of the tunnel barrier, leading to low density of the current passing through it. From the experimental point of view, in contrast to junctions with direct conductivity, this allows the experimenter to ignore nonequilibrium and thermal effects, while the low penetrance of the barrier allows the theoretician to use the "tunnel" formalism to describe the junction.

There are three base theoretical models to explain the appearance of the subharmonic gap structure: multiparticle tunneling, detection of natural radiation, multiple Andreyev reflection.

Multiparticle tunneling. The Tien–Gordon expression [37] for single-frequency tunnel current through a junction struck by rf radiation is written as

$$I(V_{dc}) = \sum_n J_n^2(\alpha)\, I_0(V_{dc} + nhf/e), \tag{1}$$

where V_{dc} is the constant component of voltage across the junction, $I_0(V_{dc})$ is the form of the current–voltage characteristic without the rf energy, f is the frequency of the incident radiation, J_n is the nth order Bessel function, $\alpha = eV_f/hf$ and V_f is the amplitude of the voltage induced across the junction by the rf radiation. $J_n(\alpha)$ is the probability that an electron will absorb energy as it passes through the tunnel barrier. The form of the current–voltage characteristic of a tunnel junction in autonomous mode, as is well known, has a singularity at $V_{dc}=2\Delta/e$.

When m electrons pass through a barrier, the probability that the first electron will absorb energy is n_1hf, that of the second is n_2hf, etc. is obviously equal to the product $J_{n_1}(\alpha)J_{n_2}(\alpha) \ldots J_{n_m}(\alpha)$. Here, $\sum_{i=1}^{m} n_i = n$. The full amplitude of the probability of absorption by these m electrons of energy nhf is obtained after summation of all such probability amplitudes. As was shown in [38], it is $J_n(m\alpha)$. Obviously, the energy acquired by the m electrons passing through the barrier and absorbing n quanta is $meV_{dc}+nhf$. When it is equal to 2Δ, a singularity should be observed in the current–voltage characteristic. Thus we explain the experimental fact of existence of the subharmonic gap structure. The value of the nth satellite arising near the mth subharmonic upon exposure to rf radiation is proportional to $J_n^2(m\alpha)$[27]. We note that the theory of process of two–frequency tunneling was developed in [39].

However, this model has serious shortcomings as an explanation for the subharmonic gap structure. For junctions with direct conductivity, where this structure is frequently observed, strictly speaking it is impossible to apply the tunnel effect formalism. Its use should lead to differences between theory and experiment. Thus, for example, the general form of the current–voltage characteristic calculated in [37] differs from the experimental results. Furthermore, in the experiment the subharmonic gap structure appears as maxima in the function $R_d(V)$, while theoretically it should appear as minima. We note yet another problem with application of multifrequency tunneling to explain the subharmonic gap structure. The probability of multifrequency tunneling is proportional to $|T|^{2m}$, where T is a tunnel matrix element. Since usually T is small, it is unusually difficult to observe even two–frequency tunneling. The use of multifrequency tunneling to explain the subharmonic gap structure must assume the possibility of 5–6–particle tunneling or even greater in some experiments, which is improbable.

Detection of natural Josephson radiation. When constant voltage V_{dc} is applied to a superconducting short, variable current $I_\omega(t)=I_c\cdot\sin\omega_0 t$ will pass through the junction. Its frequency is determined by the equation $\omega_0=4\pi\cdot e\cdot V_{dc}/h$. The junction need not necessarily be a Josephson junction. A variable current of the same frequency, though with lower amplitude, arises across bridges which are large in comparison to the coherence length, due to the periodic movement of Abrikosov eddies. Now to consider the effect of natural radiation on the form of the current–voltage characteristic we shall assume (though not absolutely) that this variable current is created at the junction by variable voltage $V_\omega\cos\omega_0 t$. Assuming that an ordinary single–frequency tunnel current also passes through the junction (which is quite realistic for a tunnel junction with a short), and using expression (1) to determine its value, we obtain in this case

$$I(V_{dc}) = \sum_n J_n^2 (V_\omega/2V_{dc}) \cdot I_0 [(2n+1) V_{dc}]. \qquad (2)$$

Since the current–voltage characteristic of a single-frequency tunnel junction has a knee-shaped singularity at $V=2\Delta/e$, expression (2) indicates that the current–voltage characteristic of a junction with a short should have singularities at voltages of $V_{dc}=2\Delta/(2n+1)e$.

The existence of even subharmonics is explained by assuming breakdown of Cooper pairs by natural radiation. Here, when $nh\omega_0/2\pi=2\Delta$, the number of quasiparticles in the junctions increases, leading to singularities in the current–voltage characteristic. This mechanism can be used to explain the existence of subharmonic satellites when external microwave radiation strikes a junction, as well as the variation in their amplitude as a function of microwave radiation power. We note that the amplitudes of even and odd subharmonics in the experiments differ little, while their mechanisms are somewhat different within the framework of self detection. This is a shortcoming of the self detection model. However, we must note that the even and odd series of gap subharmonics can be observed individually.

Multiple Andreyev reflection. An interesting mechanism of appearance of the current–voltage characteristic gap structure was suggested in [24]. The essence of this mechanism is as follows. As we know, at the boundary between a normal metal and a superconductor an electron moving from the normal metal experiences a unique sort of reflection. If its energy is less than Δ, it is reflected and moves back in the direction of a hole. The same is also true of a hole, but when it is reflected, it is converted to an electron. This phenomenon is called Andreyev reflection. If the energy of a particle striking the boundary exceeds the energy gap, it is possible for the electron to pass through the boundary.

Let us assume, as was done in [24], that a microbridge is in the normal state and the electron free path length (or short length) in it is greater than the length of the microbridge. In this case, multiple reflection of quasiparticles at the boundaries of the microbridge may occur, converting an electron to a hole and vice versa. Each time a quasiparticle passes through the microbridge its energy is increased by eV. As soon as the energy exceeds the gap in the excitation spectrum of the superconductor, it becomes possible for the electron to pass through the boundary, forming an additional quasiparticle in the superconductor. In [24], Andreyev reflection was considered using the equations of Bogolyubov. The authors were able to calculate the current–voltage characteristic, and to demonstrate that this model agrees well with the experimental data. Their model was also applicable for two different superconductors. In this case, if $2\Delta_1 \ll \Delta_2$, where Δ_1 is the value of the smaller gap, while Δ_2 is the value of the larger gap, singularities should be observed at voltages of Δ_2/e, $(\Delta_1+\Delta_2)/e$ and Δ_1/e, which previously was observed experimentally[18]. One shortcoming of the model suggested by the authors of [24] is that within its framework the subharmonic gap structure at $T=0$ should disappear, which is not observed in the experiments. In [40], dedicated to further development of the ideas suggested in [24], this shortcoming is eliminated. The authors used a superconductor–oxide–normal short–oxide–superconductor junction. In this case at the superconductor–normal metal boundary in addition to Andreyev reflection, normal reflection also occurs, consideration of which leads to the existence of a subharmonic gap structure at $T=0$. We should emphasize that the mechanism responsible for the existence of the subharmonic gap structure is Andreyev reflection.

Since a model based on multiple Andreyev reflection of quasiparticles at the boundaries of junctions is applicable for junctions which are short in comparison to the electron free path length, it is best used in the case of tunnel junctions with shorts. At the same time in order to explain the existence of a gap structure in the current–voltage characteristics of bridge junctions, particularly bridges of Nb [25] or NbN [36], the model with detection of natural Josephson radiation is more acceptable, since the condition of small dimensions of the bridge in comparison to the free path length disappears in this model.

We note that models with self detection and Andreyev reflection also explain the main singularity at $V=2\Delta/e$ quite well. Of course, this singularity can be explained most simply by the presence of normal quasiparticles, the energies of which at $V=2\Delta/e$ are sufficient to break down pairs. When such quasiparticles are present, singularities should also be observed at $V=4\Delta/e$, $6\Delta/e$, etc. (gap harmonics). Incidentally, in a model with self detection, gap subharmonics and harmonics are explained from the same point of view, which is difficult for the model with Andreyev reflection.

Certain other models have also been suggested to explain gap subharmonics (for example, cf. [41]).

5. Conclusions

In this article we have attempted to present the most complete possible review of works dedicated to gap singularities in the current–voltage characteristics of junctions with direct conductivity. The experimental data obtained to date are frequently contradictory, while the theoretical models suggested for their explanation are phenomenological in nature.

REFERENCES

1. Likharev, K. K., "Superconducting Weak Connections: Steady-State Processes," UFN, Vol. 127, pp. 185–220, 1979.

2. Riedel, E., Zum tunneleffekt bei supraleitern im mikrowellenfeld, Z. Naturforsch, Vol. 19a, pp. 1634–1635, 1964.

3. Hamilton, C. A., "Frequency Dependence of the Josephson Current," Phys. Rev. B., Vol. 5, pp. 912–913, 1972.

4. Rogovin, D., Scalapino, D. J., "Fluctuation Phenomena in Tunnel Junction," Ann. Phys., Vol. 86, pp. 1–90, 1974.

5. Eisenmenger, W., Dayem, A. H., "Quantum Generation and Detection of Incoherent Phonons in Superconductors," Phys. Rev. Lett., Vol. 18, No. 4, pp. 125–127, 1967.

6. Thomas, P. E., Rowell, J. M., "Low Level Second Harmonic Detection System," Rev. Sci. Instrum., Vol. 36, pp. 1301–1306, 1965.

7. Golovashkin, A. I., Lykov, A. N., "Some Methods and Results of Studying Bridge Junctions of Nb and Nb_3Sn," Preprint FIAN, No. 203, Moscow, 84 pp., 1977.

8. Golovashkin, A. I., Pechen', Ye. V., "Superconducting Compounds Nb_3Ge and Nb_3Si," Trudy FIAN, Vol. 151, pp. 63–136, 1984.

9. Testardi, L., Weger, M., Gol'dberg, I., Sverkhprovodyashchiye soyedineniya so strukturoy β-vol'frama [Superconducting Compounds with β-Tungsten Structure], Moscow, Mir Press, 435 pp., 1977.

10. Golovashkin, A. I., Kuznetsova, T. I., "Relationship of Critical Temperature to Physical Characteristics of Superconductors," Trudy FIAN, Vol. 151, pp. 3–62, 1984.

11. Golovashkin, A. I., Levchenko, I. S., Lykov, A. N., et al., "Measurement of Energy Gap in Nb_3Ga Superconducting Films by Tunnel Method," FTT, Vol. 24, pp. 1328–1333, 1982.

12. Vedeneyev, S. I., Golovashkin, A. I., Levchenko, I. S., et al., "Study of Tunnel Characteristics of Evaporated Superconducting Nb_3Sn Films," ZhETF, Vol. 63, pp. 1010–1019, 1972.

13. Taylor, B. N., Burstein, E., "Excess Currents in Electron Tunneling Between Superconductors," Phys. Rev. Letts., Vol. 10, No. 1, pp. 14–17, 1963.

14. Rowell, J. M., "Tunneling Between Superconductors," Rev. Mod. Phys., Vol. 36, No. 1, pp. 199–200, 1964.

15. Yanson, I. K., Svistunov, V. M., Dmitrenko, I. M., "Experimental Observation of Tunneling of Cooper Pair Between Thin Layers of Superconducting Tin," ZhETF, Vol. 47, No. 6, pp. 2091–2094, 1964.

16. Marcus, S. M., "$2\Delta/n$ Structure of Pb–PbO–Pb Tunnel Junctions," Phys. Lett., Vol. 19, No. 8, pp. 623–625, 1966.

17. Marcus, S. M., "The Magnetic Field Dependence of the $2/n$ Structure Observed in Pb–PbO–Pb Superconducting Tunneling Junctions," Phys. Lett., Vol. 20, No. 3, pp. 236–237, 1966.

18. Rochlin, G. I., Douglass, D. H., Jr., "Subharmonic Structure in Pb–Pb Superconductive Junctions," Phys. Rev. Lett., Vol. 16, No. 9, pp. 359–361, 1966.

19. Douglass, D. H., Jr., Rochlin, G. I., "Internal-Photon Assisted Tunneling in Superconducting Junctions," Bull. Am. Phys. Soc., Vol. 11, No. 1, pp. 87–88, 1966.

20. Rowell, J. M., Feldman, W. L., "Excess Currents in Superconducting Tunnel Junctions," Phys. Rev., Vol. 172, pp. 393–401, 1968.

21. Greger-Hansen, P. E., Hendrics, E., Levinsen, M. T., et al., "Subharmonic Energy–Gap Structure and Josephson Radiation Enhanced Structure in Dayem Bridges," Phys. Rev. Lett., Vol. 31, pp. 524–526, 1973.

22. Greger-Hansen, P. E., Pickett, G. R., "Subharmonic Energy Gap Structure and Gap–Enhancement by the Josephson Radiation in Small Dayem Bridges," Rev. de Phys. Appl., Vol. 9, pp. 145–152, 1974.

23. Gubankov, V. N., Koshelets, V. P., Ovsyannikov, G. A., "Nonequilibrium Phenomena in Variable–Thickness Microbridges," Fizika nizkikh temperatur, Vol. 6, pp. 50–59, 1980.

24. Klapwijk, T. M., Blonder, G. E., Tinkham, M., "Explanation of Subharmonic Energy Gap Structure in Superconducting Junctions," Physica., Vol. 109–110 B+C, pp. 1657–1664, 1982.

25. Crozat, P., Vernet, G., Adde, R., "Properties of Coherent Vortex Motion in Pb–Nb–Pb Ion Implant Variable Thickness Bridges," Proc. 2nd. Int. Conf. SQUID–80, New York, N.Y., pp. 157–163, 1980.

26. Golovashkin, A. I., Ivanenko, O. M., Mitsen, K. V., "Study of Weakly–Connected Superconducting Junctions Based on Tin Whiskers," Fizika nizkikh temperatur, Vol. 12, No. 4, pp. 368–373, 1986.

27. Hoffman Soerenson, O., Kofoed, B., Pedersen, N. F., et al., "Effects of rf Subharmonic Gap Structure in Superconducting Junctions," Rev. de Phys. Appl., Vol. 9, No. 1, pp. 153–155, 1974.

28. Aleksandrov, A. S., Yelesin, V. F., "The Nature of Anomalous Properties and Superconductivity of High-Temperature Superconductors," Vozdeystviye izlucheniy na sverkhprovodniki [Effects of Radiation on Superconductors], Collection of articles, Moscow, Energoatomizdat Press, pp. 4–18, 1983.

29. Vedeneyev, S. I., Golovashkin, A. I., Motulevich, G. P., "Low-Frequency Phonon State Density Maxima, Harmonics and Subharmonics of Nb_3Sn Energy Gap," Piz'ma v ZhETF, Vol. 16, No. 4, pp. 216–219, 1972.

30. Vedeneyev, S. I., Golovashkin, A. I., Motulevich, G. P., "Oscillations of Current in Tunnel Junctions with Nb_3Sn at Voltages Greater than the Superconducting Gap," FTT, Vol. 16, No. 3, pp. 688–671, 1974.

31. Moore, D. F., Rowell, J. M., Beasley, M. R., "Superconductive Tunneling into Niobium-Tin Thin Films," Solid. St. Comm., Vol. 20, No. 4, pp. 305–308, 1976.

32. Rowell, J. M., Schmidt, P. H., "Energy Gaps of Order and Disorder in A15 'Phases' Measured by Tunneling," Appl. Phys. Lett., Vol. 29, No. 9, pp. 622–625, 1976.

33. Suhl, H., Mattias, B. T., Walker, L. R., "Bardeen-Cooper-Schreiffer Theory of Superconductivity in the Case of Overlapping Bands," Phys. Rev. Lett., Vol. 3, No. 8, pp. 552–554, 1959.

34. McMillan, W. L., "Tunneling Model of the Superconducting Proximity Effect," Phys. Rev., Vol. 175, No. 2, pp. 537–542, 1968.

35. Golovashkin, A. I., Lykov, A. N., Pechen', Ye. V., "Study of Bridge and Tunnel Junctions Based on Nb_3Ge Films," FTT, Vol. 25, No. 5, pp. 1305–1309, 1983.

36. Lykov, A. N., Motulevich, G. P., Tikhanskiy, M. V., Sukhov, V. V., "Superconducting Microbridges Based on Niobium Nitride," FTT, Vol. 28, No. 8, pp. 2299–2305, 1986.

37. Tien, P. K., Gordon, J. P., "Multiphonon Process Observed in the Interaction of Microwave Fields with the Tunneling Between Superconductor Films," Phys. Rev., Vol. 129, No. 2, pp. 647–651, 1963.

38. Hasselberg, L. E., Levinsen, M. T., Samuelsen, M. R., "Theories of 'Subharmonic' Gap Structures in Superconducting Junctions," Rev. de Phys. Appl., Vol. 9, No. 1, pp. 157–158, 1974.

39. Schrieffer, J. R., Wilkins, J. W., "Two-Particle Tunneling Process Between Superconductors," Phys. Rev. Lett., Vol. 10, No. 1, pp. 17–20, 1963.

40. Octavio, M., Tinkham, M., Blonder, G. E., Klapwijk, T. M., "Subharmonic Energy-Gap Structure in Superconducting Constriction," Phys. Rev. B., Vol. 27, No. 11, pp. 6739–6746, 1983.

41. Ivanchenko, Yu. M., "Influence of Delay on Form of Superconductor Tunnel Characteristics," Pis'ma v ZhETF, Vol. 4, No. 9, pp. 358–361.

TUNNEL INVESTIGATIONS OF NIOBIUM NITRIDE FILMS WITH B1 STRUCTURE

D. R. Dzhurayev, G. P. Motulevich

ABSTRACT

Systematic tunnel investigations are performed on films of niobium nitride with B1 structure obtained by getter, reactive cathode sputtering of a niobium target in a glow discharge in an atmosphere of nitrogen and argon. The maximum values of superconducting parameters of the films obtained are: T_c=16.6 K (superconducting transition temperature): $\Delta(0)$=2.84 meV (superconducting energy gap at 0 K). One specimen had a double superconducting transition: the main transition at T_c=16.6 K and a high temperature transition at T_c=18.5 K. For a series of films, mean values of $\bar{T}_c$=15.4 K, $\bar{\Delta}(0)$=2.51 eV, $2\bar{\Delta}(0)/KT_c$= 3.78 were obtained. An empirical correlation was established between $\Delta(0)$ and T_c, linear in the area of $9.8 \leqslant T_c \leqslant 16.6$ K. The variation of adjusted gap $\Delta(T)/\Delta(0)$ with adjusted temperature T/T_c was measured, confirming the theoretical correlation. In the area of voltages V beyond the gap (up to 30 meV), a complex structure is observed for the first time in graphs of $-dV/dI$ and d^2I/dV^2 as a function of V (I is the current through the tunnel junction). It is well reproduced both in repeated measurements of the same tunnel junction, and in different tunnel junctions. Figures 17, Tables 7, References 16.

1. Introduction

Niobium nitride compounds with B1 structure are promising stable superconductors with high critical superconducting parameters. They are resistant to external effects, including corrosive media, and are insensitive to radiation. All of this expands the area of practical application of these compounds. One important superconducting characteristic of any superconductor is its energy superconducting gap Δ. It determines the upper frequency limit at which it can be used in various instruments and devices. Prior to [1] there were no studies of the superconducting gap of niobium nitride compounds at superconducting transition temperature $T_c > 16$ K. In this work, we have undertaken systematic tunnel studies of niobium nitride with B1 structure over a broad range of voltages V, allowing us to obtain the values of Δ, establish their correlation with T_c, determine the temperature variation $\Delta(T)$, and also discover the structure of the graph of the curve of $d^2I(V)/dV^2$ in the area of voltages V beyond the gap up to 30 meV (I is the current of the tunnel junction).

2. Films. Tunnel Junctions. Measurement Methods

Films of niobium nitride with thicknesses of ~2000 Å and ~4000 Å were manufactured by one of the authors by a technology developed by Ye. A. Antonova and V. A. Sukhov on their evaporating installation[2]. We used the method of getter reactive cathode sputtering of a niobium target in a (direct current) glow discharge in an atmosphere of nitrogen and argon. The niobium nitride was deposited on substrates of fused quartz. The film characteristics: T_c, $\Delta T_c = T_c^b - T_c^e$ and $\gamma = R_{300}/R_{25}$ depended on the partial pressures of nitrogen and argon, the pressure in the vacuum chamber, the energy of the sputtered atoms which precipitated onto the substrate, the material and temperature T_s of the substrate. Here T_c is the temperature of the superconducting transition, defined as the temperature at the center of the jump in resistance, T_c^b and T_c^e are the temperatures of the beginning and end of the superconducting transition, R_{300} and R_{25} are the resistances of the film at 300 and 25 K. γ characterizes the granularity of the film. For most of the films manufactured, T_s=700°C. T_c=14.5–16.6 K, ΔT_c=0.2–1.5 K, γ=0.87–0.94. The most homogeneous films, i.e., films with ΔT_c=0.3–0.5 K, had T_c=15.3–16.2 K.

For example, Figure 1 shows the function $R(T)$ in the area of the phase transition to the superconducting state for three niobium nitride films.

All of the resistive measurements were performed by a four-contact method, using an "Allen-Bradley" resistance thermometer, programmed with a type TSG-1 germanium thermometer. The specimens and the thermometer were mounted on a copper holder which was placed in a massive copper bomb. The accuracy of temperature measurements was ±0.05 K.

The basic element used for tunneling studies is the tunnel junction. In this work, we used two types of tunnel junctions: NbN — NbN oxide — S and NbN — NbN oxide — n. Here S and n indicate superconducting and normal electrodes. The first electrode was a freshly prepared strip of niobium nitride ~2000 Å or 4000 Å thick, deposited through a special pattern onto heated polished quartz plates measuring 1×6×12 mm^3 or 2×10×15 mm^3. The tunnel barrier between electrodes was a layer of oxide formed on the surface of the niobium nitride film in air at room temperature [1]. The oxidation time was varied from two to several hundred hours, depending on the desired tunnel junction resistance. The second (top) electrode was a strip of metal, deposited perpendicular to the first metal at room temperature. The second electrode consisted of Pb, In superconductors and the normal metals Ag, Au and Al. The area of the tunnel junction was varied from 0.07 mm^2 to 1 mm^2, the resistance R_T — from 0.01 Ohms to 10 Ohms. In order to study the temperature variation of Δ_{NbN} and for tunnel spectroscopy at voltages V up to 30 meV, it was found that the optimal junction was a tunnel junction with $R_T \gg 1$ Ohms.

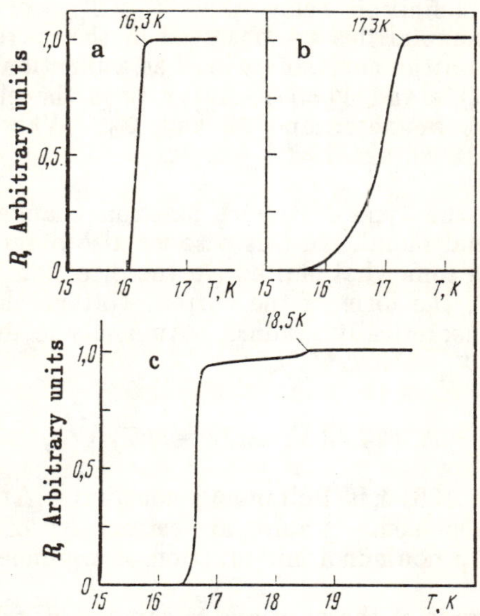

Figure 1. Graph of the Function $R(T)$ for Phase Superconducting Transitions of Three Specimens. a: No. 23. T_c=16.15 K, ΔT_c=0.3 K, specimen with "narrow" transition; b: No. 55. T_c^b=17.3 K, T_c=16.6, ΔT_c=1.5, specimen with "broad" transition; c: specimen with double transtion: main transition — T_c=16.6, plus high-temperature transition — T_c=18.5 K. The jump in resistance at the high-temperature transition is on the order of 5% of the resistance jump of the main transition.

The electronic circuits used to record the current-voltage characteristics and their first derivatives were similar to the circuits used in [3, 4], which utilized an electronic reochard[5]. Current-voltage characteristics were recorded on a two-coordinate type PDS-021M strip-chart recorder. The voltage V on the tunnel junction was monitored with an F-30 instrument, current I — with an M1104 ammeter. The variation of the first derivatives $-dV/dI$ as a function of V was studied by a modulation method. The first harmonic signal was recorded.

3. Measurement of Superconducting Energy Gap

At present, the tunnel method is among the most precise and reliable methods for measurement of the superconducting energy gap Δ[6]. The value of Δ is determined from the position of the "gap" singularities of the graph of $I(V)$ or the position of the maxima on the graph of $-dV/dI$ as a function of V. For s_1/s_2-type tunnel junctions, where s_1 and s_2 are superconductors with parameters T_{c1}, Δ_1 and T_{c2}, Δ_2, the positions of these singularities are related to Δ_1 and Δ_2. We assume for definition below that $T_{c1}>T_{c2}$. Then at temperatures[2] $T^*<T<T_{c2}$, two gap singularities are observed in the current-voltage characteristic and two maxima on the curve of $-dV/dI$ as a function of V at the following voltages: $V_1=(\Delta_1+\Delta_2)/e$ and $V_2=|\Delta_1-\Delta_2|/e$ (e is the charge of an electron). By varying V_1 and V_2 we determine Δ_1 and Δ_2. Where $0<T<T^*$, only one maximum is observed at voltage $V=(\Delta_1+\Delta_2)e$.

At $T_{c2}<T<T_{c1}$, the type of tunnel junction changes, it becomes a type s/n, where n is a normal metal. In this case we also observe one maximum, the position of which V_m is somewhat shifted (in the direction of higher values) with respect to Δ_1/e. In [7], the form of the current-voltage characteristic of an s/n tunnel junction was theoretically studied. An analytic expression was derived for the relationship of V_m with $\Delta^*(T)$:

$$\Delta^*(T)=[(V_m-akT)^h-(bkT)^h]^{1/h}, \tag{1}$$

$a=1.113$; $b=2.107$; $h=2.138$, k is Boltzman's constant. $\Delta^*(T)$ is close to $\Delta(T)$. This same work also presents graphs for estimation of the small difference $\Delta^*(T)-\Delta(T)$, which is a nonmonotonic function of argument $kT/\Delta(T)$.

Thus, measurements of the positions of gap singularities on current-voltage characteristics or the positions of the maxima on the curve $-dV/dI-V$ allow us to obtain the values of Δ for tunnel junctions of both types: s_1/s_2 and s/n.

4. Tunnel Studies of Niobium Nitride Films in the Gap Voltage Area

We undertook systematic tunnel studies of niobium nitride films in the gap voltage areas, i.e., at $V\leqslant(\Delta_{NbN}+\Delta_2)/e$, where Δ_2 is the superconducting gap

---

[2]T* is the temperature at which the concentration of quasiparticles in the superconductors becomes sufficient to observe the "difference" maximum.

of the second electrode. NbN/Pb, NbN/In, NbN/Ag, NbN/Au and NbN/Al tunnel junctions were made for this purpose. Their *I–V* characteristics were recorded. To refine the position of the gap singularities, we recorded graphs of the variation of $-dV/dI$ as a function of V. In some cases, this function was defined by numerical differentiation. The method described above was then used to determine the superconducting energy gap Δ_{NbN}. The positions of singularities produced by the different methods differed by a quantity of not over $2\cdot10^{-2}$ meV. The absolute error Δ_{NbN} was determined by the accuracy of the recording device, and was not over $9\cdot10^{-2}$ meV.

Let us begin by analyzing the NbN/Pb tunnel junctions. A large number of NbN/Pb tunnel junctions have been investigated (for example, cf. [1]). Current–voltage characteristics and $-dV/dI$ as a function of V have been recorded at various temperatures. For these tunnel junctions in the interval of 1.8 K$<T<$7.2 K two gap singularities were observed for each temperature on the current–voltage characteristic, with two corresponding maxima on the graph of $-dV/dI$ as a function of V at $V_1=(\Delta_{NbN}+\Delta_{Pb})/e$ and $V_2=|\Delta_{NbN}-\Delta_{Pb}|/e$. The value of the second maximum was significantly less than that of the first. It decreased with decreasing T. At $T<4.2$ K, the accuracy of determination of the position of the "difference" maximum decreased. At $T>4.2$ K, the two maxima were clearly observed. As an example, Figures 2 and 3 show the current–voltage characteristics of two specimens at 4.2 K. In Figure 4 we see the current–voltage characteristic of specimen 23, measured at various $T<7.2$ K. We can see from Figure 4 that as T drops, the first gap singularity, related to the sum $\Delta_{NbN}+\Delta_{Pb}$, moves into the area of greater values of V, while the second, related to the difference $|\Delta_{NbN}-\Delta_{Pb}|$, moves in the opposite direction. For all of the specimens mentioned above the leakage current as $V\to0$ is small and values of $(-dV/dI)_{V\to0}/(-dV/dI)_{V=6mV}=50$–$100$ (for example, cf. Figures 2–4).

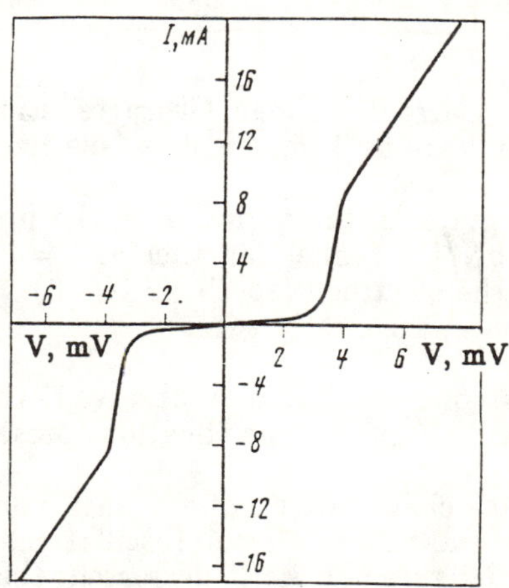

Figure 2. Current–Voltage Characteristic of NbN/Pb Tunnel Junction at 4.2 K (specimen No. 22).*

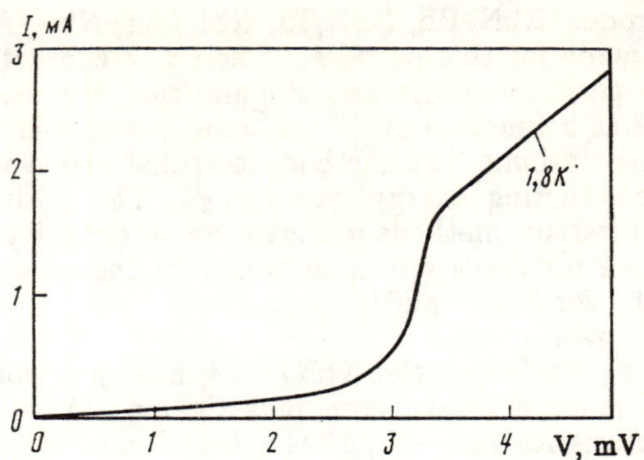

Figure 3. Current-Voltage Characteristic of NbN/Pb Tunnel Junction at 4.2 K (specimen No. 8).

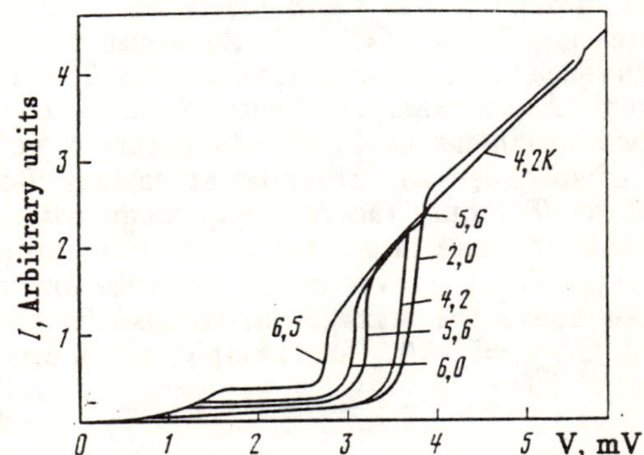

Figure 4. Current-Voltage Characteristic of NbN/Pb Tunnel Junction at Various Temperatures (specimen No. 23).

Figure 5 shows a graph of the variation in the first derivative $-dV/dI$ as a function of I for an NbN/Pb tunnel junction at T=4.2 K, obtained with two signs of the voltage on the electrode (specimen No. 8). Both maxima, related to the gap sum and difference, are clearly seen.

At T>7.2 K, just one maximum is observed, the position of which V_m, according to [7], is related to Δ_{NbN} by the equations presented above.

The current-voltage characteristic of a tunnel junction with small values of R_T<0.2 Ohms showed anomalous effects (oscillations of voltage), apparently indicating local heating (Figure 6). As T decreased, the oscillations disappeared, confirming this assumption.

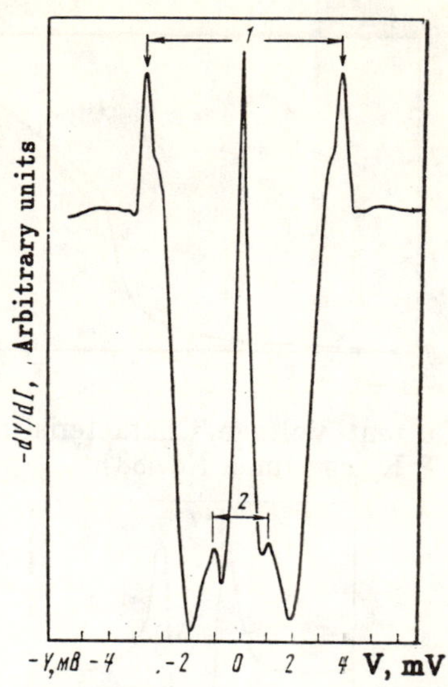

Figure 5. Variation of First Derivative $-dV/dI$ as a Function of V for NbN/Pb Current-Voltage Characteristic at T=4.2 K (specimen No. 8). Graph recorded with two signs of voltage on electrodes: 1 — $2(\Delta_{NbN}+\Delta_{Pb})/e$, 2 — $2\times(\Delta_{NbN}-\Delta_{Pb})/e$.

Up to this point we have studied the maxima on the curve of $-dV/dI$ in the area of the gap voltages, i.e., where $V \lessdot \{[\Delta_{NbN}(0)+\Delta_{Pb}(0)]/e\}$. Among the tunnel junctions used, two had maxima at V=6.15 mV, which is beyond the range of gap singularities. The position of the maximum shifted in the direction of larger V as T decreased. Its origin requires additional investigation.

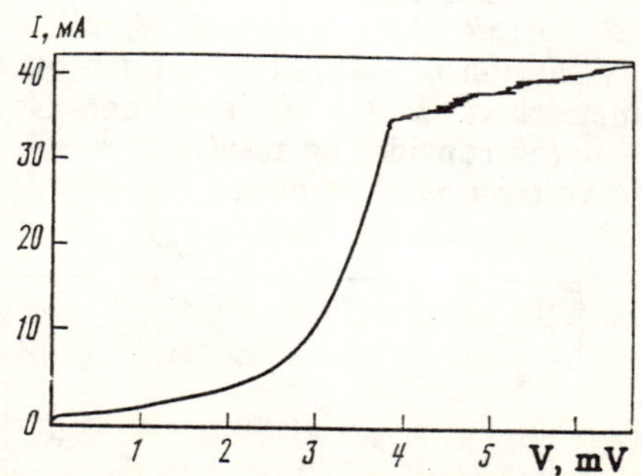

Figure 6. Anomalous Current-Voltage Characteristic of NbN/Pb Tunnel Junction with Small R_T at 4.2 K (specimen No. 75).

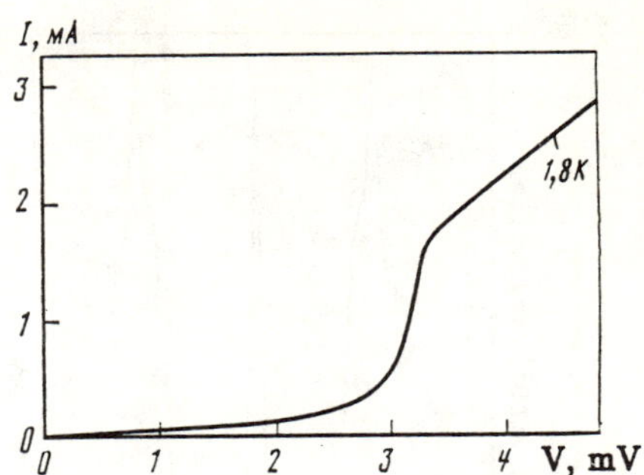

Figure 7. Current–Voltage Characteristic of NbN/In Tunnel Junction at 1.8 K (specimen No. 88).

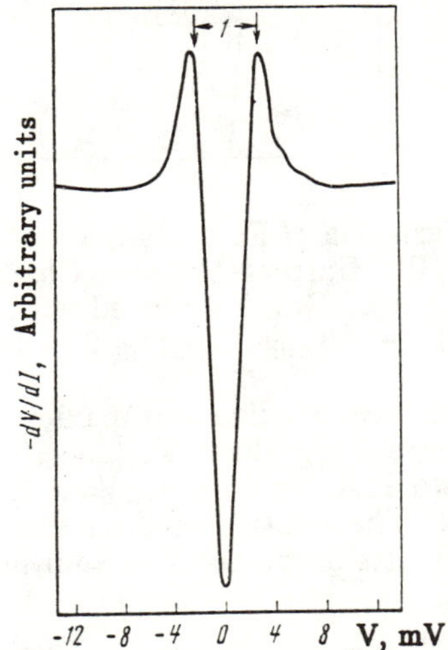

Figure 8. Variation of $-dV/dI$ as a function of V for NbN/Ag Tunnel Junction at T=1.8 K (specimen No. 57). Value of $2\Delta_{NbN}/e(1)$ noted considering results of [7]. Graph recorded with two signs of voltage on electrodes.

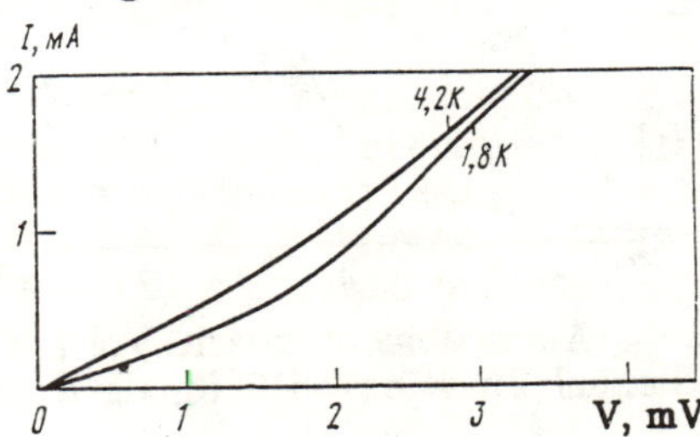

Figure 9. Current–Voltage Characteristic of NbN/Ag Tunnel Junction at 4.2 K and 1.8 K (specimen No. 57).

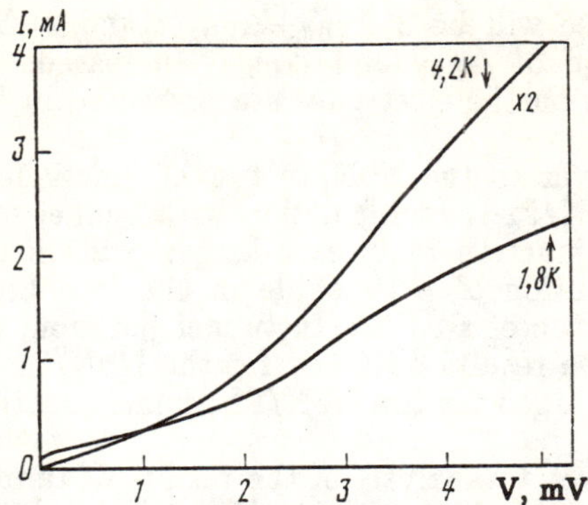

Figure 10. Current–Coltage Characteristic of NbN/Au Tunnel Junction at 4.2 and 1.8 K (specimen No. 64).

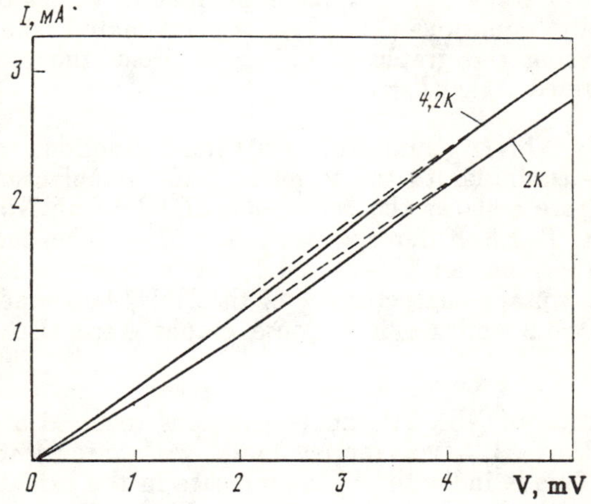

Figure 11. Current–Voltage Characteristic of NbN/Al Tunnel Junction at 4.2 and 1.8 K (specimen No. 92). Dash line shows current–voltage characteristic at temperature at which niobium nitride is in the normal state.

For unbalanced tunnel junctions with different electrodes, multifrequency effects related to the different "growth" of quasiparticles on an electron in both electrodes make the current–voltage characteristic unbalanced relatived to $V=0$. A small but noticeable addition develops to the ordinary single-frequency tunnel current[8]. Considering that the first derivatives of this addition are odd functions of V, we can eliminate its influence on the result if we write the gap singularities with two signs of the voltage on the electrodes. The distance between the gap singularities in this case will be $2(\Delta_{NbN}+\Delta_{Pb})/e$ and $2|\Delta_{NbN}-\Delta_{Pb}|/e$. The accuracy of determination of Δ_{NbN} and Δ_{Pb} is increased. The gap singularities with both voltage signs on the electrodes are presented in Figures 2, 5 and 8.

Let us now go over to the NbN/In tunnel junctions. The studies showed that to produce an NbN/In tunnel junction with tunnel resistance equal to that of an NbN/Pb tunnel junction requires a longer oxidation time. This is apparently because of dissolution of NbN oxide in the In. Figure 7 shows the current–voltage characteristic of an NbN/In tunnel junction, specimen No. 88, measured at T=1.8 K. The results obtained for the NbN/In tunnel junction are similar to the results obtained for the NbN/Pb tunnel junction.

Let us now go over to analysis of the results of tunnel studies of NbN/Ag, NbN/Au and NbN/Al tunnel junctions. We noted earlier that for s/n tunnel junctions, the function $-dV/dI(V)$ has only one maximum, related to the energy gap of the superconductor (Δ_S). Its position V_m is displaced from Δ_s/e. To produce the value of $\Delta_{NbN}(T)$, first the experimental values of $V_m(T)$ were used to find $\Delta^*(T)$ by equation (1), then we estimated the slight correction $\Delta_{NbN}(T)-\Delta^*(T)$ using the graphs of [7]. In all cases the correction was $<10^{-2}$ meV, and we assumed $\Delta_{NbN}(T)=\Delta^*(T)$.

Let us study NbN/Ag and NbN/Au tunnel junctions. The technology of their manufacture is similar to the technology for manufacture of NbN/Pb tunnel junctions. Figure 8 shows the curve $-dV/dI(V)$ for an NbN/Ag tunnel junction, measured at T=1.8 K for specimen No. 57. The current–voltage characteristic of this specimen for T=4.2 and 1.8 K is shown in Figure 9. Figure 10 shows the current–voltage characteristic of the NbN/Au tunnel junction of specimen No. 64 at T=4.2 and 1.8 K. These graphs were also used to determine $\Delta_{NbN}(0)$.

The production of NbN/Al tunnel junctions required a special technology. The difficulty is that Al, a very active metal with respect to oxygen, deoxides the NbN oxide. This is indicated by an increase in the oxidation time necessary to form a barrier with the required resistance. Figure 11 shows the current–voltage characteristic of an NbN/Al tunnel junction (specimen No. 92) for T=4.2 and 2 K. We can see that the gap singularities appear better as the temperature decreases.

5. Variation of $\Delta_{NbN}(0)$ as a Function of T_c

When niobium nitride films are manufactured, spatially heterogeneous films are sometimes produced with two superconducting phases, differing in the values of the energy gaps. Tunnel junctions based on such films have two gap singularities on the current–voltage characteristic and its derivatives at voltages V of $\Delta'_{NbN}+\Delta_2$ and $\Delta''_{NbN}+\Delta_2$, where Δ_2 is the superconducting gap of the second electrode. Such tunnel junctions will not be studied below. For a number of tunnel junctions with one high-temperature superconducting gap, measurements of Δ_{NbN} were performed at T=1.8 K. Since the niobium nitride films have high T_c, for them Δ_{NbN}(1.8 K) practically coincides with $\Delta_{NbN}(0)$. Table 1 presents the results of measurements of Δ_{NbN}(1.8 K) for 16 specimens with different T_c, obtained in the present study[3]. The same table presents T_c^b, T_c^e, T_c and $2\Delta(0)/kT_c$. For comparison, the lower portion of the table presents the data from the literature.

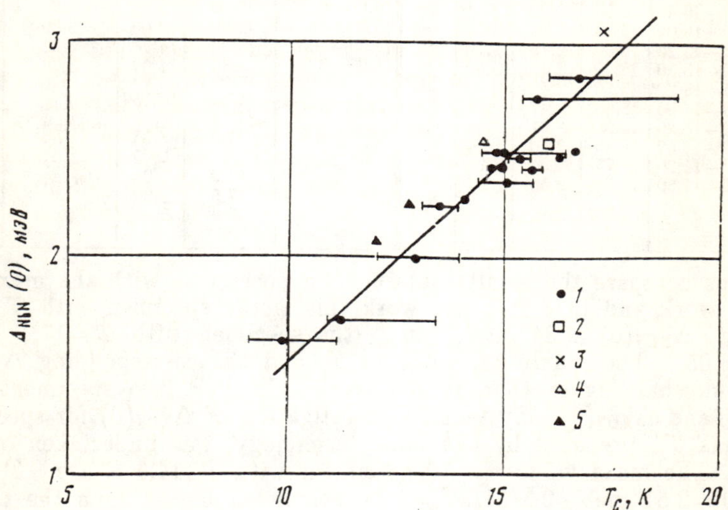

Figure 12. Empirical Correlation Between $\Delta_{NbN}(0)$ and T_c. 1 — Our data on $\Delta_{NbN}(0)$ for specimens with T_c in the interval $9.8 < T_c < 16.6$ K; horizontal lines indicate the sector of $[T_c^b, T_c^e]$; 2 — mean value for specimens with $T_c > 14.6$ K $\overline{\Delta}_{NbN}(0)$, related to mean value T_c; 3, 4, 5 — results of [9, 10] and [11], respectively. Line produced by averaging experimental points by least squares method.

Let us study the specimens with $T_c > 14.6$ K. We can see from the table that we studied 12 such specimens and two specimens with $T_c < 17.1$ K were studied in [9].

---

[3]Our data in the table are directly measured values of $\Delta_{NbN}(0)$ with no corrections for proximity effect.

Table 1
Superconducting Parameters of Niobium Nitride Films Obtained in This Work and by Other Authors

$T_c^b - T_c^e$, K	T_c, K	$\Delta(0)$ meV	$2\Delta(0)/kT_c$	Reference	$T_c^b - T_c^e$, K	T_c, K	$\Delta(0)$ meV	$2\Delta(0)/kT_c$	Reference
17.25—15.75	16.60	2.84	3.97	This Work	16.30—14.00	14.60	2.40	3.81	
16.60—16.40	16.50	2.50	3.51		14.00—12.75	13.40	2.25	3.82	
16.30—16.00	16.15	2.46	3.53		13.80—12.00	12.90	2.00	3.59	
18.80—15.25	15.60	2.75	4.16		13.30—10.90	11.20	1.70	3.52	
15.60—15.30	15.45	2.40	3.60		11.00—9.00	9.80	1.60	3.78	
15.45—15.20	15.30	2.46	3.73						
15.40—14.80	15.05	2.47	3.81			17.1	3.05	4.14	[9]
15.50—14.20	14.90	2.37	3.69			<17.1	—	4.03±4.37	[9]
14.90—14.60	14.80	2.40	3.76			14.5	2.55	4.08	[10]
16.60—14.20	14.75	2.50	3.93			12.8	2.25	4.08	[11]
14.80—14.50	14.70	2.50	3.94			11.9	2.08	4.05	[11]
						7.5	2.00	6	[12]

Let us compare the results obtained for specimens with the maximum T_c in our work and in [9]. In our work this is the specimen with T_c=16.6 K. For it, $\Delta_{NbN}(0)$=2.84 meV. In [9], the specimen with T_c=17.1 K has $\Delta_{NbN}(0)$=3.05. The maximum value of T_c and the corresponding $\Delta_{NbN}(0)$ in [9] is somewhat higher than in our work. In [10-12], the specimens have smaller T_c and $\Delta_{NbN}(0)$. Systematic investigation of $\Delta_{NbN}(0)$ for specimens with various T_c, prepared by the same technology, was undertaken only in our work. The mean values for 12 specimens with $T_c \geqslant 14.6$ K are: $\bar{T}_c$=15.4 K, $\bar{\Delta}_{NbN}(0)$=2.51 meV, $2\bar{\Delta}(0)/kT_c$=3.78, somewhat higher than the theoretical value.

To establish the correlation between $\Delta_{NbN}(0)$ and T_c, we used our values of $\Delta_{NbN}(0)$ for specimens with T_c in the range $9.8 \leqslant T_c \leqslant 16.6$, as well as the results of [9-11]. The empirical correlation between Δ_{NbN} and T_c is shown in Figure 12[4]. We can see from the figure that the mass of points can be approximated by a straight line, i.e., the variation of $\Delta_{NbN}(0)$ as a function of T_c in the interval of T_c from 10 K to 17 K is approximately linear. The data from the literature are shifted slightly relative to our data in the direction of higher values of $\Delta_{NbN}(0)$, by approximately 0.2 meV, within the limits of error of the measurement of Δ.

[4]The results of [12] are not presented on the figure, since they are so strongly divergent with all other results that the question arises whether they refer to niobium nitride with B1 structure.

6. Temperature Variation of Superconducting Energy Gap of Niobium Nitride Films[5]

Three NbN/Pb and one NbN/In tunnel junctions were used to study the temperature variation of energy gap $\Delta_{NbN}(T)$. The superconducting characteristics of these NbN films are presented in Table 2, from which we see that different NbN films having T_c from 14.7 K to 16.6 K were used.

The greatest interval of adjusted temperatures T/T_c was obtained in the study of an NbN/Pb tunnel junction (specimen No. 66) with R_T=2.2 Ohms. Using this tunnel junction, we measured $\Delta_{NbN}(T)$ up to 14.5 K (the corresponding T/T_c=0.93). We recorded the current-voltage characteristics of the NbN/Pb and NbN/In tunnel junctions at various T (cf. Figures 4 and 14). Numerical differentiation was used to compute the function $dI(V)/dV$. In Figure 14 for example we present the variation of dI/dV as a function of V for the NbN/Pb tunnel junction of specimen No. 66.

Where $T<T_{c2}$ (where T_{c2} is the superconducting transition temperature of the second electrode), we found the position of the sum and difference maxima, which allowed us to determine the values of Δ_{NbN} and Δ_2 (Δ_2 is the superconducting energy gap of the second electrode).

Table 2
Superconducting Characteristics of Niobium Nitride Films

Characteristics	Specimen No. 23	No. 55	No. 66	No. 88
$T_c^b - T_c^e$, K	16.00-16.30	17.25-15.75	17.00-15.25	15.10-14.50
T_c, K	16.15	16.6	15.6	14.7
$\Delta(0)$, meV	2.46	2.84	2.75	2.50
$2\Delta(0)/kT_c$	3.53	3.97	4.15	3.94
R_{300}/R_{25}	0.97	0.97	0.93	0.97
R_T, Ohms	0.27	0.80	2.2	1.5
Tunnel Junction	NbN/Pb	NbN/Pb	NbN/Pb	NbN/In

For $T>T_{c2}$, the tunnel junction was of NbN/n type. For this type of tunnel junction, there is just one maximum on the curve of $dI/dV(V)$ at voltage V_m, the position of which was used to find Δ_{NbN} by an algorithm based on the results of [7] and described in Section 3.

---

[5]Niobium nitride films with just one superconducting gap were used to study the temperature variation of the superconducting energy gap in Δ_{NbN}(T)

Table 3
Experimental Values of Δ_{NbN} at Various Temperatures T in Various Tunnel Junctions

$\Delta_{NbN}/\Delta(0)$	T/T_c	T, K	Δ_{NbN}, meV	$\Delta_{NbN}/\Delta(0)$	T/T_c	T, K	Δ_{NbN}, meV
ТП NbN/Pb, R_T = 2,20 Ohms				ТП NbN/Pb, R_T = 0,27 Ohms			
1	0,13	1,8	2,75	1	0,11	1,8	2,46
0,990	0,27	4,2	2,71	0,984	0,26	4,2	2,42
0,980	0,37	5,8	2,70	0,935	0,35	5,6	2,30
0,980	0,43	6,7	2,70	0,911	0,37	6,0	2,24
0,950	0,47	7,3	2,60	0,911	0,40	6,5	2,24
0,950	0,54	8,5	2,60	0,907	0,50	8,0	2,23
0,890	0,63	9,8	2,44	0,907	0,55	8,9	2,23
0,860	0,70	11	2,37	ТП NbN/Pb, R_T = 0,80 Ohms			
0,700	0,81	12,6	1,93	0,990	0,25	4,2	2,80
0,400	0,93	14,5	1,10	0,980	0,51	8,5	2,78
				ТП NbN/In, R_T = 1,50 Ohms			
				1	0,12	1,8	2,50
				0,990	0,15	2,2	2,47
				0,990	0,16	2,4	2,47
				0,980	0,19	2,8	2,46
				0,970	0,29	4,2	2,44

We compared the experimental values of $\Delta_{NbN}(T)$ with the theoretical values using the equation

$$\Delta(T)/\Delta(0)=\text{th}[(T/T_c)\cdot\Delta(T)/\Delta(0)], \qquad (2)$$

presented in [13].

Our study of current-voltage characteristics and their first derivatives of NbN/Pb and NbN/In type tunnel junctions allowed us to determine the temperature variations $\Delta_{NbN}(T)$, $\Delta_{Pb}(T)$ and $\Delta_{In}(T)$. The results are presented in Tables 3-5[14].

Figure 15 shows the variation of $\Delta(T)/\Delta(0)$ as a function of T/T_c for niobium nitride, lead and indium films. For comparison, the same figure shows the theoretical curve.

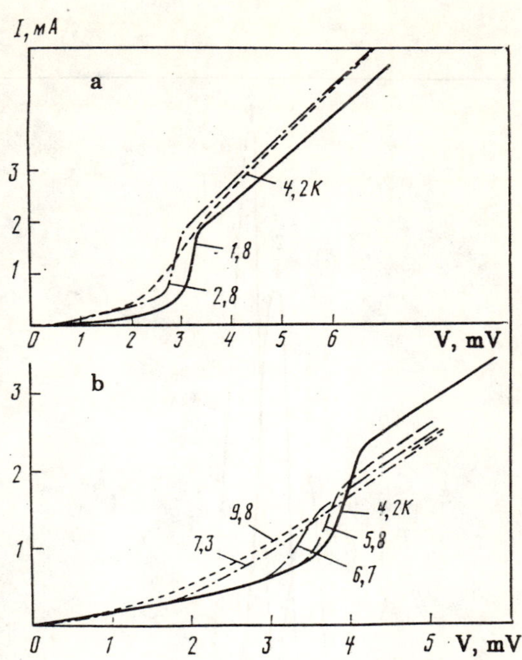

Figure 13. Current–Voltage Characteristics Measured at Various Temperatures. a — NbN/In (specimen No. 88); b — NbN/Pb (specimen No. 66).

We can see from this figure that for tunnel junctions with $R_T>1$ Ohms, the experimental points agree well with the theoretical curve. For tunnel junctions with $R_T<1$ Ohms, the points corresponding to $T>4.2$ K, i.e., when the specimen was in helium vapor, are somewhat lower than the theoretical curve. This is apparently related to local heating of the specimen caused by the high current flowing through the tunnel junction. As T approaches T_c, the gap decreases, the current through the tunnel junction also decreases and the experimental points once more fall on the theoretical curve. In our case, there was just one such specimen ($R_T=0.27$ Ohms), the results for which are shown in Figure 15, *2*. The experimental points for lead and indium used in our tunnel junctions fall on the theoretical curve within the limits of error.

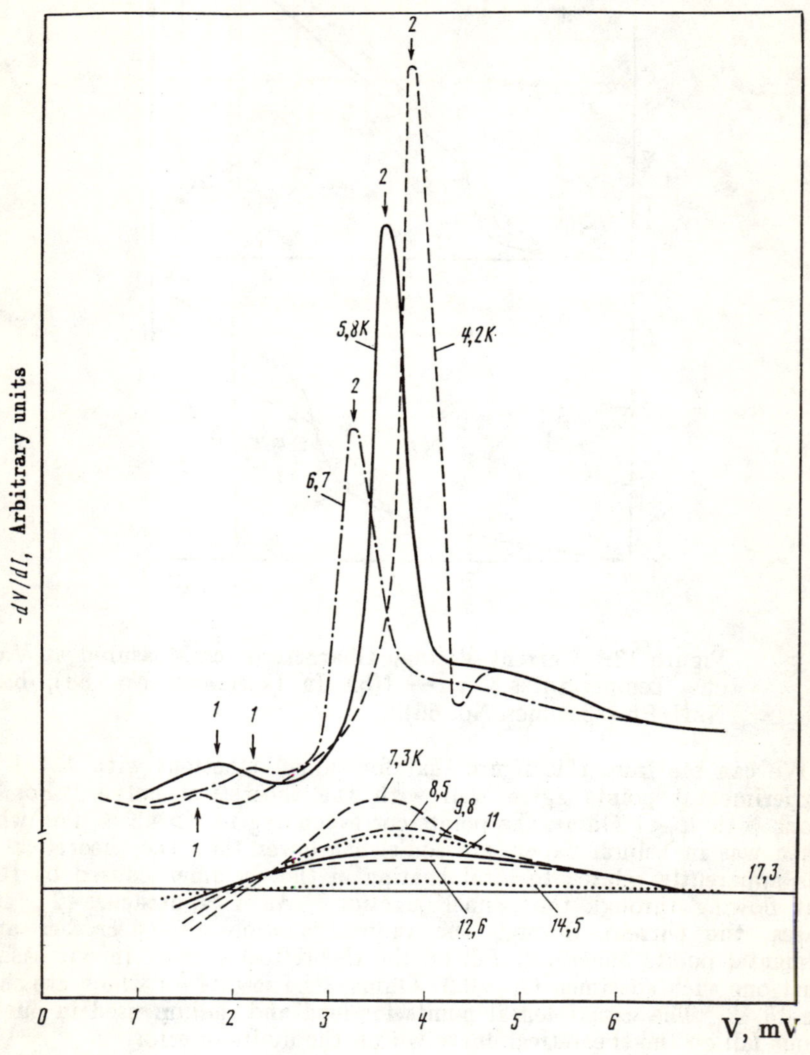

Figure 14. *dI/dV* as a Function of *V* for NbN/Pb Tunnel Junction at Various Temperatures (specimen No. 66). 1 — $|\Delta_{NbN}-\Delta_{Pb}|/e$; 2 — $(\Delta_{NbN}+\Delta_{Pb})/e$.

Table 4
Temperature Variation of Δ_{Pb}

Δ_{Pb}/Δ_0	T/T_c	T, K	Δ_{Pb}, meV	Δ_{Pb}/Δ_0	T/T_c	T, K	Δ_{Pb}, meV
0.978	0.25	1.8	1.35	0.645	0.83	6.0	0.89
0.906	0.58	4.2	1.25	0.449	0.90	6.5	0.62
0.725	0.78	5.6	1.00	0.420	0.93	6.7	0.58
0.688	0.81	5.8	0.95				

Table 5
Values of Δ_{In} at Various Temperatures

Δ_{In}/Δ_0	T/T_c	T, K	Δ_{In}, meV
0.880	0.53	1.8	0.55
0.784	0.60	2.2	0.49
0.656	0.71	2.4	0.41
0.624	0.83	2.8	0.39

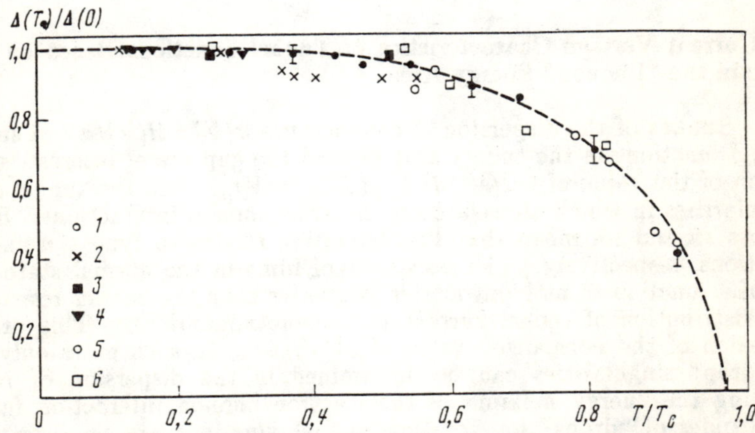

Figure 15. Experimental Values of Adjusted Gap $\Delta(T)/\Delta(0)$ Corresponding to Adjusted Temperature T/T_c for Films of NbN, Pb and In. 1–4 relate to NbN films: 1 — tunnel junction 66; 2 — 23; 3 — 55; 4 — 88; 5,6 relate to the second electrodes: 5 — Pb; 6 — In; dash line shows theoretical data. Errors on vertical scale shown by vertical line sectors, errors on horizontal scale fall within experimental points.

Thus, considering local heating we can consider that $\Delta_{NbN}(T)/\Delta_{NbN}(0)$ for one-phase specimens agrees with the theoretical curve.

The temperature variation $\Delta_{NbN}(T)$ was also studied in [9] and [11]. In [11], R_T=5 Ohms. Agreement with theoretical curve is obtained. Unfortunately, measurements were performed for specimens with relatively low T_c (T_c=11.9 and 12.8 K) and in the temperature interval 2–6.9 K, which corresponds to the interval of adjusted temperatures of 0.17–0.58.

In [9], the temperature variation $\Delta_{NbN}(T)/\Delta_{NbN}(0)$ as a function of T/T_c was shown for two specimens. In both cases deviations were observed from the theoretical curve. As T approached T_c, the experimental point once more fell on the theoretical curve.

The authors relate the deviation observed to the presence of a surface layer, the temperature of the superconducting transition of which T_c'=12–12.8 K, determined by the proximity effect. Resistive measurements of T_c cannot detect T_c, since the surface layer is shorted by the base of the film, which has higher T_c (~17 K). However, tunnel studies with sufficient sensitivity should show the presence of two gaps. Nothing is said of this. The resistance of the tunnel junction is not presented, and therefore the significance of local heating cannot be evaluated.

7. Current–Voltage Characteristics of Tunnel Junctions Based on Niobium Nitride in the "Phonon" Energy Area

Studies of the dispersion of conductance $\sigma(V)=dI(V)/dV$ of superconductor tunnel junctions in the energy area beyond the gap are of interest, since normalization of the value of $(dI(V)/dV)_{sn}/(dI(V)/dV)_{nn}$ yield the tunnel state density, singularities in which are related to electron–phonon interactions. Here, the subscripts sn and nn mean that the derivative relates to type s/n and n/n tunnel junctions, respectively. The resistance of films in the normal state in the vicinity of a junction in niobium nitride is greater than the barrier resistance, causing the distribution of tunnel current to become nonuniform. This prevents determination of the normalized value of $dI(V)/dV$. In such cases only the position of "drop" singularities can be determined in the dispersion of $(dI(V)/dV)_{sn}$, yielding the energy maxima of the electron–phonon interaction function. The amplitudes of "drops" for s/n-type tunnel junctions are small, much less than for s_1/s_2-type tunnel junctions. The latter junctions are therefore much easier to study. Singularities appear in the dispersion $dI(V)/dV$ of s_1/s_2 tunnel junctions, related to electron–phonon interaction of s_1 and s_2. If the maxima of the electron–phonon interaction functions in s_1 and s_2 are displaced, they have practically no influence on each other.

In the present work we studied NbN/Pb, NbN/In and NbN/Ag tunnel junctions. A modulation method was used (cf. Section 2). A current source was used and the variation of $-dV/dI$ as a function of V was recorded. Measurements were performed at a temperature of 1.8 K. The positions of the "drops" for the variation of $-dV/dI$ as a function of V and dI/dV as a function of V are practically the same. Figure 16*a* shows the variation of $-dV/dI$ as a function of V for an NbN/Pb tunnel junction. Both the "drops" related to Pb and the "drops" related to NbN are clearly seen. The variation of d^2I/dV^2 as a function of V was also measured to allow more precise determination of the position of the "drops." The values of $-dV/dI$ were inverted and numerical differentiation performed. All computations were performed on a WANG-2200 computer.

For some tunnel junctions measurements were performed with two polarities of voltage on the electrodes. The reading on the abscissa for each polarity was performed for the sum of the gaps $(\Delta_{NbN}(0)+\Delta_{Pb}(0))$ or $(\Delta_{NbN}(0)+\Delta_{In}(0))$. For NbN/Ag tunnel junctions, the reading on the abscissa for each polarity was performed from the $\Delta_{NbN}(0)$ gap. A correction was introduced for the displacement of the position of the maximum of $-dV/dI$ with respect to $\Delta_{NbN}(0)$, calculated by the method of [7].

Figure 16*b* shows the variation of d^2I/dV^2 as a function of V in an NbN/Pb tunnel junction. The two main minima are related to the Pb. Their positions, at 4.5 and 8.5 meV, agree well with the values from the literature. Other minima are also observed, related to niobium nitride.

Initially [15], we interpreted the positions of the minima of the function $d\sigma(V)/dV$ in the area $eV<24$ meV as energies corresponding to the maxima of the electron–phonon interaction function. However, the variation of d^2V/d^2I as a function of V in the interval $5<eV<29$ meV (energy read from zero) is presented in [16], where a complex structure was also found, which disappeared in a magnetic field ($H=0.3\,T$). The position of the singularities was not reproduced in different tunnel junctions, or for repeated measurements in the same tunnel junction. The author concluded that the singularities observed were not related to the electron–phonon interaction, but rather the radiation of Josephson eddies, the development of which results from fluctuations, as is indicated by the inability to repeat the positions of singularities. The disappearance of the structure in the field of 0.3 T makes our initial interpretation doubtful, but, in contrast to [16], the positions of singularities in our experiments were reproduced both for repeated measurements of the same tunnel junction (within an accuracy of 0.1–0.3 meV), and for different tunnel junctions (within an accuracy of 0.3–1.0 meV).

The reproducibility of the results with repeated measurements on the same tunnel junction is illustrated by Table 6 and Figure 17. Curves *1–3* were recorded at the same gain. They are displaced on the ordinate in the figure. Curve *4* was recorded with the sensitivity doubled, revealing a small singularity at 14.2 meV. The positions of the "drops" on the curve $-dV/dI$–V coincide with the positions of the minima on the curve of d^2I/dV^2–V with an accuracy of better than 0.2 meV. The results of processing curves *1–4* (Figure 17) are shown in Table 6.

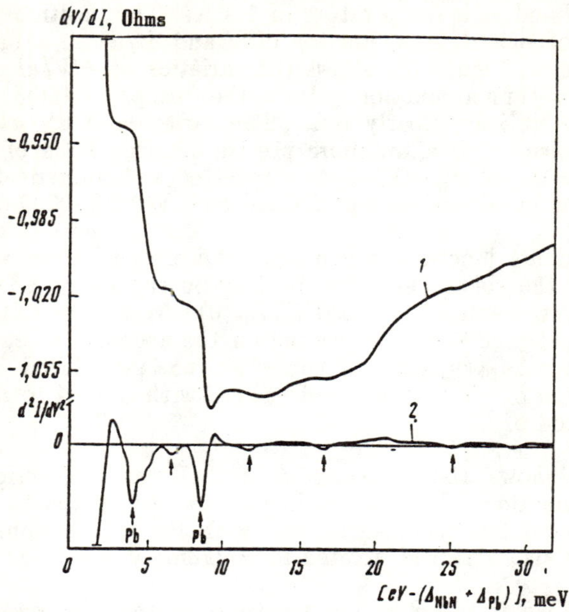

Figure 16. $-dV/dI$ as a Function of V *(1)* and d^2I/dV^2 as a Function of V *(2)* for NbN/Pb Tunnel Junction at T=1.8 K. Abscissa read from sum of gaps $\Delta_{NbN}(0)+\Delta_{Pb}(0)$. Two minima related with Pb so marked. Other arrows show positions of minima of function d^2I/dV^2.

The niobium nitride film used for Figure 17 had the following parameters: T_c=15.6±0.1 K; $\Delta_{NbN}(0)$=2.75±0.09 meV and $2\Delta_{NbN}(0)/kT_c$=4.16±0.15.

The reproducibility of the results for four NbN/Pb junctions, two NbN/In junctions and one NbN/Ag junction is shown in Table 7, which presents data for each type of tunnel junction studied. The error was determined by the spread of positions of the minima in different tunnel junctions of the same type. Only those minima which appeared in at least six of the seven tunnel junctions were considered. Minima related to Pb and In were also eliminated. The weighted mean for all types of tunnel junction is presented, with the ratio of the number of tunnel junctions of a given type in the total number of tunnel junctions used as the weight. The maximum error includes both the error within each type of tunnel junction and the error related to the deviation from the mean. We can see from the table that in the 0–26 meV range there are six minima, reproduced with the accuracy noted.

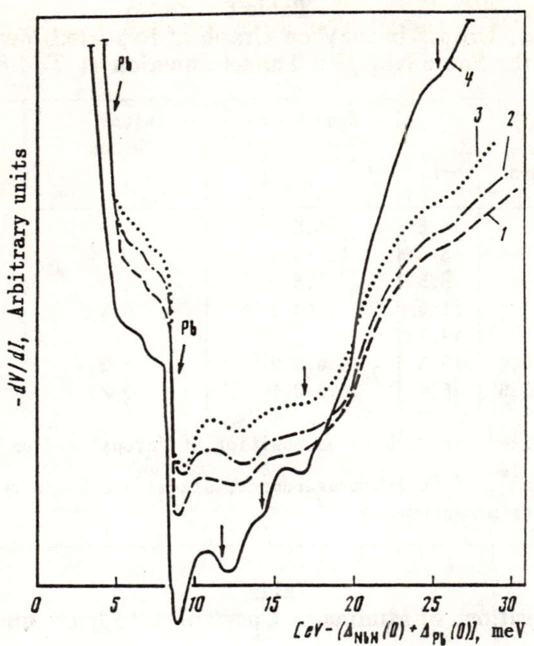

Figure 17. $-dV/dI$ as a Function of V for Repeated Measurements of One NbN/Pb Tunnel Junction at T=1.8 K. For clarity, curves shifted on ordinate. Arrows note positions of "drops," related both to niobium nitride and to lead. On curve *4* ordinate gain double the gain used in all other cases. Readings on abscissa measured from sum of gaps ($\Delta_{NbN}(0)+\Delta_{Pb}(0)$).

Comparing the corresponding values for positions of the minima in our work with those of [16], we conclude that the results of [16] agree with our own results with this same accuracy. We must note that the technology used in [16] for preparation of specimens was not the same as we used.

Measurements performed on NbN/Pb, NbN/In and NbN/Al tunnel junctions allowed us to determine the energies of the singularities of interest to us throughout the entire interval (0–26) meV. Actually, in NbN/Pb tunnel junctions the "strong" singularities related to Pb do not allow us to detect singularities related to niobium nitride in the vicinities of 4 and 8 meV. In these areas they can be detected by the use NbN/In and NbN/Al tunnel junctions. In those areas where singularities related to In might interfere (6 and 14 meV), NbN/Pb and NbN/Al tunnel junctions work well.

Table 6
Positions of "Drops" in meV on Graph of Repeated Measurements on the Same NbN/Pb Tunnel Junction at T=1.8 K

Measurement*				Mean Value	Values of "Drops"	Source of Singularity
1st	2nd	3rd	4th			
4.5	4.5	4.5	4.5	4.5	7.8	Pb
6.5	6.5	6.6	6.5	6.5±0.1	0.74	Nb/N
8.5	8.5	8.5	8.5	8.5	4.8	Pb
11.5	11.6	11.6	11.5	11.5±0.1	0.4	Nb/N
			14.2			
16.7	16.8	16.6	16.6	16.6±0.2	0.2	Nb/N
25.3	25.4	25.3	25.4	25.3±0.1	0.2	Nb/N

Note. Data obtained for Pb both on position of "drops" and on their values agree with[16]. * In 4th measurement, signal amplitude twice as great as in other measurements.

Table 7
Positions of Minima in Curve of $d\sigma(V)/dV$, meV

Present Work							Data of [16]	
NbN/Pb	Source of Singularity	NbN/In	Source of Singularity	NbN/Ag	Source of Singularity	Weighted Mean for All Types	NbN/Pb	Source of Singularity
4.3±0.3	Pb						4.5	Pb
6.2±0.5	NbN	6.4±0.4	In	6.5±0.6	NbN	6.3±0.5	6.5	NbN
8.5±0.3	Pb	8.8±0.8	NbN	8.9±0.3	NbN	8.8±0.6	8.5	Pb
11.3±0.8	NbN	11.7±0.3	NbN	11.4±0.6	NbN	11.4±0.6	11.5	NbN
14.2±0.6	NbN	14.0±0.3	In	15.0±0.2	NbN	14.4±0.6	15	NbN
17.0±0.4	NbN	18.5±0.8	NbN	17.0±0.3	NbN	17.5±1.0	18	NbN
							22.5	NbN
25.2±0.6	NbN	25.8±0.5	NbN	25.8±0.4	NbN	25.5±0.6	24	NbN

We have thus discovered a complex structure in the functions $\sigma(V)$ and $d\sigma(V)/dV$, consisting of a number of "drops" on the first curve and a number of minima on the second curve. The positions of the minima for seven different tunnel junctions are reproduced with an accuracy of 0.3–0.8 meV. This means that the structure is not random, but rather is related to the niobium nitride film.

Conclusions

Systematic tunnel studies of niobium nitride films with B1 structure lead to the following conclusions.

1. For series consisting oof 12 films with $T_c > 14.6$ K, mean values were obtained as follows: $T_c = 15.4$, $\overline{\Delta}(0) = 2.51$ meV, $2\overline{\Delta}(0)/kT_c = 3.78$ (cf. Table 1)[6]. The high T_c allows these films to be used as superconducting elements in modern cryoengineering products operating at $T \sim 10$ K. The value of $\overline{\Delta}(0) = 2.5$ meV determines the high-frequency limit ($6 \cdot 10^{11}$ Hz) for various electronic devices utilizing these films. In NbN it is 1.7 times higher than in Nb. In combination with good stability and resistance to external actions, the excellent superconducting parameters make niobium nitride a processing material for various practical applications.

2. The maximum values of parameters obtained in our work and in [9] are as follows: $T_c = 16.6$ K, $\Delta(0) = 2.84$ meV and $T_c = 17.1$ K, $\Delta(0) = 3.05$ meV. Furthermore, in our work there was one specimen with a double superconducting transition: the main transition at $T_c \sim 16.6$ and a higher-temperature transition at $T_c \sim 18.5$ K (cf. Figure 1). This means that the maximum capabilities of niobium nitride films have not yet been reached.

3. For a series consisting of six specimens a correlation was established between $\Delta(0)$ and T_c of niobium nitride films. In the sector $9.8 \leqslant T_c \leqslant 16.6$ it was found to be linear (cf. Figure 12). The results of [9-11] agree with this conclusion.

4. Based on the results of our studies of three NbN/Pb tunnel junctions and one NbN/In tunnel junction for niobium nitride films, the experimental variation of corrected gap $\Delta(T)/\Delta(0)$ as a function of corrected temperature is presented. Throughout the entire interval studied $0.11 \leqslant T/T_c \leqslant 0.93$, the experimental points fall on the theoretical curve (Figure 15). It was shown that to eliminate the influence of overheating of the tunnel junction on the position of the experimental point ($\Delta(T)/\Delta_0$, T/T_c) requires tunnel junctions with $R_T > 1$ Ohms. For tunnel junctions with small R_T, the experimental points fall below the thoeretical curve (cf. Figure 15, specimen No. 23 with $R_T = 0.27$ Ohms). The influence of overheating is slight where $T < 4.2$ K, when a specimen is kept in liquid helium and at T such that the superconducting gap $\Delta(T)$ and corresponding value of V are clearly decreased.

5. The tunnel studies of four NbN/Pb tunnel junctions, two NbN/In tunnel junctions and one NbN/Al tunnel junction in the gap area of voltages V $0 < V - (\Delta_{NbN} + \Delta_2) < 26$ meV (Δ_2 is the gap of the second electrode) allowed us for

---

[6]The measurements of $\Delta(0)$ for films with T_c of over 16 K were performed by us for the first time.

the first time to observe a complex structure in the graphs of $-dV/dI$ as a function of V and $d^2I(V)/dV^2$, consisting of a number of "drops" on the first graph and a number of corresponding minima on the second graph[15]. Repeated measurements on the same tunnel junction reproduced the positions of the singularities with an accuracy of 0.1–0.2 meV (cf. Table 6 and Figure 17). When measurements were performed on seven different tunnel junctions, the positions of the singularities were reproduced with an accuracy of 0.3–0.8 meV (cf. Table 7). This means that the structure obtained is not random, but rather is related to the niobium nitride film. The data of [16] agree with our data.

In the area $0 < V - (\Delta_{NbN} + \Delta_2) < 26$ meV there are six singularities, as indicated above.

REFERENCES

1. Antonova, Ye. A., Dzhurayev, D. R., Motulevich, G. P., Sukhov, V. A., "Superconducting Energy Gap in Niobium Nitride," ZhETF, Vol. 8, pp. 2426–2429, 1981.

2. Antonova, Ye. A., Sukhov, V. A., "Relationship Between Structure and Electrophysical Properties of Niobium Nitride Films," Fizika nizkikh temperatur, Vol. 17, pp. 1002–1009, 1981.

3. Thomas, D. E., Rowell, J. M., "Low L Second Harmonic Detection System," Rev. Sci. Instr., Vol. 36, pp. 1301–1306, 1965.

4. Vedeneyev, S. I., Golovashkin, A. I., Motulevich, G. P., "Study of Phonon Spectrum of Nb_3Sn Alloy," Preprint FIAN, No. 116, Moscow, 23 pp., 1973.

5. Sidel'nikov, V. I., Tyurikov, D. A., "Bias Current Generators for Studies of Superconducting Junctions and Tunnel Experiments," Preprint, FIAN, No. 116, Moscow, 23 pp., 1978.

6. Tunnel'nye yavleniya v tverdykh telakh [Tunnel Phenomena in Solids], Edited by E. Burshteyn and S. Lundquist, Moscow, Mir Press, 421 pp., 1973.

7. Pike, G. E., "Superconductor Energy Gaps from Tunneling," J. Appl. Phys., Vol. 42, pp. 883–886, 1971.

8. Ivanchenko, Yu. M., Medvedev, Yu. V., "Spectroscopy of Quasiparticle Excitations in Normal Metals by Multifrequency and Inelastic Tunnel Effect," Fizika nizkikh temperatur, Vol. 2, pp. 141–168, 1976.

9. Pan, V. M., Goishnyak, V. P., Rudenco, E. M., et al., "Investigation of the Properties of Superconducting Niobium Nitride Films," J. Criogenics, Vol. 23, pp. 258–262, 1983.

10. Shinoki Fujitoshi, Takada Susumi, Kosaka Shin and Hayakawa Hisao, "Fabrication of High Quality NbN/Pb Josephson Junction," Jpn. J. Appl. Phys., Vol. 19, Suppl. 19-1, pp. 591–594, 1980.

11. Komenou, K., Yamashita, T., Onodera, Y., "Energy Gap Measurement of Niobium Nitride," Phys. Lett., Vol. 28A, pp. 335–336, 1968.

12. Jones, H., Fischer, O., Bongi, G., "Some Properties of Reactively Sputtered NbN Thin Films with Low T_c," Colloq. Intern. C.N.R.S. (Grenobls), Vol. 242, pp. 107–110, 1974.

13. Thouless, D. J., "Strong-Coupling Limit in the Theory of Superconductivity," Phys. Rev., Vol. 117, pp. 1256–1260, 1960.

14. Dzhurayev, D. R., "Temperature Variation of Superconducting Energy Gap in Niobium Nitride Films," Kratkiye soobshcheniya po dizike, No. 3, pp. 58–62, 1984.

15. Motulevich, G. P., Dzhurayev, D. R., Antonova, E. A., Sukhov, V. A., "Tunneling of Charge Carriers in Niobium-Nitride-Based Junctions," Pis'ma v ZhETF, Vol. 36, pp. 313–316, 1982.

16. Shaternik, V. Ye., "Proximity Effect and Tunneling of Electrons in Superconducting Multilayer Structures," Dissertation, Candidate of Physical-Mathematical Sciences, Kiev, Institute of Metal Physics, 1984.

TUNNEL STUDIES OF SPATIALLY HETEROGENEOUS NIOBIUM–GALLIUM FILMS WITH A15 STRUCTURE

D. R. Dzhurayev, I. S. Levchenko, G. P. Motulevich

ABSTRACT

Films of Nb–Ga compounds are prepared with A15 structure with a gradient of composition and tunnel junctions are made from the films. It is found that under certain conditions when the composition of the film is near stoichiometric, instead of a smooth change in superconducting gap Δ and superconducting transition temperature T_c along the concentration gradient, the superconducting phase is divided into areas ("domains") differing in superconducting parameters T_c and Δ. An empirical correlation is observed between Δ and T_c, which in the range T_c (8.0–16 K) is near linear, not passing through 0. For a specimen with $T_c \approx 16.5$ K a superconducting gap $\Delta = 2.79$ meV and ratio $2\Delta/kT_c \approx 3.92$ are obtained. This is the first time such a large value of Δ has been obtained for Nb–Ga compounds.

1. Introduction

Studies of spatially-heterogeneous superconductors are of great interest, since spatial heterogeneity can increase T_c and Δ (cf. [1-4]) (T_c is the temperature of transition to the superconducting state, Δ is the superconducting energy gap.) In Nb–Ga compounds a composition gradient can be used to produce spatially heterogeneous superconductors containing various superconducting phases with A15 structure. Tunnel measurements of Δ, supplemented by measurements of T_c of the various superconducting phases allow us to determine the correlation between Δ and T_c.

Nb–Ga compounds have high critical superconducting parameters and are of interest from both the scientific and the practical points of view. However, in spite of this fact, prior to our earlier work[5] no detailed studies of the superconducting energy gap of these alloys were conducted. We know of only two works[6, 7] in which the gap in the alloy Nb_3Ga was measured. In [6], the measured values of Δ_{Nb_3Ga} were very small (0.23 and 0.45 meV). However, after special theoretical processing the authors predicted a value of Δ_{Nb_3Ga}=2.25 meV. In [7], values of $2\Delta_{Nb_3Ga}/kT_c$ indicate the existence of two gaps with ratios of $2\Delta_{Nb_3Ga}/kT_c$ equal to 1 and 3–4. However, the corresponding values of Nb_3Ga and T_c are not presented in this work. In the present work, we studied the energy gaps Δ of Nb–Ga alloys with various T_c. For the tunnel junctions which we studied, we measured the current–voltage characteristics, their first derivatives and the variation of resistance as a function of temperature $R(T)$ in the area of the superconducting transition.

2. Films. Tunnel Junctions. Measurement Methods

We used an improved technology in comparison to [8] to manufacture films of superconducting Nb–Ga compounds with A15 structure[9]. Films 0.2–0.8 μm thick were obtained by simultaneous precipitation of Nb and Ga in a vacuum of $(1.3–4)\cdot 10^{-3}$Pa on a vacuum installation with an oil-vapor pump with a nitrogen trap. The substrates were niobium foil or ruby and sapphire polished plates heated to 750–900°C. The niobium was evaporated by an electron beam, the gallium was evaporated thermically from a tantalum or molybdenum boat. The precipitated films were subjected to mulststage processed, including high-temperature annealing and low-temperature annealing, which increased the quantity of A15 phase. The optimal parameters: temperature (T_{HTA}, T_{LTA}) and time (t_{HTA} and t_{LTA}) were selected experimentally.

High-temperature annealing was performed in two stages. First Nb–Ga films with a thickness of ≅1000Å were formed and subjected to high-temperature annealing at T_{HTA}=1200–1500°C and t_{HTA}=1–5 s. Then, after deposition of the film was completed, it was once more subjected to high-temperature annealing at the same T_{HTA} and t_{HTA}=2–20 s in the vacuum chamber which was still sealed at a pressure of $P=(4–5)\cdot 10^{-4}$ Pa. Low-temperature annealing was performed in a vacuum of $(1.3–2.6)\cdot 10^{-4}$ Pa under the following conditions: T_{LTA}=750–800°C, t_{LTA}=(10–100) min and T_{LTA}=850–900°C, t_{LTA}=(10–20) min. Low-temperature annealing usually increased T_c by 1–2.5 K. For five specimens of different compositions immediate analysis of the composition was performed right after deposition on a "Kameka" installation. We can see from Table 1 that as t_{HTA} increased, the atomic %Ga in the compounds rapidly decreased.

Table 1
Change in Atomic Percent Ga in Nb–Ga Compounds at Various t_{HTA} for T_{HTA}=1500°C

Specimen	at.%Ga	t_{HTA}, s	Specimen	at.%Ga	t_{HTA}, s
No. 23.3	7-11	20	No. 42	46-48	3
No. 7	20-23	15	No. 30	48-56	1
No. 40	39-40	8			

The manufacturing conditions of Nb–Ga films which we used yielded specimens with a gradient in composition primarily in the direction of the measurement current. We made Nb–Ga–oxide–Pb tunnel junctions by a method developed at the FIAN Optical Laboratory by S. I. Vedeneyev[7, 10]. First, a strip of Nb–Ga was laid down on a "hot" substrate. It was oxidized in air. Then a strip of Pb was laid down through a special mask perpendicular to the first strip. When a foil substrate was used, after deposition and oxidation of the Nb–Ga strip, the foil was glued to a polished quartz plate, the edges of the working surface were covered with BF–4 glue, after which the Pb strip was applied. The areas of the tunnel junctions were 0.1–0.5 mm^2, resistance 0.03–0.70 Ohms.

All information obtained in our tunnel studies was contained in the nonlinearities of the tunnel junction current-voltage characteristics. The recordings of the current-voltage characteristics were made on a type PDS–021M two–coordinate recorder. The voltage across the tunnel junctions was monitored by a F–30 instrument, the current — by an M104 ammeter. In order to increase sensitivity, the derivatives of the current-voltage characteristics were studied first. A modulation method was used. The first harmonic was recorded. A bias current generator developed in [11] was used as an electronic rheochord.

In certain cases, the derivative $dI(V)/dV$ was determined by numerical differentiation of the smoothed $I-V$ characteristic (I being the current, V the voltage across the tunnel junction). Tunnel measurements were performed at T=4.2 K.

Resistance measurements were performed by a four–contact method.

3. Results of Measurements and Their Discussion

In certain specimens, the function $R(T)$ of the Nb–Ga film had multiple steps (Figure 1), indicating a series of enclosed film sectors with different T_c. The values of T_c were determined from the middle of each resistance jump. The results are shown in Table 2, where we also see the temperatures of the beginning and end of the corresponding transition to the superconducting state.

Table 2
Parameters of Nb–Ga Compound Films

Specimen	Substrate	Interval	$T_c^b - T_c^e$, K	T_c, K	$\Delta_{Nb\text{-}Ga}$, meV	$\frac{2\Delta_{Nb\text{-}Ga}}{kT_c}$	at.%Ga
No. 1	Ruby	1	16.7-16.3	16.5	2.79	3.92	
		2	14.7-14.3	14.5	2.19	3.50	20-30
		3	12.8-9.3	11.3	1.47	3.00	
No. 2	Ruby	1	17.2-15.2	16.2	2.75	3.94	
		2	15.2-13.5	14.2	2.27	3.66	---
		3	13.5-7.6	9.5	1.47	3.60	
No. 3	Foil	---	12.2-11.5	11.8	1.79	3.52	---
No. 4	Sapphire	---	8.9-8.5	8.7	1.03	2.75	46-48
No. 5	Sapphire	---	8.4-8.2	8.3	0.99	2.77	39-40

Note: $T_c^b - T_c^e$ represent the temperatures of the beginning and end of the superconducting transition, k is Boltzman's constant.

Several "gap" singularities were observed for tunnel junctions based on these films in the graph of the function $I(V)$ (cf. Figure 2*b*). Their positions are noted by the arrows. In Figure 2*a* we show a graph of the function $dI(V)/dV$ obtained by numerical differentiation of the function $I(V)$. Comparing the two graphs, we see that the three "gap" singularities shown in the graph of the function $I(V)$ correspond to the three main maxima on the graph of the function $dI(V)/dV$. Their values are greater than the measurement errors.

In addition to the main maxima, there was also a finer structure in the area of 1.0–1.5 mV, the amplitude of which is so small that it cannot be seen in the graph of the function $I(V)$. In this area are the overlapping "difference" and small gaps. We utilized only singularities which are clearly seen on the graph of $I(V)$[7]. The three main maxima of the function $dI(V)/dV$ are related to the sums $\Delta_1+\Delta_{Pb}$, $\Delta_2+\Delta_{Pb}$ and $\Delta_3+\Delta_{Pb}$, where Δ_1, Δ_2 and Δ_3 are the three superconducting energy gaps of the Ni–Ga compounds. The value of Δ_{Pb} was accepted as the value from the literature Δ_{Pb}=1.25 meV[12]. The values of $\Delta_{Nb\text{-}Ga}$ obtained are presented in Table 2. The error in determination of Δ_i consisted of the error for Δ_{Pb} plus the error of position $\Delta_i+\Delta_{Pb}$ (i=1, 2, 3). The resultant error was ±0.09 meV.

---

[7]In [5], in contrast to ths work, additional results are presented which relate to the "slot" singularities, seen only on the graph of dI(V)/dV. The numbering schemes used in this work and [5] are different.

For tunnel junctions in which the function $R(T)$ of the Nb–Ga film indicated the presence of just one superconducting transition (for example, cf. Figure 3), the graphs of $I(V)$ and $dI(V)/dV$ contained just one summary gap singularity (for example, cf. Figure 4), indicating the presence of just one superconducting gap. The other specimens were processed similarly. The results are also shown in Table 2.

The centers of the film areas used for measurement of $R(T)$ and Δ_i virtually coincided. However, the distance between potential contacts was greater than the width of the Pb film. In spite of this, we assumed that the temperature intervals of the superconducting junctions could be compared to the corresponding superconducting energy gaps. It was considered that higher T_c corresponded to higher Δ_i.

We can see from Table 2 that for two specimens we observed simultaneously a multistep function $R/(T)$ and several "gap" singularities on the graph of $I(V)$, meaning that the specimen was divided into areas with different T_c and Δ. For one of the specimens (No. 1) a composition was measured which was found to be close to the stoichiometric. For specimens, the composition of which was far from stoichiometric (No. 4 and 5) this subdivision was not observed. We can assume that in the case of subdivision of a specimen into areas with different superconducting parameters (T_c and Δ), these areas correspond to different lattice parameters. The basis for this assumption is the results of [13] and [14].

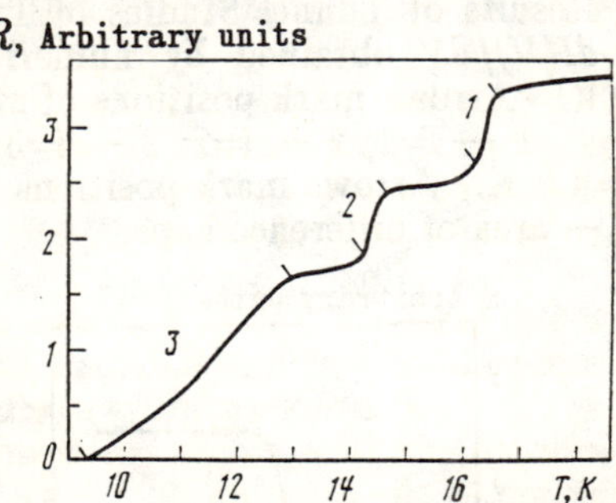

Figure 1. Temperature Variation of Electrical Resistance $R(T)$ of Nb–Ga Compound Film. 1–3, Temperature intervals of superconducting transitions, diagonal lines mark boundaries of these intervals.

A study of the variation in lattice parameter a_0 of the A15 phase of compounds Nb–Ga as a function of Ga content was performed in [13]. The resutls of this work are shown in Figure 5*a*, from which we see that in the range of Ga content from 19–24 at.% there is a sharp change in lattice parameter from 5.168 to 5.181Å.

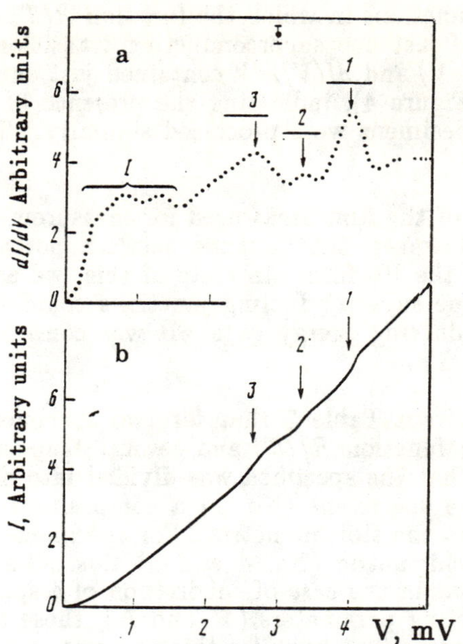

Figure 2. Results of Tunnel Studies of Tunnel Junction 1. a: Graph of $dI(V)/dV$ obtained by numerical differentiation of function $I(V)$. Arrows mark positions of maxima of corresponding $\Delta_i+\Delta_{Pb}$: 1 — i=1; 2 — i=2; 3 — i=3; b: graph of function $I(V)$ at T=4.2 K. Arrows mark positions of "gap" singularities (jumps); I — area of difference gaps.

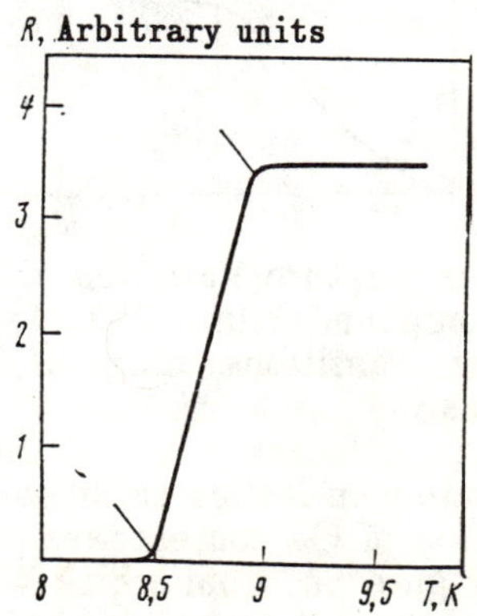

Figure 3. Temperature Variation of Resistance $R(T)$ for Nb–Ga Film for Tunnel Junction 4. Diagonal lines mark limits of superconducting transition.

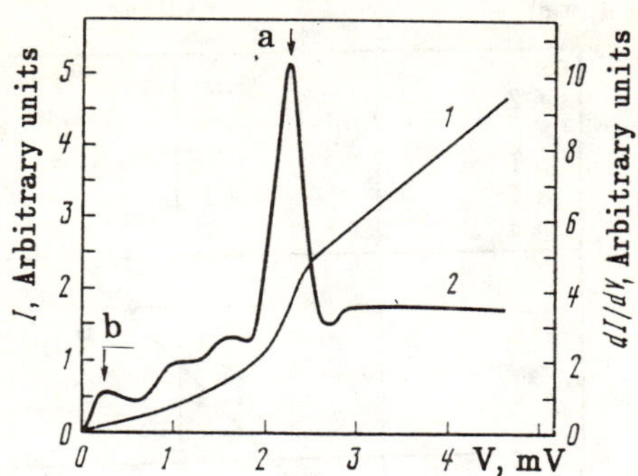

Figure 4. Results of Tunnel Studies of Tunnel Junction 4. 1 — Graph of function $I(V)$; 2 — graph of function $dI(V)/dV$. Arrows mark positions of maxima corresponding to $\Delta+\Delta_{Pb}$ (a) and $|\Delta-\Delta_{Pb}|$ (b).

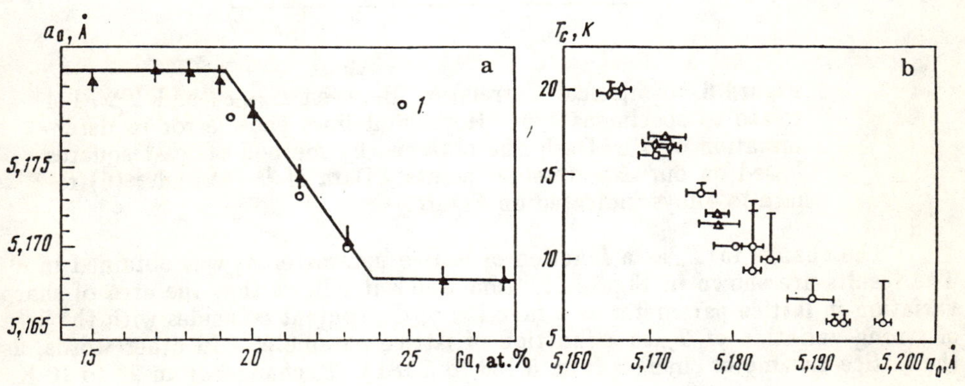

Figure 5. Lattice Parameter a_0 as a Function of Atomic Percent Ga[13] (a) and T_c as a function of a_0 (b) for Nb–Ga Compounds with A15 Structure. a: Vertical lines show errors in determination of a_0, 1 — data of [14]; b: horizontal lines indicate range of change of a_0, vertical lines show range of T_c and T_c^b[14]. Methods of preparation of Nb–Ga compounds: ◇ hardening and annealing[15]; Δ chemical precipitation of vapors[14]; ▯ arc melting and annealing[14]; 0 iodine transportation[14].

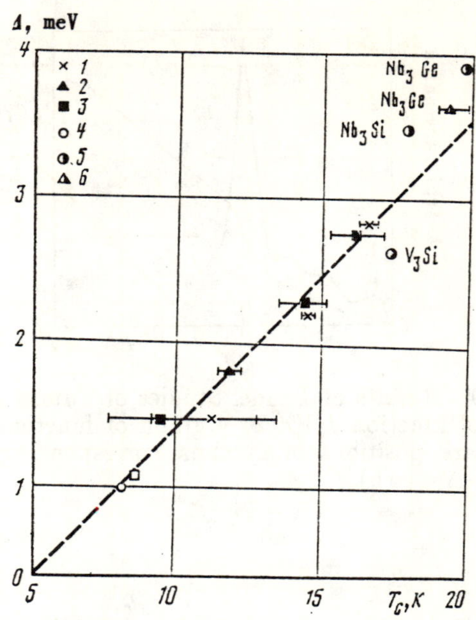

Figure 6. Empirical Correlation Between $\Delta_{Nb\text{-}Ga}$ and T_c. 1–4 relate to specimens 1–5. Horizontal lines show error in determination of T_c. Dash line obtained by method of least squares based on our experimental points. Data of [16] (5) and [4] (6) relate to alloys indicated on Figure.

The change in T_c as a function of lattice parameter a_0 was obtained in [14] The results are shown in Figure 5*b*, from which it follows that the area of sharp variation in lattice parameter as a function of Ga content coincides with the area of strong variation of T_c as a function of lattice parameter. In other words, as the lattice parameter changes from 5.168 to 5.181 Å, T_c changes from 20 to 10 K.

For specimen No. 1 an x-ray diagram was measured, indicating the presence of phase A15, as well as other phases. The lattice parameter of the A15 phase was d_0=5.170 Å. We can see from Figures 5*a* and 5*b* that the specimen falls in the area of strong variation of a_0 as a function of atomic percent Ga and simultaneously in the area of strong variation of T_c as a function of a_0. Thus, slight changes in a_0 in various areas, related to slight changes in A15 phase composition, should lead to significant changes in superconducting parameters.

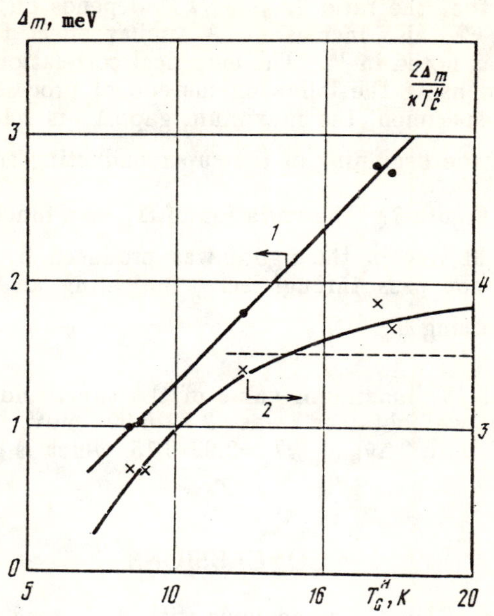

Figure 7. Variation in Maximum Gap Δ_m (1) and Ratio $2\Delta_m/kT_c^b$ (2) as a Function of Temperature of Beginning of Superconducting Transition T_c^b. Line 1 obtained by averaging experimental points using method of least squares. Curve 2 constructed based on data of line 1. Dash line shows value according to BCS theory.

Let us return to Table 2. Five specimens were studied in all. Of these, four had multistep transition to the superconducting state, three had single-step transition. The results in Table 2 indicate an empirical correlation between Δ and T_c in Nb–Ga compounds with A15 structure, as shown in Figure 6. This correlation in the interval $8 \leqslant T_c \leqslant 16.5$ was near linear. The straight line in the figure was calculated by the method of least squares based on our experimental points. The error for $\Delta_{Nb\text{-}Ga}$, as was indicated above, was 0.09 meV. It is identical for all experimental points and is not shown in Figure 6. The error in determination of T_c was accepted to be the difference $T_c^b - T_c^e$, since the energy gap could relate to any part of this temperature interval. It is noted by the horizontal lines. The figure also presents data for Nb_3Ge, Nb_3Sn and V_3Si. Points relating to compounds with A15 structure and $T_c \geqslant 18$ K are slightly above our straight line.

The linear variation represented by the straight line does not pass through zero. This means that the ratio $2\Delta_{Nb\text{-}Ga}/kT_c$ depends on T_c. As T_c increases, the ratio $2\Delta_{Nb\text{-}Ga}/kT_c$ also increases. A similar effect for the alloys Nb_3Ge, Nb_3Sn and V_3Si was noted in [16]. The empirical correlation between $\Delta_{Nb\text{-}Ga}$ and T_c can be obtained using the following method of processing the experimental results. For each specimen, the maximum gap Δ_m is selected, compared with the temperature of the beginning of the superconducting transition T_c^b. The results are shown in Figure 7. The variation of Δ_m as a function of T_c^b is also near linear. The straight line in the figure was produced by the method of least squares. It does not pass through zero, indicating an increase in the ratio $2\Delta_m/kT_c^b$ with increasing T_c.

We note that the maximum value of the superconducting gap of Nb–Ga compounds which we obtained was 2.79±0.09 meV. It corresponds to T_c=16.5±0.1 K and ratio $2\Delta_{Nb\text{-}Ga}/kT_c$=3.92±0.15, which is greater than the theoretical value of BCS.

CONCLUSIONS

1. The films of Nb–Ga compounds with A15 structure, with composition gradient primarily in a predetermined direction and tunneling junctions based on them which we produced allowed us to detect a new phenomenon. It was found that under certain conditions when the film composition is near stoichiometric (with a slight shortage of gallium), instead of a smooth variation of $\Delta_{Nb\text{-}Ga}$ and T_c along the concentration gradient, the superconducting phase is divided into individual areas ("domains") differing in superconducting parameters T_c and $\Delta_{Nb\text{-}Ga}$. The presence of different values of T_c was detected by measuring the function *R(T)*. A multistep transition to the superconducting state was observed (cf. Figure 1). The presence of different $\Delta_{Nb\text{-}Ga}$ was detected by the tunneling method. A number of "gap" singularities (of jump type) were observed on the graph of the function *I(V)* and corresponding maxima on the graph of the function $dI(V)/dV$ (cf. Figure 2). For films with compositions far from stoichiometric in the direction of increased atomic percent Ga, this phenomenon was not observed.

We should note that the condition of subdivision of an Nb–Ga film into "domains" with different superconducting parameters coincide with the conditions under which the film parameters fall in the area of strong variation of lattice parameter a_0 as a function of atomic percent Nb, and simultaneously in the area of strong variation of T_c with a_0 (cf. Figure 5).

2. The empirical correlation between $\Delta_{Nb\text{-}Ga}$ and T_c where T_c varies from 8.0–16 K was near linear, not passing through zero (cf. Figure 6). With this relationship between Δ and T_c, the ratio $2\Delta/kT_c$ is is not constant, but increases with increasing T_c. The rise rate decreases with increasing T_c, and a tendency toward saturation is observed.

3. For the specimen with T_c=16.5±0.1 K a superconducting gap $\Delta_{Nb\text{-}Ga}$=2.79±0.09 meV was observed, with $2\Delta_{Nb\text{-}Ga}/kT_c$=3.92±0.15. This is the first time such a large value of superconducting gap has been observed in Nb–Ga compounds.

REFERENCES

1. Nabutovskiy, V. M., Shapiro, B. Ya., "Localized Superconducting States in Spatially Heterogenous Superconductors," Fizika nizkikh temperatur, Vol. 7, pp. 855–862, 1981.

2. Averin, V. V., Buzdin, A. I., Bulayevskiy, L. N., "Theory of Localized Superconductivity," ZhETF, Vol. 84, pp. 737–748, 1983.

3. Khlyustikov, I. N., Khaykin, M. S., "Proximity Effect and Phase Diagram of Superconductivity of Twinning Plane of Metallic Crystals," Pis'ma v ZhETF, Vol. 36, pp. 132–135, 1982.

4. Golovashkin, A. I., Lykov, A. N., Pechen', Ye. V., "Study of Bridge and Tunnel Junctions Based on Nb_3Ge Films," FTT, Vol. 25, pp. 1305–1309, 1983.

5. Dzhurayev, D. R., Levchenko, I. S., Motulevich, G. P., "Energy Gap of Superconducting Nb–Ga Alloys," Fizika nizkikh temperatur, Vol. 8, pp. 498–503, 1982.

6. Nordman, J. E., Honck, L. L., "Excess Current in Nb and Nb_3Ga Josephson Junction," Proc. on Future Trends in Superconductive Electronic, Charlottesville, Virginia, pp. 354–358, 1978.

7. Vedeneyev, S. I., "Tunneling Studies of Electron–Phonon Interaction in Transition Metals and A15–Type Structures," Preprint FIAN, No. 193, Moscow, 61 pp., 1977.

8. Golovashkin, A. I., Levchenko, I. S., Motulevich, G. P., "Characteristics of Superconducting Alloys with A15 Lattice Obtained by Evaporation in a Vacuum," Elektronnyye kharakteristiki i elektron–fononnyye vzaimodeystviya sverkhprovodyashchikh metallov i splavov [Electronic Characteristics and Electron–Phonon Interactions of Superconducting Metals and Alloys] (Trudy FIAN, Vol. 82), Moscow, Nauka Press, pp. 72–108, 1975.

9. Dzhurayev, D. R., "Superconducting Nb–Ga Alloys on Dielectric Substrates," Preprint FIAN, No. 105, Moscow, 23 pp., 1983.

10. Vedeneyev, S. I., Golovashkin, A. I., Motulevich, G. P., "Study of Phonon Spectrum of Nb_3Sn Alloys," Preprint FIAN, No. 116, Moscow, 23 pp., 1973.

11. Sidel'nikov, V. I., Tyurikov, D. A., "Bias Current Generators for Studies of Superconducting Connections and Tunnel Experiments," Preprint FIAN, No. 118, Moscow, 23 pp., 1978.

12. Franck, J. P., Keeler, W. J., "Pressure Dependence of the Energy Gap of Superconducting Pb," Phys. Rev. Lett., Vol. 20, pp. 379–381, 1968.

13. Jorda, J. L., Flukiger, R., Muller, J., "Effects des details du diagramme de phases sur la temperature sapraconductric du systeme Nb–Ga ($T_{c\max}$=20.7 K)," Recents progr. connais. propr. Phys. et chim. gallium et. composes. Colloq. Marscile, Marscile, pp. C6/1–C6/11, 1976.

14. Webb, G. W., Engelhardt, J. J., "Superconducting, Metallurgical and Synthesis Properties of Nb_3Ga," Proc. 1974 Appl. Superconductivity Conf., IEEE Trans. on Magn., Mag–11, No. 2, pp. 208–213, 1975.

15. Webb, G. W., Vieland, L. J., Miller, R. E., Wicklund, A., "Superconductivity above 20 K in Stoichiometric Nb_3Ga," Solid State Commun., Vol. 9, pp. 1769–1773, 1971.

16. Moore, D. F., Zubeck, R. B., Rowell, J. M., Beasley, M. R., "Energy Gaps of the A15 Superconductors Nb_3Sn, V_3Si and Nb_3Ge Measured by Tunneling," Phys. Rev. B., Vol. 20, pp. 2721–2738, 1979.

SUBJECT INDEX